# Breaking Ocean Waves

Geometry, Structure, and Remote Sensing

Eugene A. Sharkov

# Breaking Ocean Waves

Geometry, Structure, and Remote Sensing

 Springer

Published in association with
**Praxis Publishing**
Chichester, UK

Professor Eugene A. Sharkov
Space Research Institute
Russian Academy of Sciences
Moscow
Russia

SPRINGER–PRAXIS BOOKS IN GEOPHYSICAL SCIENCES
SUBJECT *ADVISORY EDITOR*: Philippe Blondel, C.Geol., F.G.S., Ph.D., M.Sc., Senior Scientist, Department of Physics, University of Bath, Bath, UK

ISBN 978-3-540-29827-4 Springer Berlin Heidelberg New York

Springer is part of Springer-Science + Business Media (springer.com)

Library of Congress Control Number: 2007928830

Cover design: Jim Wilkie
Project management: Originator Publishing Services Ltd, Gt Yarmouth, Norfolk, UK

Printed on acid-free paper

# Contents

# Preface

The surface of the ocean has drawn the most intent attention of human beings from the very beginning of mankind's history. The breaking of large oceanic waves has an especially hypnotic effect on any ordinary man or woman observing this magnificent phenomenon. These natural processes have repeatedly been the themes in painting (see Plate 1), literature, and music. Very few—if any—people have remained indifferent to this manifestation of the greatness of nature. No less important are the oceanic wave breaking processes for scientific concepts and views in the study of the World Ocean as well.

The study of the physical and dynamic characteristics of gravity waves on the sea surface while they break and the subsequent foam activity and formation of drop-spray clouds are amongst the major problems facing modern satellite oceanology, physics of the ocean–atmosphere interaction, and oceanic engineering (see Plate 2). In particular, the contribution of foam and drop-spray systems of various types to mean values and spatiotemporal variations in microwave (radio emission and backscattering), infrared, and optical parameters of a rough sea surface is significant. Knowledge of the detailed statistical characteristics of breaking wave fields is also important in the study of the dynamics of sea waves (generation of waves, nonlinear interactions, dissipation, etc.).

Despite the external accessibility and seeming simplicity of visual and instrumental observation of the oceanic wave breaking process, detailed scientific data on spatial breaking fields in many different areas of the World Ocean obtained in field remote-sensing experiments are still not yet available. This is principally due to the high spatiotemporal variability in the process of breaking gravity oceanic waves under rough sea conditions and high values of wind speeds over the oceanic surface. Under such complicated aerial navigation conditions, there arise both field experimental methodological complications in remote sensing of these rapid processes by ships and airplanes, and complications of administrative character (e.g., banning of air flights under complex meteorological conditions). And space-borne instruments still do not

**Plate 1.** Thirty-six views of Mt. Fuji: View through waves off the coast of Kanagawa, by Katsushika Hokusai (18th century).

possess the spatial resolution sufficient for remote recording of oceanic wave breaking processes.

Note also that—depending on the scientific approaches and particular tasks of remote sensing and sea hydrodynamics—the basic characteristics of the breaking process should now be considered in two principally different aspects at least: in the individual aspect (i.e., in the form of a temporal set of individual breakings) and in plural representation (i.e., in the form of a spatial field of breaking oceanic waves). It should further be noted that—depending on the geometrical characteristics of a remote-sensing system (flight altitude, instantaneous field of view, and time constant of signal storage)—the contribution from the same disperse structure or from a set of structures can in principle be different.

This book represents the first detailed analytical description of the state of remote investigations (in the optical and microwave ranges of electromagnetic waves) of one of the major nonlinear elements of sea dynamics: the process of breaking gravity waves and their subsequent evolution and the dynamics of dispersed foam systems of various classes and the drop-spray phase. Issues in the methodology of multi-scale optical and microwave remote measurements are considered; and the techniques used to study individual breakings and meso-scale, discrete, breaking, random point fields are described. The results of field investigations are presented. The advantages and

**Plate 2.** Oil platform in rough seas (the North Sea basin).

limitations of various remote complexes, used to reveal the spatiotemporal features of the fields of gravity wave breaking and disperse systems from air carriers of various classes, are also considered in the book. The latest achievements in the field of electrodynamics of emission and scattering of electromagnetic waves by polydisperse close-packed polyhedral media, as well as in the field of electrodynamics of dense flows of the spherical particles of water, are fully described in the book.

The principal feature of the book consists in an integrated description of the spatiotemporal and structural properties of breaking oceanic waves along with its electrodynamics. Emphasis is placed on the physical aspects of breaking processes necessary to judge the possibilities and limitations of remote-sensing methods in specific cases of oceanic surface observation. Numerous practical applications and illustrations, based on air-borne, ship-borne, and up-to-date laboratory experiments, are given in the book.

The book is based on scientific findings from several Russian scientific airborne remote-sensing expeditions to the Far East (the Pacific Ocean), the Black Sea, the Caspian Sea, and the Barents Sea, as well as from several scientific marine expeditions to the tropics and, once more, to the Far East as part of a number of major research projects of the Russian Academy of Sciences. These findings were presented in lectures by the author at the Moscow Physical and Technical Institute (in Dolgoprudnyi, near

Moscow) and at the Moscow University of Geodesy, Mapping, and Aero-Photo Surveying to students of physics and geophysics.

In the field of application of remote observations of the oceanic surface, a book is needed that would represent a systematic and unified statement of the fundamental concepts and issues of the theory of breaking gravity waves, the electrodynamic properties of disperse systems arising during the breaking process, as well as of the various instrumental and methodological issues of microwave and optical remote measurements. In addition, it would be useful to provide a unified and systematic description of the latest achievements in the field of microwave sensing of a rough sea surface (i.e., one that is easily accessible to undergraduate students, post-graduate students, researchers, engineers, and instrument operators). The present book was conceived to give as large a systematized idea of the possibilities and modern achievements of methods of the remote sensing of a rough sea surface as possible to a wide range of specialists and interested readers.

The format of the book is constructed in such a way that the reader could acquire the necessary knowledge of the physical mechanisms of breaking gravity waves that he or she needs, in addition to the most complete information available on the modern level of development of microwave and optical remote diagnostics of a rough sea surface. The content of the present book is essentially broader than the requirements that are usually set out in a handbook for students. Much of it contains detailed information and can be used as a reference book to the many special issues of the microwave and optical remote diagnostics of the oceanic surface.

The first chapter of the book considers the scientific and applied aspects of remotely sensing the sea surface, the role and place of optical and microwave methods and instruments in the study of breaking waves, the basic concepts of the modern theory of breaking gravity waves, the possibilities of passive and active methods of microwave diagnostics of a rough sea surface. The second chapter is devoted to the results of airborne sensing of spatial fields of breaking sea waves in two modes: limited fetch and fully developed sea state. On the basis of experimental data, important modeling ideas are proposed for the spatial field of breaking sea waves that occur as a result of the formation of Poisson's point field of non-interacting centers. The third chapter presents the results of experimental investigation into the geometrical characteristics (linear and two-dimensional sizes) of the process of individual breaking gravity wave (whitecapping) and foam fields of various types. On the basis of experimental data, the statistical models of breaking processes are constructed. Critical analysis of existing theoretical concepts of wave breaking as a result of the threshold mechanism for a random Gaussian three-dimensional field (breaking criteria, threshold mechanism restrictions, etc.) is carried out. The fourth chapter gives the results of experimental investigations into the lifetime of the disperse phase of a whitecapped gravity wave: in particular, revealing the exponential character of the temporal evolution of a whitecapping crest and patch foam structures, and detection of a specific group of gravity wave breaking (microbreaking). The fifth chapter is devoted to studying the nature of formation of the disperse structure and the contribution of a drop-spray phase—formed as a result of breaking—to the mass and moisture exchange in the ocean–atmosphere system. The sixth chapter contains a detailed analysis of the electrodynamics of absorption and emission of close-packed media of colloid-type foam. The generally colloidal, physical, and disperse properties of close-packed foam structures are considered in detail. Great attention is given in this chapter to the methods of describing the electromagnetic properties of rarefied and close-packed disperse structures; also the results of detailed experimental investigations are presented in which two types of colloidal structures were found that essentially differ in their emissive characteristics (viz., a monolayer of multiple emulsion and a foam layer of polyhedral structure). An entire spectrum of electromagnetic models of foam systems is analyzed in the chapter, and a model is found that agrees well with experimental data—namely, a model of the inhomogeneous dielectric layer that involves the scattering of hollow spheres and a smooth transitional phase boundary. The seventh chapter is devoted to studying the electrodynamics of a drop-spray phase as a flow of highly concentrated drop medium. Optical models for rarefied flows in the radiative transfer theory and their restrictions are considered. Then, the results of specialized experiments on studying the electromagnetic properties of dense drop flows and the possibilities of their use for forming the electromagnetic models of a drop-spray phase of breaking waves are analyzed in detail. Chapter 8 is devoted to detailed analysis of remote field investigations of the transition zone in the ocean–atmosphere system by means of optical, IR, and microwave air–space missions, beginning with the first successful Russian missions on the "Cosmos-243" and "Cosmos-384" satellites carrying microwave multi-frequency instruments. The results of field experiments carried out onboard research vessels by means of microwave active–passive instruments in the Indian Ocean are outlined in detail. Prominence is given in this chapter to the description of modern models of the state of the ocean–atmosphere system under storm conditions (models, hypotheses, preliminary experiments, etc.). The modern situation in the instrument field of potential microwave remote missions, the ways of developing observation techniques and methods, and the exploiting of new frequency ranges for detailed studying of the state of the oceanic surface are all fully considered.

A detailed bibliography is given at the end of the book that should be useful both for undergraduate students and post-graduate students of applicable specialties, as well as for researchers.

The book is aimed at researchers, university teachers, and undergraduate and postgraduate students working in geography, meteorology, climatology, atmospheric physics, geophysics, oceanography, and in the environmental science areas of remote sensing and geophysics.

Many of the experimental and full-scale results, used in preparing the book, were obtained by the author during his work at the Space Research Institute of the Russian Academy of Sciences (SRI RAS). The full-scale laboratory experiments performed during 1974–1993 by SRI co-workers using highly sensitive optical and microwave instruments, as well as the unique results obtained with their help, have determined in many respects the design of future air–space microwave instruments for studying the state of the oceanic surface.

# Acknowledgments

The author is grateful to his colleagues, without whose support the unique full-scale laboratory experiments given in this book could not have been fulfilled. He especially thanks I. V. Pokrovskaya, M. D. Raev, V. M. Veselov, I. V. Chernyi, and V. Yu. Raizer.

The author is thankful to Yu. Preobrazhenskii for his constructive approach to translating the manuscript into English. The typing by Nataly Komarova of such a complex manuscript is appreciated.

The author also wishes to express his thanks, for support and encouragement received, to Clive Horwood of Praxis.

The author also wants to acknowledge the advice and recommendations of his colleagues during preparation of the manuscript.

Fully realizing the complexity and responsibility of the present publishing project, the author would like to thank readers in advance for any constructive criticism and remarks (email: *easharkov@iki.rssi.ru*).

# Figures

# Tables

# Abbreviations and acronyms

| | |
|---|---|
| ADP | Antenna Directional Pattern |
| AFA-100 | Type of Russian aerosurveying camera |
| AFA-TE-100 | Type of Russian aerosurveying camera |
| AN-30 | Antonov-30 (Russian research aircraft) |
| DSP | Drop-Spray Phase |
| EBS | Effective Backscattering Surface |
| ESA | Effective Scattering Area |
| HF | High-Frequency |
| IL-14 | ILyushin-14 (Russian aircraft) |
| IR | InfraRed |
| IW | Internal Wave |
| JONSWAP | Joint North Sea Wave Project |
| LF | Low-Frequency |
| MB | MesoBreaking |
| MGPI | Moscow State Pedagogical Institute |
| MKF-6 | Type of Russian space-borne multiband photographic camera |
| OI | Optical Image |
| PSD | Polar Scattering Diagram |
| RAS | Russian Academy of Sciences |
| RCS | Radar Cross-Section |
| RFT | Rapid Fourier Transformation |
| RHV | Research Hydrographical Vessel |
| RTI | Range–Time–Intensity (diagram) |
| RV | Research Vessel |
| SAS | Surface-Active Substance |
| SRI | Space Research Institute |
| VH and HV | Cross-polarization |

# 1

# Introduction and rationale

This chapter is devoted to consideration of: (1) scientific and applied aspects of remote sensing of the sea surface under conditions in which its continuity is violated; (2) the role and place of optical and microwave means and methods in studying the spatiotemporal characteristics of breaking wave fields; (3) basics of the modern theory of breaking gravity waves; (4) possibilities for passive and active methods of microwave diagnostics of a rough sea surface.

## 1.1 BREAKING OCEAN WAVES IN THE ATMOSPHERE– OCEAN SYSTEM

Under moderate and extremely rough sea conditions the sea surface ceases to be a single-bounded surface with a distinct water–air boundary—an intermediate density zone arises, which represents a polydisperse mixture of finite volumes of air and water with a particular and highly fluctuating (in space and time) transition gradient. The existence of such a two-phase medium drastically changes both the character of how the two media interact (including energy exchange, gas exchange, heat exchange, mass exchange parameters), and the dynamic regime of a rough sea surface.

### 1.1.1 Wave dynamics at wave breaking

In spite of considerable effort undertaken by the scientific community in studying wave dynamics and wave modeling of rough seas for the last 60 years, there exists a lot of physical and methodological problems, which have been the subject of serious criticism and discussion (Cavaleri, 2006; Papadimitrakis, 2005a, b).

One of the most complicated wave dynamics problems is solution of the issue of wave energy dissipation at the moment of gravity wave breaking. Investigations have taken various directions: spatiotemporal estimations of breaking parameters have been carried out by means of dimensional considerations (Banner and Phillips, 1974; Phillips, 1977; Kitaigorodskii, 1997, 2001) and simplified numerical models have been used (Vinje and Brevig, 1981; Banner and Peregrine, 1993; Krasilnikov, 1987). There exists a series of works on wave breaking, performed under laboratory conditions (Deane and Stokes, 2002), which certainly cannot be correlated fully to conditions of gravity sea waves breaking in natural situations (in "deep" water, in particular). A detailed analysis of the modern situation of studies investigating gravity wave breaking, based on spectral methodology, is presented in a paper by Young and Babanin (2006).

A very important structural component of gravity wave breaking is the disperse phase (the foam, drop-spray structures, and the aerated layer), which arises in the breaking process and substantially (from 10 to 50%, according to the conclusions of some models) "incorporates" in itself the dissipated energy of a wave. However, as was rightly noted in a paper by Young and Babanin (2006), the methodology of accounting for dissipated wave energy at the moment of wave breaking should be strictly correlated with those spatiotemporal scales in which the wave field is studied in itself. This circumstance puts on the agenda the necessity for a statement of theory and the performance of successive field experiments, under high-sea conditions, by studying the temporal and spatial characteristics of breaking wave fields and the disperse phase formed when an individual gravity wave breaks over the water surface (foam structures and drop-spray clouds) and inside the water volume (the aerated layer).

### 1.1.2  Energy exchange at wave breaking

It has been believed for a long time that to determine the turbulent heat and water fluxes over the sea it is sufficient to multiply wind velocity by water–air temperature drop and the heat exchange coefficient, or by the humidity drop and the water exchange coefficient. The dimensionless coefficients of heat exchange ($C_H$) and water vapor exchange ($C_E$) (the Stanton number and Dalton numbers, respectively) were assumed to be constant, not dependent on wind velocity and stratification conditions and equal to each other and to the sea surface resistance coefficient $C_U$ (Zilitinkevich et al., 1978). However, the gradual accumulation of experimental data—both in situ and remotely sensed—and laboratory data has resulted in serious reconstruction of the mentioned scheme. According to modern data, coefficient $C_U$ grows as the wind strengthens, and the distinctions between it and $C_H$ and $C_E$ coefficients grow in this case. Long before obtaining these results it was supposed that coefficient $C_U$ could essentially grow during a storm. In addition, it was emphasized that spray clouds, which fill the near-water layer of air during a storm, play a considerable part in the momentum transfer. However, only recently the interest in studying the thermodynamic interaction of the ocean and atmosphere under storm conditions has noticeably grown. This is mainly associated with revelations about the key role of the heat and water exchange of the ocean and atmosphere in global atmospheric processes (Lappo et al., 1990). Even rough estimations, which take into account the possible increase of $C_H$, $C_E$ coefficients during a storm, have shown that during a short period of stormy conditions the ocean could impart to the atmosphere, in addition, huge quantities of heat and water, which, in their turn, are capable of noticeably changing the state of the atmosphere over extensive areas. In World Ocean regions where the occurrence of storms is considerable, the contribution of storms to the average annual or seasonal heat and water transfer is significant. The regions with a high incidence of storms mainly coincide with the energy-active regions of the ocean that have greatest effect on atmospheric processes. This circumstance makes it clear that the intensification of energy and mass transfer through the ocean surface during a storm should be taken into account when constructing the models of climate and models of general circulation of the atmosphere and ocean, as well as when developing techniques for long-term weather forecasting.

Estimates presented earlier by Bortkovskiy (1983) have shown that tropical cyclones highly influence the heat and water balance in the boundary layers of the atmosphere and ocean within the scale of the whole tropical zone. So, the above statement concerning the zones in which the ocean has an active effect on atmospheric processes is valid also for tropical cyclogenesis regions (Sharkov, 1998, 2000). Moreover, analysing and forecasting the evolution and motion of tropical cyclones are practically impossible without knowledge of the spatial distribution of fluxes of heat, water, and momentum inside a stormy zone. However, in the modern models of tropical cyclones some arbitrary values are often assigned to fluxes, which are far from the real ones. The correct parameterization of energy and mass transfer during a storm requires studying the physics of transfer processes in the near-surface layers, in which two-phase currents are formed during the storm: spray enters the lower layer of air, and bubbles fill the upper layer of water. As preliminary calculations have shown (Bortkovskiy, 1983), even at a wind velocity of $20\,\mathrm{m\,s^{-1}}$ the heat and water transfer by spray becomes commensurable with the corresponding turbulent fluxes. Therefore, it follows that the heat and water fluxes under storm and tropical cyclone conditions cannot be determined reliably enough if the role of spray is disregarded. Studying this transfer mechanism is key to solution of various problems. In their turn the spray field parameters are related with such variable factors as the degree and stage of rough-sea development, the state of the interface surface (the degree to which they are covered with whitecaps and foam), surface contamination by oil products, detergents, and natural (biogenic) substances. Laboratory and field investigations have shown that spray generation is associated with wave crests breaking as a result of wind, and that the basic generation mechanism consists in air bubbles which burst and float up after the passage of whitecaps—that is, the breaking crests of waves.

An important point of energy exchange in the ocean–atmosphere system is the attempts to determine the degree of the foam coverage effect on global and regional radiation budgets, keeping in mind the change of effective reflectance of the ocean surface in the optical and IR ranges in the presence of foam systems. However, because of the absence of reliable experimental data on foam coverage on regional and global scales, the corresponding estimates are rather contradictory (see, e.g., Frouin *et al.*, 2001).

Thus, the analysis of heat and water exchange in the ocean and atmosphere under storm conditions, as well as of momentum and gas fluxes closely related with these conditions, requires consideration of a wide scope of processes and phenomena, which mainly concern the state of the interface surface and the disperse phase arising from it.

### 1.1.3   Gas exchange in the ocean–atmosphere system

As is known, owing to its considerable extension over the Earth's surface area, the World Ocean acts as a kind of regulator of the content of greenhouse gases: carbon dioxide, water vapor, methane, and some other minor gas constituents. The current greenhouse effect is considered to be produced by the first two gases. And, principally because of the possible anthropogenic load of industrial activity, the role of carbon dioxide and variation of its content in the atmosphere are important. The World Ocean's role in regulating the carbon dioxide content in the Earth's atmosphere seems to be very important and is determined by three factors.

The first factor is related with the high solubility of carbon dioxide in sea water. The second factor consists in the fact that, unlike the majority of gases, carbon dioxide is chemically active at dissolution in water and disturbs the dynamic balance of an ocean's carbon cycle. Namely, it causes a series of successive reactions, beginning with the formation of carbonic acid at $CO_2$ dissolution in sea water and finishing with precipitation of insoluble calcium and magnesium carbonates, which gradually turn into sedimentary rocks as limestones and dolomites. The third factor is photosynthesis. Under the effect of solar energy, inorganic carbon dioxide transforms into organic carbon. Since sunlight is highly absorbed by water, photosynthesis occurs in the ocean's surface layer only, at a depth of no more than 70–80 m. Nevertheless, 20 billion tons of organic carbon are produced in the ocean annually.

The mechanisms of gas transportation from the atmosphere into the ocean are diverse (Bortkovskiy, 2006). They include: the direct molecular diffusion of gases into liquid, the involvement of gases from the atmosphere in the deeper layers of the ocean in the form of bubbles when gravity waves break under strong wind conditions and in storm zones, as well as the intermixing of gases in surface layers under strong storm conditions and in tropical cyclone zones. But there also exists a more delicate mechanism—internal micro-convection—which is caused when a thin surface layer of the ocean cools due to the "molecular, turbulent" diffusion and evaporation of water.

Carbon dioxide absorption by the World Ocean is mainly concentrated at tropical latitudes, which occupy almost one-half of its area. At mid-latitudes the fluxes change their direction many times during a year. On average, they compensate each other, but in some areas the flux from the atmosphere into the ocean still prevails (such as in the Gulf Stream area). And, virtually, only high latitudes serve as a source of carbon dioxide transfer into the Earth atmosphere throughout a year.

Note that as the wind velocity changes over the sea surface the character of gas exchange itself essentially changes—so, at weak wind it has a diffusive character, and at strong wind the gas exchange is almost completely controlled by the disperse medium arising when gravity waves break (Bortkovskiy, 2006).

## 1.2  BREAKING OCEAN WAVES AND MICROWAVE REMOTE SENSING

Microwave images of the ocean–atmosphere system and direct observations of the ocean surface, obtained from aircraft and satellites, reveal unique information on various phenomena occurring both directly in the near-water layer of the atmosphere, the near-surface layer of the ocean, and, somewhat surprisingly, the ocean depths. Certainly, the electromagnetic waves of the microwave range can only penetrate the water by no more than a few millimeters or centimeters (depending on the wavelength range), and the processes taking place in the ocean depth can be visualized as a result of their surface manifestations. Here, we note that information, obtained in active and passive modes of microwave sensing, can principally differ due to the different character of physical mechanisms of natural radiation and backscattering of objects in the ocean–atmosphere system (Sharkov, 1998, 2003; Bulatov et al., 2003; Kanevsky, 2004). So, on radar images one can clearly identify the traces of flows and oceanic vortices, as well as the hydrological fronts representing the boundaries that separate sea water according to various properties (temperature, salinity, density, color, various suspensions, and organic admixtures). High-resolution radar records the surface of a rough sea or, more precisely, the large-scale component of a rough sea with characteristic wavelengths exceeding 20–40 m. Of special interest are surface manifestations of internal waves in the ocean (Sabinin and Serebryany, 2005). In addition, radar is capable of observing any smoothing of the surface—that is, slicks, including those caused by oil pollution and surface-active substances.

Not only intra-oceanic processes but also a lot of atmospheric processes occurring in the near-water layer of the atmosphere are reflected in the ocean surface (Bulatov et al., 2003). In particular, one can see on radar images the manifestations of near-surface wind, which influences the centimeter component of a wave's spectrum on the sea surface (gravitation–capillary waves) and, thereby, influences radiowave scattering. The small-scale swell of the ocean surface, excited by wind, can be used to visualize some other atmospheric motions: atmospheric fronts, atmospheric internal waves, as well as atmospheric convective cells, which develop over the ocean under unstable stratification conditions (Bulatov et al., 2003).

Capillary waves excited by the near-surface wind reflect both atmospheric and intra-oceanic processes, which modulate in one way or another the short gravitation–capillary waves on the ocean surface, which, in turn, are revealed by modulations in the backscattered signal and natural radiation of the sea surface. Thus, the radar and radiothermal images of the ocean surface visualize motions both in the ocean itself and in the atmosphere, as well as the gaseous and disperse structure of the atmosphere itself. The complex analysis of fields of radiothermal and radar signals, currently being developed, will undoubtedly become a serious advance in studying the ocean–atmosphere system.

An important advantage of microwave-range waves, as compared with electromagnetic waves of optical and IR ranges, consists in the fact that they can penetrate through cloud cover, thus providing round-the-clock and all-weather observations of the ocean. One more advantage of the microwave range that is important for remote diagnostics of the ocean consists in the fact that these waves interact with surface disturbances in a diffractive manner. And, thereby, they visualize such motions in the ocean and atmosphere that are principally inaccessible to observation in other ranges of the electromagnetic spectrum and, principally, in the optical range.

The resonance (Bragg) theory of radiowave scattering on the sea surface—which is generally accepted now—is based on using the small-perturbation method. Within the framework of the disturbance method, the height of surface roughnesses (capillary waves) is considered to be small compared with the electromagnetic wavelength, so that in Maxwell's equations a small parameter arises, in which the expansion is just carried out.

Various structures on the ocean surface become visible as a result of these swell modulation mechanisms: mainly, flows and surface-active substances (SASs), which influence the surface tension of a liquid.

The greater part of the phenomena relating to the ocean surface observed by means of radar can be explained within the framework of resonance scattering (Moor and Fung, 1979; Alpers and Hasselmann, 1982; Phillips, 1988; Phillips et al., 2001). At the same time, the wide range of phenomena testifies to the existence of so-called non-resonance scattering mechanisms. It was experimentally found (Kalmykov et al., 1976; Lewis and Olin, 1980) that at small sliding angles a considerable contribution, along with scattering by a swell, is made by reflections arising when a wave breaks. Here, at the horizontal polarization, splashes are sometimes observed, whose amplitude exceeds by 10–15 dB the mean level of scattering by a swell. In this case, one can consider as firmly established the fact that there exists a correlation between speckle and large gravity waves breaking as shown by a sensing signal in the spatial resolution element (Kalmykov et al., 1976; Lewis and Olin, 1980; Cherny and Sharkov, 1988). The breaking of large surface waves—which for expedience we call macro-breakings—should be distinguished from meso- and micro-breakings of waves in the meso-scale wave spectrum (see Chapters 3 and 4 for more details). The breaking of large waves represents a complicated dynamical process that is a subject of intensive investigations (see, e.g., Longuet-Higgins and Turner, 1974). The mechanisms that control large waves breaking in the open ocean and in coastal zones are still insufficiently clear and are described as a rule only statistically.

The effect of various wave-breaking-related factors on the formation of radar reflections has been studied in a series of works (see the review by Bulatov et al., 2003). The following phenomena have been considered in such a way: the diffraction of electromagnetic waves on the sharp crests of surface waves, the "corner" model of radiowave scattering in the region of hydraulic jump resulted from a spilling breaker, the mirror reflection of radiowaves from the crests of surface waves at the instant of formation of a "plunging" breaker. The principally different (namely, of non-diffraction type) non-Bragg mechanism of scattering on two-phase media—the "water drops in air" (drop-spray clouds) and the "air bubbles in water" (the aeration layer)—was offered and developed in papers by Kalmykov et al. (1976) and Cherny and Sharkov (1988).

In recent years a series of radiophysical experiments was carried out in a special wave pool (Sletten et al., 2003). These experiments, however, have not fully revealed the physical backscattering mechanism. Naturally, the results of these investigations cannot be fully correlated to the real situation in the open ocean.

Of essential importance for elucidating the physical reasons behind formation of the backscattering field is the statement of theory of the detailed radiophysical and optical experiments by detailed filming and photo-recording of individual large gravity waves as they break. These experiments, carried out under field sea conditions, should be similar in their geometrical parameters (observation angle, resolution element, etc.) to the situation arising at observations from air- and space-based carriers. The classification of the investigation and methodology of such experiments are described in the next section.

## 1.3 CLASSIFICATION OF INVESTIGATION TECHNIQUES: METHODOLOGY OF EXPERIMENTS

Despite certain progress, achieved within the last 20 years, in studying the spatial-statistical properties of foam fields and the small-scale structure of foam-spray systems, the physical properties and structure of sea foam and drop-spray clouds are still insufficiently known. This is principally explained by the complexity of *in situ* measurements under sea conditions—especially, under high seas. Nevertheless, one can say a lot about how to develop a general approach to studying the breaking process and foam-spray structures—mainly, by means of optical methods. These methods, along with acoustic ones, will allow us in the near future to obtain the necessary and sufficient information on aerated and disperse layers on the sea surface and in the near-water layer by means of microwave remote investigation.

It seems expedient to devise a specific classification of optical methods for studying the processes of gravity wave breaking to which we shall adhere in presenting the material in this book:

- Type I—aerial photography onboard a high-altitude airborne platform for revealing the regularities in distribution of the centers of dissipation of rough seas (gravity wave breaking) over large areas of sea basins (of about several km$^2$) without differentiating the type of foam systems (the experimental technique was presented and described in detail in papers by Pokrovskaya and Sharkov, 1986, 1987a, b, 1994—see Chapter 2).
- Type II—aerial photography onboard a low-velocity airborne platform and perspective surveying onboard a research vessel for studying the spatial-statistical characteristics of the area and linear geometry of foam structures of various types, but without revealing their temporal dynamics (the method was offered and developed in papers by Bondur and Sharkov, 1982, 1990—see Chapter 3).
- Type III—investigation of the temporal dynamics of the process of individual gravity sea waves as they break, of the temporal evolution of geometrical properties of separate disperse formations on the sea surface by means of their repeated surveying onboard a research vessel, as well as filming and photography using

long-focal lenses (the method was developed in Bortkovskiy, 1983, and Cherny and Sharkov, 1988—see Chapter 8).

- Type IV—investigation of the disperse structure of an aerated layer and surface disperse systems by macro-photography of foam mass samples taken from the surface (Raizer and Sharkov, 1980) or directly in the near-surface layer of the sea by means of hermetically sealed boxes (see Chapter 4).

By virtue of the specificity imposed by optical instrumentation platforms on the observation process, each of these types of remote observations of short-living disperse systems possesses certain limitations. And, principally, one should note here that observations according to type I and II do not reveal the temporal dynamics of foam structures, since in these types of observations the individual foam structure is recorded at a certain instant of its own "life", unknown to the researcher. And, thus, the statistics of time of "life" of foam structures is associated with the statistics of the general geometry of foam systems determined from observations of type II. The latter feature substantially "smears" the statistical densities of the distributions of geometrical parameters of foam systems found in works by Bondur and Sharkov (1982, 1990) (see Section 3.5 for more details).

## 1.4   CONCLUSIONS

**1**   A brief review of the spectrum of oceanological tasks and remote-sensing problems results in the necessity of statement of theory and performance of successive field experiments—under high-sea conditions—in studying the temporal and spatial characteristics of breaking wave fields and the disperse phase formed as individual gravity waves break over the water surface (foam structures and drop-spray clouds) and inside the volume of water (the aerated layer).

**2**   Of essential importance for elucidating the physical reasons behind formation of the backscattering field and natural radiation is the statement of theory of detailed radiophysical and optical experiments by means of detailed filming and photo-recording the process of individual large gravity waves breaking. These experiments, carried out under field sea conditions, should be similar in their geometrical parameters (observation angle, resolution element, etc.) to the situation arising at observations from air- and space-based carriers.

# 2

# Spatial stochastic breaking wave fields in the atmosphere–ocean system

This chapter presents the results of experimental remote investigations into the paterns of the statistical spatial structure and the elements of stochastic dynamics of chaotic natural disperse media, formed in the ocean–atmosphere system during moderate sea and breaking gravity sea wave conditions. On the basis of the data obtained, a spatial stochastic model of breaking wave fields is formed that represents a basis for revealing the physics of the interaction between a turbulent wind flow and gravity waves.

## 2.1 STATEMENT OF THE PROBLEM OF STUDYING THE SPATIAL STOCHASTIC STRUCTURE OF BREAKING WAVES

As already noted in Chapter 1, the most detailed investigations of breaking wave fields can be performed by analyzing optical images. The experimental techniques and instruments for optically surveying a disturbed sea surface are diverse: both small-scale surveys from ships, and large-scale ones from air- and space-based carriers. The experimental data, obtained as a result of ship-based investigations, possess some natural limitations (spatial locality, strong perspective distortions, incomplete accounting for the static independence of ranked sets, etc.). In light of these limitations, the experimental data are not suitable for getting information on the large-scale spatial characteristics of foam fields. On the other hand, application of an aerial view from high-altitude airplanes and, even more so, from space carriers does not allow high spatial resolution (i.e., in the dm and cm scale) to be obtained, which is required for detailed analysis of foam structures. Analysis of the possibilities of the means of remote sensing required for experimental solution of these problems has led us to the conclusion of the expediency of using two types of optical surveying: the large-scale, multizonal photo-instrument MKF-6 and the aerial view system

AFA-100. These instruments were installed, respectively, onboard the Russian airplane laboratories AN-30 and IL-14. Synchronous surveys from both have been performed from essentially different ranges of altitudes: 5100 and 400 m, respectively. In this way, on the one hand, a large spatial field of the overall picture was achieved allowing the full breaking field pattern to be viewed, and, on the other hand, satisfactory spatial resolution was provided for analyzing individual breaking processes.

## 2.2   TECHNIQUE AND CONDITIONS FOR PERFORMING FIELD EXPERIMENTS

Large-scale optical sensing of the sea surface in sea wave fetch mode and in a zone of developed seas, whose results are analyzed in this chapter, have been carried out in the water area of the Caspian Sea within the framework of a complex of works of the USSR Academy of Sciences called "Kaspii-81" ("Caspian Sea-81") under scientific guidance of the author of this book (Bespalova et al., 1982; Grushin et al., 1990). Special attention while performing the experiment was given to choosing meteorological conditions that would allow the "net" fetch mode to be accomplished and, thus, to trace the evolution of the spatial–angular spectrum of sea waves (including wave breaking mode)—from the initial mode of a steady wind flow interacting with a weakly disturbed sea surface up to the completely developed mode. Note that similar meteorological conditions are extremely rarely met in open ocean conditions, and, in addition, they are difficult to check instrumentally. In light of these circumstances, the planned experiment was carried out in a rather limited (from an oceanic scale viewpoint) water area of the Caspian Sea. In performing the experiment, both conditions of principal character were satisfied (namely, the appropriate steady atmospheric pressure situation and its monitoring, the presence and recording of the initial stage of wind flux interaction with the surface), and conditions of technological character were met (the presence of an oil installation in the fetch zone, the possibility of accurate geographical attribution of means). This field experiment was carried out under controllable conditions—at least from the viewpoint of knowledge of the spatial–spectral characteristics of sea waves, turbulent wind regime, thermal stratification, and atmospheric pressure field. This is the reason we have used below the materials of the aforementioned experiment for studying the spatial–statistical characteristics of a breaking gravity wave field.

Remote measurements of sea waves in the limited fetch zone were carried out onboard airplane laboratories IL-14 and AN-30 in the water area of Caspian Sea using the standard aerial view system AFA-TE-100 (IL-14) (Lavrova and Stetsenko, 1981) and the multizonal survey instruments MKF-6M (AN-30) (Sagdeev et al., 1980). The measurements were accompanied by detailed visual observations of skilled onboard observers. The flights were carried out on October 31, 1981, in the first half of a day. The Sun altitude was 35–40°.

### 2.2.1  Meteorological conditions during the experiment and the flight performance technique

Large-scale meteorological conditions during the "Kaspii-81" complex of experiments are represented by the near-surface weather maps (Figure 2.1) for October 31, 1981, at 00 hr 00 min and 09 hr 00 min, Moscow time (USSR Hydrometeocenter archive data), as well as by the data of measurement of the magnitude and direction of a wind flux velocity vector during weather observation flights using air navigation instruments. Synoptic analysis of atmospheric pressure field maps indicates that a steady southern wind flux from the Earth surface up to altitudes of 5 km—which lasted nearly three weeks (the last 10 days of October to the first 10 days of November, 1981)—was caused by the presence of a low atmospheric pressure gradient and a pressure crest, directed from southern latitudes toward Transcaucasia and the Caspian Sea; this caused warm and dry air to be carried away (Figure 2.1).

During altitude flight of the airplane laboratories, optical surveying was performed over a rectilinear crossing of extension 250 km along the steady wind flux vector and general sea wave propagation. Surveying was performed on the IL-14 laboratory from 10 hr 36 min to 11 hr 19 min at an altitude of 400 m at a speed of 270 km/h, and on the AN-30 laboratory from 08 hr 58 min to 09 hr 46 min (track I) and from 10 hr 35 min to 11 hr 17 min (track II) at an altitude of 1500 and 5100 m, respectively, at a speed of 375 km/h. The scheme of crossings is shown by segment B

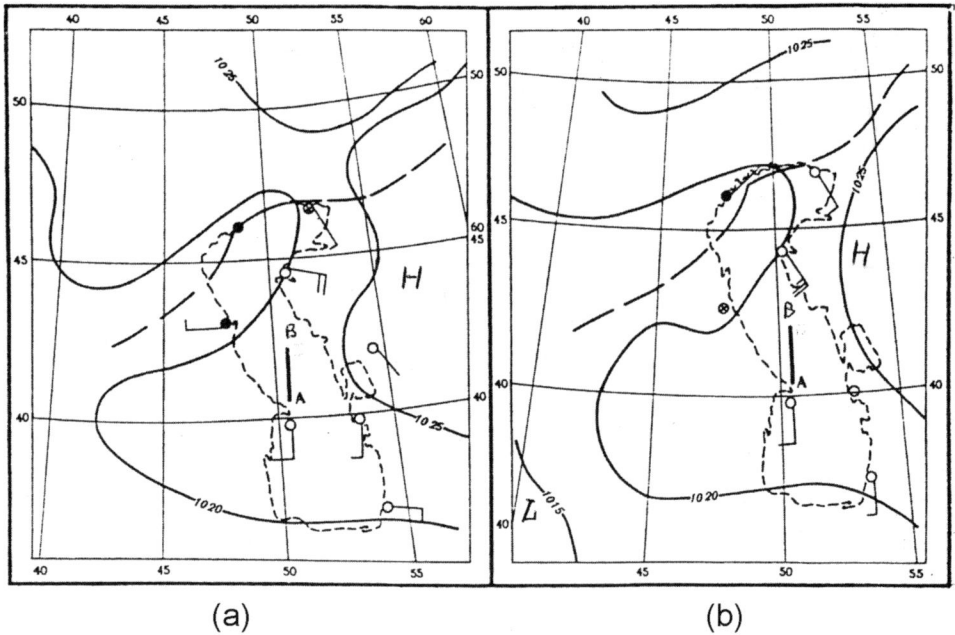

(a)                                          (b)

**Figure 2.1.** The near-surface synoptic situation over the Caspian Sea on October 31, 1981: (a) at 00:00 Moscow time; (b) at 09:00. Pressure is given in millibars.

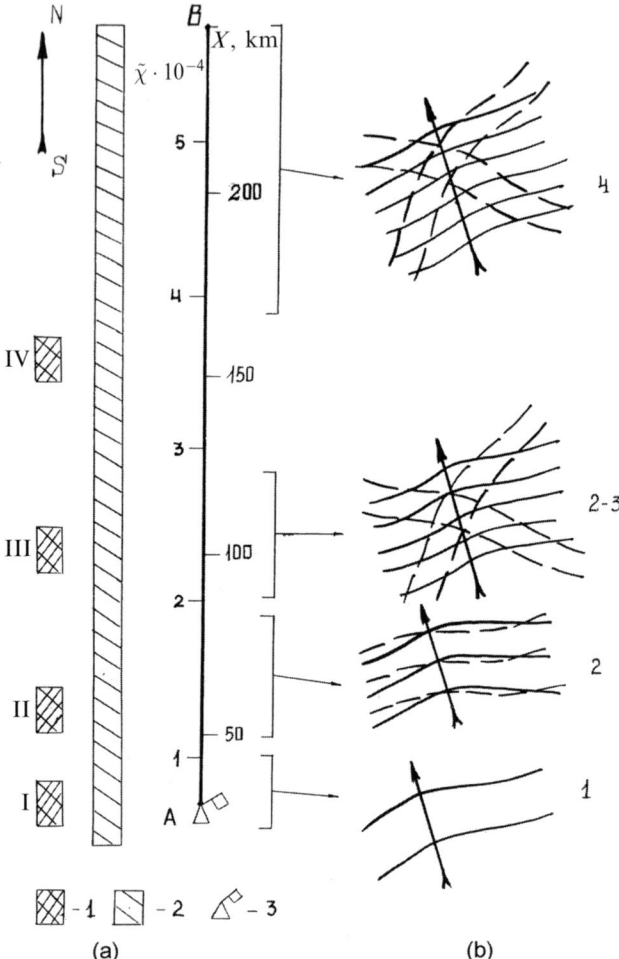

**Figure 2.2.** The schematic superposition of observing data by visual observation and by aerial photography: (a) areas with photo instruments AFA-TE-100 (1) and MKF-6M (2); oil platform with meteorological instruments and wave recorders (3). (b) Sea surface visual estimation by the onboard meteorologist. Fat arrows show the main direction of sea wind wave propagation. Solid and dashed lines show fronts for wind waves and swells correspondingly. Figures show the Beaufort notation (balls). $X$ is fetch ( km); $\tilde{\chi} = Xgu_{10}^{-2}$ is nondimensional fetch. I–IV are areas for the aerial photography. The geographical location of the airway AB is pointed at in Figure 2.1.

in Figure 2.1. Moreover, Figure 2.2 schematically presents the zones of synchronous optical survey by the AFA-TE-100 and KF-6 instruments (Figure 2.2a), as well as the data of visual observations of sea surface state, the direction of the prevailing wind, and the presence of swell fronts at various stages of sea wave development (Figure 2.2b).

According to the primary goal of the experiment, each track was flown over the oil installation by the weather station, where contact measurements of sea waves, as well as measurements of the air- and thermodynamic characteristics of a near-surface layer of the atmosphere and sea surface were carried out. Southward of the oil installation (at a distance of about 24 km) there was an extensive shallow 0.5 m deep (a sandbank called "Oil Rocks"), on which large gravity sea waves, formed in the water area of the Caspian Sea southward of the shallows, have been almost fully destroyed. The initial fetch point ($X = 0$) in Figure 2.2 corresponds to the geographical position of an extended shallow.

### 2.2.2   Technique for performing the contact part of the experiment and data processing

Wave oscillations of the sea surface were recorded from the oil installation by means of a two-string wavemeter, which allowed sea surface elevations to be recorded within the following limits: a wave height from 0.005 to 2 m and a period of time from 0.05 to 20 s. The error in wave height determination in the low-frequency range (0.05–2 Hz) was no greater than 1.5 cm, and in the high-frequency range (2–15 Hz) no greater than 0.5 cm. Processing of a wind wave time series for the purpose of obtaining sea wave spectra was performed by the computer. The significant interval of spectra estimations with a probability of 0.8 was equal to:

$$0.72S(\omega) < \hat{S}(\omega) < 1.8S(\omega) \tag{2.1}$$

where $\hat{S}(\omega)$ and $S(\omega)$ are, respectively, the estimate and the true power spectrum of sea waves; and $\omega$ is the circular frequency of gravity waves.

Thus, once observation point frequency spectra were obtained—for comparison with spatial spectra obtained by processing the optical spatial images of the sea surface—they were recalculated as one-dimensional spatial spectra using the isotropic dispersion relationship:

$$S(\omega) = \left. \frac{kS(k)}{\partial \omega(k)/\partial k} \right|_{k=\omega^2/g} \tag{2.2}$$

where $S(k)$ is the one-dimensional spatial spectrum; $\omega(k) = (gk)^{1/2}$; and $k$ is the wave number magnitude.

Simultaneously with measuring sea wave characteristics, a set of aerodynamic and thermal measurements of the characteristics of a near-surface layer of the atmosphere and sea surface were carried out from the oil installation (see Section 2.2.3).

### 2.2.3   The stratification state and the turbulent mode of a near-surface layer. The "net" fetch conditions

From analysis of the spatial scales of a synoptic atmospheric pressure situation (Figure 2.1) it can easily be seen that the size of the experimental range (200–250 km) is essentially smaller than the characteristic scales for the spatial variability of synoptic processes (2000 km). Thus, the near-surface layer and the Ekman boundary layer of the atmosphere can be considered as statistically stationary and horizontally homogeneous, and the single-point (from the oil installation) measurements of meteorological fields can be (to a certain approximation) spread into the entire flight trajectory.

To determine the state of a turbulent near-surface layer, the results of gradient measurements have been processed within the framework of the technique based on the semi-empirical Monin–Obukhov theory of a near-surface layer (Zilitinkevich *et al.*, 1978). A statistically stationary and horizontally homogeneous near-surface layer possesses the following characteristics: the vertical turbulent fluxes of a momentum (the stress of friction) $\tau$, of heat $H$, of vapor $E$, and the Monin–Obukhov scale $L$ are equal to the following numerical values: $\tau = 0.05 N\,\mathrm{m}^{-2}$; $H = -22.2 W\,\mathrm{m}^{-2}$; $E = 2 \pm 2.4 W\,\mathrm{m}^{-2}$. Vertical development of the near-surface layer is characterized by the Monin–Obukhov length scale $L$, which is equal under given experiment conditions to $L = 30\,\mathrm{m}$. Estimates of the value of dynamic velocity $u_*$ and of the scale of temperature pulsations $T_*$ were equal: $u_* = 0.23\,\mathrm{m\,s}^{-1}$ and $T_* = 0.37\,\mathrm{K}$. One should pay attention to a peculiar thermal regime where the turbulent flux of vapor is very small (in magnitude) and, moreover, strongly fluctuates both in time and in space (i.e., intermittent turbulence).

To estimate the degree of stability of the stratification mode, we shall evaluate the dynamic Richardson number $R_f$ by using the measured (average) values of wind velocity at the levels of 10 and 2 m, as well as the values of $\tau$, $H$, and $E$:

$$R_f = \frac{BoH/c_p + 0.61gE}{\tau(\partial u/\partial z)} \simeq 0.03 \qquad (2.3)$$

This relation indicates that the experiment was carried out under conditions of a prominent temperature inversion of a near-surface layer ($T_{10} - T_S > 0$), of a steady temperature stratification ($R_f > 0$), with a weak effect on vapor stratification (the Bowen relation value $Bo > 1$), which causes a peculiar turbulent mode of a near-surface layer with rather weak vertical development ($L \approx 30\,\mathrm{m}$). Turbulent pulsations are essentially suppressed by hydrostatic stability: dispersions of the pulsation components of velocities equal a value of about $u_*$, and the dispersion of temperature of about $T_*$; the fluxes of vapor and heat are directed from the atmosphere toward the sea surface, the thermal flux being nearly an order of magnitude greater than the flux of vapor. To describe the altitude profiles of temperature and mean wind velocity we can make use of the linear–logarithmic approximation. To estimate the degree of correlation of sea waves with the wind flux, we shall make use of the technique developed on the basis of processing a series of contact experiments including the international JONSWAP experiment. This technique includes consideration of two inequalities (Carter, 1982):

$$D > 1.167\frac{X^{0.7}}{u_{10}^{0.4}}; \qquad X < 2.32u_{10}^2 \qquad (2.4)$$

where $D$ (hr) and $X$ (km) are the time and spatial dimension of the wind flux interaction with the surface; and $u_{10}$ is the wind velocity at an altitude of 10 m from

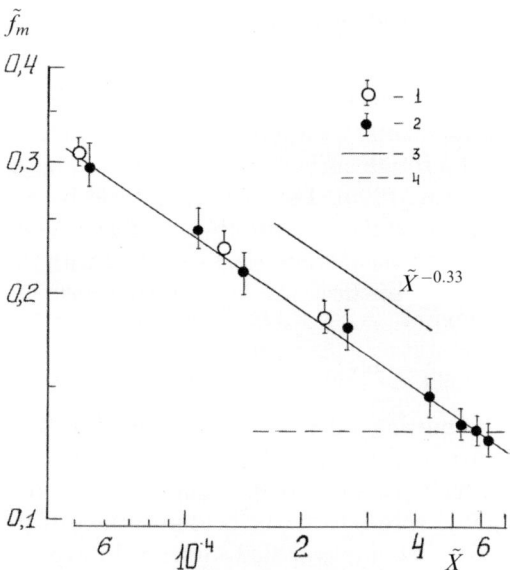

**Figure 2.3.** The experimental dependence of nondimensional frequency for maximum sea wave spectra $\tilde{f}_m = f_m u_{10} g^{-1}$ from nondimensional fetch $\tilde{X}$: (1) data of computer processing using restored elevation spectra; (2) data of computer processing of aerial photography frames using photo instument MKF-6M; (3) approximation $\tilde{f}_m = 5.02 \tilde{\chi}^{-0.33}$; (4) nondimensional frequency for the maximum of the Pierson–Moskowitz spectrum of a developed sea state.

the mean sea level. Satisfaction of the first inequality (which is really satisfied in our case: $D > 20$ hr, $u_{10} = 6.2$ m s$^{-1}$, $X > 250$ km) implies that the wind flux interaction with the surface has occurred in the mode of "spatial" (rather than "temporal") fetch. In this case—up to $X > 200$ km—limited fetch conditions were met (i.e., the second inequality was satisfied) with subsequent transition into the "full" fetch mode (i.e., mature wave fields) at $X > 250$ km. The statement of theory of spatial evolution of windy sea waves is proven by the experimental dependence of the dimensionless frequency of the one-dimensional spectrum maximum ($\tilde{f}_m$) on the dimensionless fetch (Figure 2.3). This dependence, obtained by Grushin et al. (1990), is well approximated by the exponential dependence $\tilde{\chi}^{-0.33}$, which is well known from the data of some experiments (Carter, 1982). In the given case the numerical coefficient is about 40% higher than that accepted in early approximations.

Of interest is the fact that under full fetch conditions ($X = 250$ km) the $\tilde{f}_m$ value almost precisely corresponds to the dimensionless frequency value obtained from the Pirson–Moskowitz spectrum of a developed sea state ($\tilde{f}_m = 0.13$) (Pierson and Moskowitz, 1964). It can easily be seen that the wave age relations $c/u_* \geq 30$ are also satisfied for $X > 200$ km (Papadimitrakis, 2005a,b) (where $c$ is the phase velocity of a gravity wave corresponding to the spectrum maximum frequency), and for the dimensionless fetch the inequality $Xg/u_*^2 > 4 \cdot 10^7$ is met, which characterizes a fully developed sea state according to the data by Volkov (1968).

### 2.2.4   Restoration of the spectral characteristics of the sea surface from its optical images

To restore the spatial spectral characteristics of the large-scale sea wave field through-out the crossing of measurements, we applied the special technique for restoring the spectral characteristics of a rough surface from its optical images obtained by aerial survey (Lupyan and Sharkov, 1990). The blackening field by observation point of a rough surface's optical image (OI) on a photo-carrier $D(x, y)$ can be considered, in the most general case, as a stochastic function associated with the field of altitudes of the surface under study. This function has the following form (Kazevich *et al.*, 1972; Lupyan and Sharkov, 1990):

$$D(x, y) = P[\eta(x, y)] = P_F P_I P_A[I[\eta(x, y)]] \qquad (2.5)$$

where $P$ is the full transmitting operator consisting of $P_F$, $P_I$, $P_A$—transmitting operators of a photo-carrier (film), camera, and atmosphere, respectively; $I$ is the optical flux intensity, which passes into the chamber from the surface point with coordinates $x, y$; and $\eta(x, y)$ is the stochastic field of surface displacements. For some approximations (which can be implemented under field experiment conditions) we can find the restoring operator $P_S$, such that

$$\frac{P_S\{S_D[D(x, y)]\} - S_Z[\eta(x, y)]}{S_Z[\eta(x, y)]} < \varepsilon \qquad (2.6)$$

under the condition that $k \in (k_{\min}, k_{\max})$. Here $S_Z$ is the power spectrum of sea surface displacements, $S_D$ is the power spectrum of the field of brightness, $\varepsilon$ is the *a priori* specified number characterizing the restoration error. Naturally, this inequality is not met for any $\varepsilon$; however, as model calculations have shown, for various illumination situations we can find such conditions of surveying from the fixed altitude and that the given condition will be satisfied for $\varepsilon \sim 0.05$–$0.1$ for the range of wavenumbers we are interested in ($0.5$–$30\,\mathrm{m}^{-1}$). The scheme, according to which the given estimates were obtained, is as follows. We supposed that the following conditions were satisfied: (1) atmospheric distortions are absent; (2) the scheduled survey is performed with $x$ and $y < H$, where $H$ is the survey altitude; (3) the photo-carrier blackening range lies in the region of normal exposures; (4) the facet model is valid. Then the initial operator can be presented in the form of

$$P = A \log[I(\alpha, \beta)] + C \qquad (2.7)$$

where $\alpha = \partial \eta(x, y)/\partial x$; $\beta = \partial \eta(x, y)/\partial y$; $A$ and $C$ are constants; and $I$ can be presented as

$$I = R(\gamma) L_H(\theta, \phi) + [1 - R(\gamma)] L_M \qquad (2.8)$$

where $R(\gamma)$ is the Fresnel reflection coefficient (in power); $L_H(\theta, \phi)$ is the sky illumination function; $L_M$ is the under-surface flux background; angles $\gamma$, $\theta$, $\phi$ are related to the small slopes $\alpha$, $\beta$ by well-known relationships (Kazevich *et al.*, 1972). To estimate the basic factors influencing restoration accuracy, we shall make use of empirical expressions $L_H(\theta, \phi)$ for two limiting cases: a clear cloudless sky and dense homogeneous cloudiness. Then, after appropriate simplifications, we can show that $D(x, y)$ is representable in the following canonical form (Lupyan and Sharkov, 1990):

$$P = \sum\sum A_{nm}(x, y)(\partial/\partial x)^n(\partial/\partial y)^m \tag{2.9}$$

In this case it can be shown that by neglecting the quadratic and higher-order terms in $P$, as well as imposing the constancy condition (i.e., independence of $x, y$) on coefficients $A_i$ ($i = 0, \alpha, \beta$), we can narrow the range of wavenumbers ($k_{min}, k_{max}$) of studied surface roughness, in which the spectrum of altitudes can be correctly restored from the optical image. On the other hand, these simplifications allow us to find the analytical relationship between the spectrum of sea surface displacements $S_Z$ and the field of brightness $S_D$

$$S_Z = P_S[S_D(k_1, k_2)] \tag{2.10}$$

where

$$P_S = [S_D(k_1, k_2)]/A_\alpha k_1^2 + A_\beta k_2^2 \tag{2.11}$$

where $\mathbf{k}(k_1, k_2)$ is the wavenumber. These relations are valid within the range of spatial frequencies $k \in (k_{min}, k_{max})$, whose boundary values are related to the specific values of errors in restoring the spectrum of altitudes, which (the errors) in their turn inevitably arise as a result of using the restoring operator $P_S$. A more rigorous derivation of optimum conditions for restoring the spectral characteristics of a rough surface from its optical image is given in the paper by Lupyan and Sharkov (1990).

The technique for processing negative optical pictures of the sea surface with the purpose of obtaining two-dimensional spatial spectra of the sea surface was implemented in a paper by Grushin *et al.* (1990). This technique consists of the following stages:

(1) The choice of fragment on a negative picture of the sea surface should satisfy the following conditions: (a) at all its points the blackening of negative $D$ should lie in the normal exposure area (i.e., at places, where $D \sim \lg I$); (b) the maximum geometrical size of the fragment ($L$) can be obtained from the estimation relation $L \sim 5(2\pi/k_{min})$, where the second multiplier determines the maximum spatial harmonic of the rough surface studied, restored with the specified relative error $\varepsilon$; in our case $\varepsilon \sim 10\%$, which just corresponds to $k_{min} \sim 0.3\,\mathrm{m}^{-1}$. The minimum spatial harmonic of the surface $2\pi/k_{max}$ is determined (under the given surveying conditions) by the spatial resolution $\Delta x$ of the "airplane–camera–film" system. In our case, because "smearing" was incompletely compensated for, $\Delta x$ was equal to 0.4–0.5 m, which finally determined the value of $k_{max} \sim 10$–$11\,\mathrm{m}^{-1}$; (c) the geometrical center of the fragment under study was situated on the line of the surface plane intersection with the vertical plane making an azimuth angle $\pm\pi/4$ to the plane of observation—that is, to the plane in which observation points (airplane) and the source (Sun) are situated. By virtue of problematic symmetry in the image fragment, situated as described above, the approximate equality $A_\alpha \approx A_\beta \approx A = \mathrm{const.}$ is fulfilled, and in this case

$$S_Z(k_1, k_2) \sim AS_D(k_1, k_2)/k^2 \tag{2.12}$$

where $k^2 = k_1^2 + k_2^2$, and the angular distribution of energy in $S_D$ and $S_Z$ coincides, and characteristic features of this distribution can be identified directly on $S_D$ without transforming it.

(2) To get the optical picture fragment ready for the computer, it was digitized and processing was performed by means of software for spectral processing of two-dimensional signals.

(3) The spectrum obtained underwent the following operations:

(a) averaging over the azimuth angle, which corresponded to the integrated operation of obtaining a single-point spatial spectrum

$$S_Z(|\bar{k}|) = 2 \int_0^\pi S_Z(|\mathbf{k}|, \theta) \, d\theta \tag{2.13}$$

(b) averaging of the initial spectrum $S_Z(|\mathbf{k}|, \theta)$ by means of a sliding window of dimension $5 \times 5$ pixels, which ensured improvement of spectrum estimate up to a relative error lower than 20%;

(c) for detailed analysis of spectral features the linear and circular sections of initial spectra were made and printed out. This operation corresponded to obtaining the sections of a spatial spectrum at fixed angle $S_Z(|\mathbf{k}|, \theta)$ and of angular spectra at fixed spatial frequencies $S_Z(|\mathbf{k}|, \theta)$. As a rule, two linear sections were fulfilled (along the main and perpendicular directions of propagation of a dominating wave component) and three to four circular sections at various radii (i.e., for various wavenumbers).

(4) To increase the range of spatial frequencies, the initial image was compressed both twice and four times. Thus, the final Wiener spectrum covered the range of spatial frequencies from $0.15\,\mathrm{m}^{-1}$ to $10\,\mathrm{m}^{-1}$, which corresponded to the range from 0.7 to 40 m in the lengths of surface waves.

As a result of full processing, the visualized two-dimensional spectra of sea surface altitudes were obtained, on which these features could be emphasized by artificial colors and, thus, made suitable for visual perception. Moreover, by obtaining one-dimensional sections of spectra as well as spectra averaged over the angle, we were able to compare them with the results of contact (single-point) experiments, as well as with the conclusions of theoretical analysis, which in the overwhelming majority of modern works was carried out for one-dimensional spectra.

### 2.2.5   Spatial–spectral structure of sea waves

Four fragments of negative images—for one of the four surveying sets I–IV (Figure 2.2)—obtained by the AFA-TE-100 instrument were subjected to digital processing. The results of digital restoration of the $S_Z(|\mathbf{k}|)$ fragment of an optical image from setI, carried out directly over the oil installation (at $\tilde{X} = 5.3 \cdot 10^3$), will be analyzed below. This makes it possible to compare the restoration of $S_Z(|\mathbf{k}|)$ with contact measurements of $S_Z(\omega)$, which is key to the given problem. Both types of measurements—contact and remote—were performed with a high (for such experi-

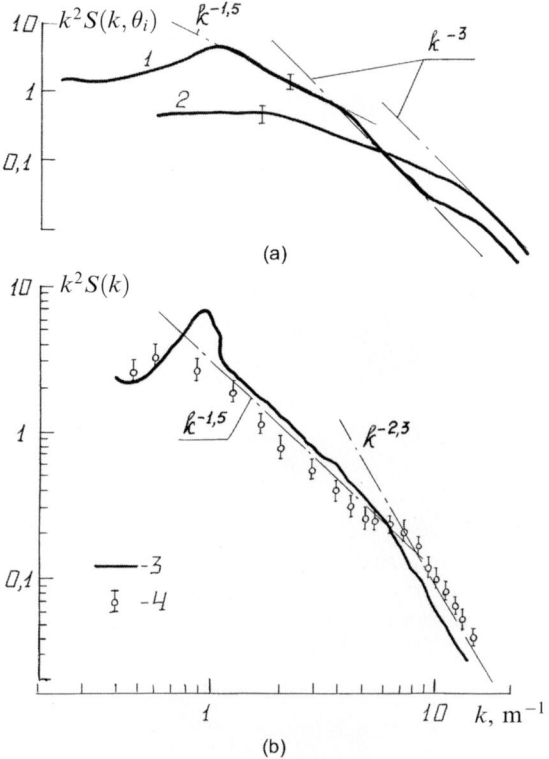

**Figure 2.4.** One-dimensional spectra of sea surface displacement using *in situ* and optical measurements: (a) recovered one-dimensional spectra $S(k, \theta_i)$ with $\theta_i = 0°$ (1) and with $\theta_i = 90°$ (2) relative to the general direction of the propagation of waves; (b) one-dimensional spectrum $S(k)$ (3) and single-point spectrum (4), measured by *in situ* instruments. Spectrum $S(k)$ was fixed to *in situ* measurements at $k = 6.5 \, \text{m}^{-1}$. Confidence intervals with 80% probability are marked by vertical line segments.

ments) degree of time synchronism (i.e., to an accuracy of a few minutes). Such a time divergence could not naturally cause any noticeable change in the sea disturbance structure.

It can easily be seen (Figure 2.4) that the azimuth characteristics of the spectrum of the sea surface elevation field $S_Z(|\mathbf{k}|, \theta)$ have a complicated structure in the installation area—the prominent azimuth anisotropy of a dominating wave component is changed by the noticeable isotropy of short-wave components. And, moreover, analysis of linear sections of spatial spectra throughout the range of wavenumbers studied after the "mosaicking" operation (Figure 2.4a) shows that high-frequency components ($k > 6 \, \text{m}^{-1}$), propagating in the direction perpendicular to the wind flux, dominate over components collinear to the wind velocity vector. All these spatial features cannot be detected by in situ measurements, and comparison can be carried out only for a single-point spectrum averaged over the azimuth angle

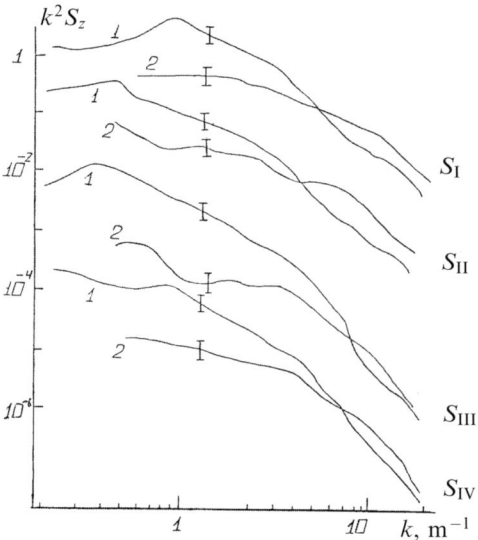

**Figure 2.5.** Sections along (1) and across (2) the main direction of two-dimensional spatial spectra $S_I$, $S_{II}$, $S_{III}$, $S_{IV}$, obtained as sea state increases. Symbols I–IV correspond to a series of photographs by AFA-TE instruments (see Figure 2.2). Spectra $S_{II}$–$S_{IV}$ were normalized on the following basis: $S_{II}(150)^{-1}$; $S_{III}(150)^{-2}$; $S_{IV}(150)^{-3}$.

$S_Z(|\mathbf{k}|)$. Analysis of Figure 2.4—which presents the results of contact measurements, transformed into the space of wavenumbers, and the single-point spectrum of altitudes $S_Z(|\mathbf{k}|, \theta)$, restored from optical measurements (at an accuracy of a constant multiplier)—shows that the latter not only correctly reflects the general spectral dependence throughout the range of wavenumbers studied of about six octaves, but also coincides with the contact spectrum in details. For example, the break in spectral dependence for $k = 6$–$7\,\mathrm{m}^{-1}$ is present on both compared spectra.

The results of processing all sets of optical images of the sea surface are presented in Figure 2.5 in the form of spatial spectra sections along and across the wavevector of the sea disturbance direction. Analysis of this figure indicates that, as the fetch increases, sea waves develop rapidly. And in this case, along with decreasing of $k_m$ (the spectral maximum frequency), spatial spectrum asymmetry grows in the spectral maximum area.

Of doubtless interest is the fact that—from the spatial spectra obtained—we can distinguish two characteristic sub-ranges of wavenumbers, in which spectra possess very certain regularities:

- Region I—for $k > k_1 = 4$–5 (or, in dimensionless frequencies, $\tilde{k}_1 = (k_1 u^2/g) = 15$–19), $S_Z \sim k^{-(4-4.2)}$—the spectrum possesses azimuth isotropy during the whole development interval (i.e., Phillips' interval) (Phillips, 1977).
- Region II—for $k_1 > k > k_m - S_Z \sim k^{-(3-3.3)}$—the region with a prominent azimuth anisotropy and generated due to weak nonlinear interactions (the so-called spectrum with a constant flux of action) (Zacharov and Zaslavskii, 1982).

The turbulent ideology of forming the spectrum (Zacharov and Zaslavskii, 1982) of the sea surface gives rise to estimation of the lower boundary of the pumping region as $\tilde{k} \sim 4$–6. It follows from the data considered that where energy is pumped from the wind field, the disturbance spectrum is anisotropic even at the earliest stages of development, the growth of anisotropy being intensively continued for $\tilde{k} \sim 4$–5 as well.

Thus, detailed analysis of the atmospheric pressure situation and of the state of turbulence of the near-surface layer—presented in Section 2.2.3—shows that wind flux interaction with the water surface has occurred under net fetch conditions with a strongly prominent, stable temperature stratification of a near-surface layer and, respectively, with turbulent pulsations suppressed by hydrostatic stability.

Under conditions of developed sea disturbance—characterized by the following values of interaction parameters: dimensionless fetch $\tilde{X} > 6 \cdot 10^4$, dimensionless frequency of the sea disturbance spectrum maximum $\tilde{f}_m \approx 0.13$, the wavelength of an energy-carrying component $\Lambda = 42$ m—the set of surveys with the MKF-6 instrument was carried out at a carrier (AN-30) altitude of 5100 m and with the surveying scale of M = 1:40800. Surveying was also carried out from the airplane laboratory IL-14 at an altitude of 400 m and with a surveying scale of M = 1:40800. The parameters characterizing the degree of sea disturbance and the length of fetch at performing each of the four surveying sets (I–IV) (see Figure 2.2) are presented in Table 2.3.

The spatial spectra used when surveying zones I–IV, restored by the technique of Section 2.2.4, possess a rather complicated structure related to features of the nonlinear interactions of waves in various spectral ranges (see Section 2.2.5). However, to an accuracy satisfactory for the purposes of our investigation (i.e., that of a constant coefficient), they can be approximated by the following expressions:

$$S_Z(k, \varphi, X) = \begin{cases} 0 & k < k_m \\ BXk^{-10/3} \cos^4 \varphi & k_m < k < k_1 \\ k^{-4} & k_1 < k < k_2 \\ 0 & k_2 < k \end{cases} \tag{2.14}$$

where $k$ is the magnitude of a wavevector of the stochastic displacements of the surface; $\varphi$ is the azimuth angle measured from the general direction of an energy-carrying component $(k_m = k u_{10}^2 g^{-1})$; $k_1$ corresponds to the sea disturbance component, above which the spectrum acquires a nearly isotropic character $(k_1 \sim 6 \text{ m}^{-1})$; $k_2$ is the high-frequency cutoff of a spectrum $(k_2 \sim 1 \text{ cm}^{-1})$.

## 2.3  SPATIAL–STATISTICAL PROPERTIES OF A BREAKING WAVE FIELD IN DEVELOPED SEAS

This section describes the techniques used in processing and the results of natural optical observations with the purpose of obtaining information on the laws of distribution of the specific spatial density of breaking centers $N(t)$ and on the spatial correlation features of the same point field (the independence of centers, the homogeneity of the field) (Pokrovskaya and Sharkov, 1986, 1987a).

### 2.3.1  The technique of formation and processing of a random-point field

Aerial pictures, taken in field experiments, have been grouped into separate ranked sets, where—for the purpose of providing statistical independence of sets—only non-overlapping image pictures were analyzed. For the purposes of effective detection and identification of foam structures the highest contrast channel of the MKF-6 instrument was used, whose negative pictures were magnified by as much as five times. In view of the presence of a strong brightness background on pictures from solar patches of light (zone B of the picture in Figure 2.6), only those parts of working pictures have been analyzed that have definitely contained foam structures (zone A of the picture in Figure 2.6). Disregarding the detailed structure of an individual wave breaking, we shall depict the foam structure apparent from breaking by a point on the plane of a paper transparency. In this way the limited section of the sea surface is presented on a transparency by a set of points (centers) representing the arrangement

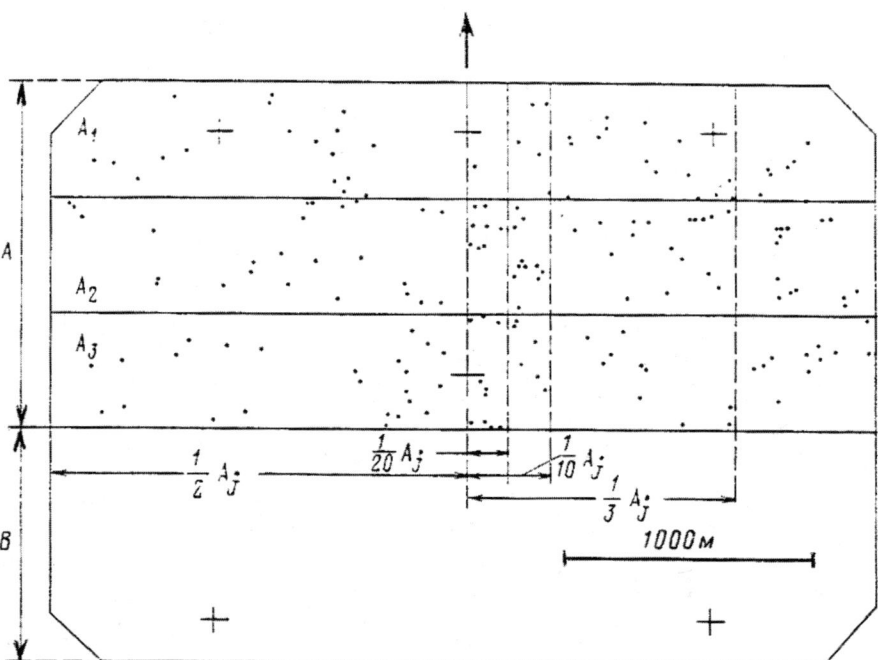

**Figure 2.6.** Aerial photography frame #4319 of instrument MKF-6 on October 31, 1981 involves a random field of breaking centers. The working zone of this frame is marked by $A$. The sub-frames $A_1, A_2, A_3$ are the main spatial elements of the sample (zone #1). The zone is the sun glitter area. The zones $(1/m)A_i$ ($m = 2, 3, 10, 20$) are the $m$th part of main zone #1 with area $S_0 = 1.49\,\text{km}^2$. The big arrow shows the geographic orientation (N–S).

of breaking waves of the water area at a fixed instant of time. Analysis has shown that foam structures with linear dimensions lower than $2\,\text{m}$ have been detected on negatives with a low degree of reliability and, for this reason, have not been transferred to a transparency.

The spatial working field A of size $3264 \times 457\,\text{m}^2$ (Figure 5.1) was simply chosen in our investigations as an initial spatial element ($S_0$ with an area of $1.49\,\text{km}^2$) of the basic ranked set consisting of 39 elements (13 non-overlapping images taken by the MKF-6 instrument). On the basic spatial field (zone $k = 1$) the zones were distinguished ($k$ is the number of a studied zone, $k = 1, \ldots, 11$) by areas that were reduced with respect to the basic field ($k = 1$) by a factor of 2, 3, 10, 20, 40, 80, 160, 320, 640, and 1280 (see Table 5.1). Further, for all studied zones the quantities of individual foam structures ($N$) were calculated, and the ranked sets $N_{ij}^k$ were formed for the studied sea water area in the form of a matrix field, where subscript $i$ (which means the "line", conventionally) corresponds to the spatial direction "west–east" across the general direction of sea waves; subscript $j$ (which means the "column", conventionally) corresponds to the "south–north" direction along the general direction of sea waves; $k$ is the number of the zone studied ($k = 1, \ldots, 5$), where $i$ runs through the values from 1 to $k$ for $k = 1, 2, 3$, and from 1 to 10, 20 for $k = 4, 5$. The transparency example, with the random field of centers and geometry of the picture subdivided into basic spatial elements (zones), is presented in Figure 2.6. We shall consider $N_{ij}^k(x_j)$ to be a whole-number random flux—that is, the random number of indiscernible centers that have appeared in a spatial window of the $k$th zone with a spatial coordinate $x_j = \Delta x_j$, where $\Delta x = 457\,\text{m}$ and $0 < x < x_1 = 17.8\,\text{km}$. Proceeding from this concept, we shall consider the spatial characteristics of the matrix field $N_{ij}^k(x_j)$. According to the terminology of stochastic process theory (e.g., Bharucha-Reid, 1960; Feller, 1971), the field $N_{ij}^k(x_j)$ is in essence the field of increments of a whole-number stochastic process.

### 2.3.2   Laws of specific density distribution

For each rank-order set, histograms were constructed that represent the statistical analogs of probability densities in accordance with well-known rules (Bendat and Piersol, 1966)

$$P^k(N^k) = \frac{1}{Q} \sum_{m=1}^{M} \frac{n_m(N^k)}{h_m} \tag{2.15}$$

where $n_m(N^k) = n_m(N^k)$ for $N_m^k < N^k < N_{m+1}^k$ and $n_m(N^k) = 0$ for $N_{m+1}^k < N^k < N_m^k$; $h = N_{m+1}^k - N_m^k$ is the grouping interval; $Q = i \cdot j$ is the full volume of a rank-order set for each zone; $M$ is the number of grouping intervals. For each ranked set the following quantities were calculated: ranked mean values $\bar{N}^K$, variances (unbiased estimates) $D = (\sigma^K)^2$, the coefficients of asymmetry $\gamma_1$ and of excess $\gamma_2$. The results of processing are presented in Table 2.1 and Figure 2.7a–d.

The experimental histograms of zones $k = 4$–11 differ highly from the Gaussian distribution—especially, in the asymmetry coefficient value—which is clearly seen in

**Table 2.1.** Distribution parameters for the specific density of breaking centers.

| Zone number $k$ | Sampling area | Sampling volume | Numerical parameters of distributions | | | | Approximate law type | Main parameters of approximate law |
|---|---|---|---|---|---|---|---|---|
| | | | Mean | Variance | Asymmetry coefficient | Excess coefficient | | |
| 1 | $S_0$ | 39 | 149 | 2916 | −0.13 | −1.07 | ND | $\bar{N} = 149;\ \sigma = 54$ |
| 2 | $(1/2)S_0$ | 39 | 73 | 676 | −0.06 | −0.95 | ND | $\bar{N} = 73;\ \sigma = 26$ |
| 3 | $(1/3)S_0$ | 39 | 52 | 441 | 0.17 | −1.0 | ND | $\bar{N} = 52;\ \sigma = 21$ |
| 4 | $(1/10)S_0$ | 390 | 15.2 | 75.69 | 0.65 | 0.017 | BD | $p = 1.02 \cdot 10^{-1};\ N_0 = 149$ |
| 5 | $(1/20)S_0$ | 780 | 7.6 | 25 | 0.79 | 0.45 | BD | $p = 5.1 \cdot 10^{-2};\ N_0 = 149$ |
| 6 | $(1/40)S_0$ | 840 | 3.5 | 9 | 1.07 | 1.01 | BD | $p = 2.3 \cdot 10^{-2};\ N_0 = 149$ |
| 7 | $(1/80)S_0$ | 920 | 1.7 | 3.5 | 1.2 | 1.89 | NBD | $p = 0.49;\ n = 1.65$ |
| 8 | $(1/160)S_0$ | 1920 | 0.86 | 1.7 | 1.09 | 2.5 | GD | $p = 0.575$ |
| 10 | $(1/320)S_0$ | 960 | 0.396 | 0.62 | 1.8 | 3.1 | GD | $p = 0.71$ |
| 11 | $(1/640)S_0$ | 1920 | 0.168 | 0.199 | 3.9 | 17 | GD | $p = 0.857$ |
| 12 | $(1/1280)S_0$ | 3940 | 0.0862 | 0.093 | 3.4 | 15.0 | GD | $p = 0.92$ |

*Comments*: ND is the normal distribution; BD is the binomial distribution; NBD is the negative binomial distribution; GD is the geometric distribution; and $S_0 = 1.49\ \mathrm{m}^2$.

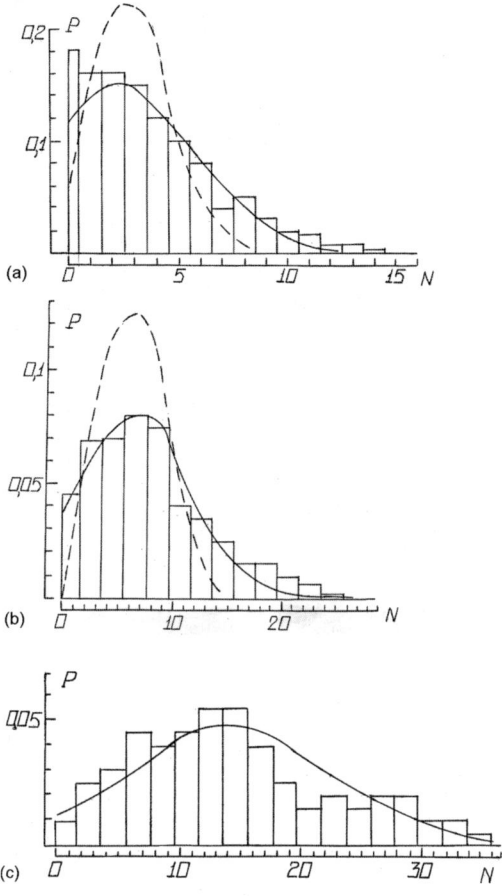

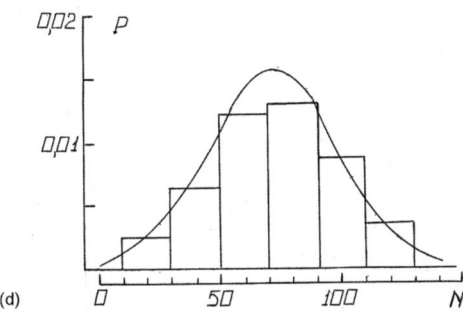

**Figure 2.7.** Experimental histograms of foam structure densities and theoretical distributions within spatial windows: (a) $\frac{1}{40}S_0$; (b) $\frac{1}{20}S_0$; (c) $\frac{1}{10}S_0$; (d) $\frac{1}{2}S_0$. The solid curves show the binomial distribution on diagrams (a) and (b) and the normal distribution on diagrams (c) and (d); the dashed lines on diagrams (a) and (b) show the Poisson distribution with parameters obtained during processing corresponding to the histograms (see Table 2.1).

Figure 2.7a, b. As the area of the picture studied (i.e., zone $k = 6$–$1$) grows, the experimental histograms approach the Gaussian distribution. The process of transition to the normal distribution is clearly demonstrated by a set of diagrams in Figure 2.7a–d. Analysis of Table 2.1 and Figure 2.7 indicates that, as the area of a studied zone of the picture increases, the Gaussian component begins to dominate in the character of distributions of $N^k$. This component is first revealed as a binomial distribution ($k = 6, 5, 4$) and then ($k = 3, 2, 1$) as a purely normal distribution, which corresponds in the given case to Wiener's stochastic process with independent increments (Feller, 1971). In accordance with our viewpoint on the proximity of a spatial field of foam activity to random point processes, we shall consider how experimental histograms accord with two fundamental distributions of the theory of random point processes (i.e., with the binomial and Poisson types).

The Bernoulli trials, used to describe the binomial process, can be interpreted in our case (Pokrovskaya and Sharkov, 1986) as follows: by "event A" we mean the appearance of an individual foam structure—at constant probability $p$ in a studied spatial picture—that occupies a partial area of the basic element $S_0$; and by independent trials we mean $N_0$—the total number of foam structures on the water area of the basic spatial element $S_0$. Then, the total number of foam structures $N$ (i.e., the events with outcome A in the studied picture $N^k$) represents a stochastic whole-number quantity $\mu$, which obeys the binomial distribution:

$$P^k\{\mu = N|N_0, p\} = C_{N_0}^N p^N (1-p)^{N_0-N} \tag{2.16}$$

where $N = 0, 1, \ldots$; with parameters $N_0$ and $p$, where $N = N_0 p$; $\sigma^2 = N_0 p(1-p)$, and the numerical values of $N_0$ and $p$ are determined by the physical state of surface and meteorological conditions (the number of sea waves, the value of fetch, the velocity of wind, etc.), $C_{N_0}^N$ is the binomial coefficient. Since in our case $N_0 \gg 1$, we have chosen the approximate Moivre–Laplace formula as the reference estimate used to construct theoretical distributions:

$$P^k\{\mu = N|N_0, p\} = \frac{1}{\sqrt{N_0 p(1-p)}} \Phi\left\{ \frac{N - \bar{N} + 0.5}{\sqrt{N_0 p(1-p)}} \right\} \tag{2.17}$$

for $N \geq 0$ and $P^k = 0$ at $N < 0$, where $\Phi$ is the normal distribution density. The results of calculations for $k = 4, 5, 6$ (where the approximating law is the binomial distribution), using the experimental parameters given in Table 2.1, are presented in Figure 2.7a, b. For the picture areas studied from $\frac{1}{3}S_0$ to $S_0$ (the numbers of zones are $k = 1, 2, 3$) the most acceptable approximating law was, as expected, the normal distribution (see Table 2.1 and Figure 2.7c, d). In accordance with Pierson's criterion $\chi^2$ (Bendat and Piersol, 1966): for zone $k = 6$ (Figure 2.7a) the value $X^2 = 20$; for zone $k = 5$ (Figure 2.7b) $X^2 = 16.38$, whereas the 1st percentile critical value of a unilateral criterion $\chi^2(9; 0.99) = 21.67$ in this case; for zone $k = 4 - X^2 = 11.1$ and $\chi^2(15; 0.95) = 25.0$; and for zone $k = 2$, $X^2 = 0.2$ and $\chi^2(3; 0.95) = 7.8$. Thus, for all considered zones, hypotheses on the proximity to the binomial ($k = 6, 5, 4$) and to the normal ($k = 1, 2, 3$) distribution can be accepted in accordance with Pierson's criterion.

In the theory of random point processes the Poissonian process (the event flow) plays a fundamental part, being a limiting one for the binomial process with $N_0 p = $ const. and $N_0 \to \infty$. It is important to find the degree of proximity of Poissonian distributions to experimental histograms. However, even a simple external comparison (see Figure 2.7a, b) indicates that for zones with $k = 5, 6$ the hypothesis on the Poissonian character of events cannot be accepted. The non-Poissonian character of a breaking field for the spatial dimensions studied (areas from 0.037 to 1.49 km$^2$) is most likely associated with the fact that the probability of appearance of an individual formation $p$ is constant in some part of region $S_0$, where $p \approx S_0^K / S_0$, and, as the intensity of a process (e.g., sea disturbance state) increases, the mean density $N$ grows in proportion to $N_0 (N \sim N_0)$.

The basic physical question consists, in the given case, in finding the approximating law of distribution for zones with a small ($k = 7$–11) area, for which the character of the general statistics of the process reveals itself in the sharpest form. Figure 2.8 presents the experimental histograms of whole-number values of a specific density of breaking waves for zones with $k = 7, 8$—with picture areas and linear dimensions of

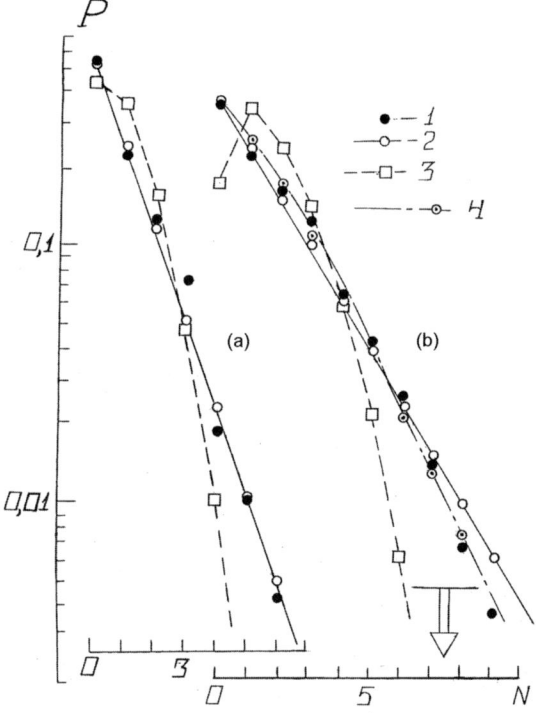

**Figure 2.8.** Experimental histograms of breaking wave densities (1) for zones $\frac{1}{160} S_0$ (a) and $\frac{1}{80} S_0$ (b). Approximated theoretical distributions in the form the whole-number values and their envelopes: 2—geometrical distribution with parameters $p = 0.575$ (a) and 0.35 (b); 3—Poisson distributions with parameters $\lambda = 0.86$ (a) and 1.7 (b); 4—negative binomial distribution with parameters $p = 0.49$, $n = 1.65$ (b).

$0.018 \, \text{km}^2$ (136 m) and $0.0093 \, \text{km}^2$ (96 m), respectively—as well as the approximating theoretical distributions of several types: Poisson, binomial, negative binomial, and geometrical (for the convenience of analysis the histograms are presented in a semilogarithmic scale). Even a simple external comparison (Figure 2.8) is enough to convince us that the hypothesis about the purely Poisson and binomial character of processes for zones with a small area (spatial pictures with $k = 7$–11) disappears. And, in this case, with 0.95 probability we may accept the hypothesis about a gamma-density-randomized Poisson distribution in the form of a negative binomial distribution (the mixed Poisson distribution) for $k = 7$. And for $k = 8$–11 a purely geometrical distribution (the Yule–Farrey distribution) (Feller, 1971; Johnson and Leone, 1977) can be accepted. Physically, this can be interpreted as follows: the mean value of the Poisson distribution (in other words, the number of wave breakings in the given frame of observation) is a stochastic, rather than constant quantity, which possesses, in its turn, a gamma-type statistical distribution. Thus, the Bernoulli scheme for interpreting the results of small frames needs to be complicated by inclusion of additional conditions—namely, the condition that experiments continue until event A occurs $n$ times exactly. And, if we take $n = 1$, then we obtain a particular case of negative binomial distribution known as the geometrical distribution (Johnson and Leone, 1977). In the case under investigation this fact implies that the event (the breaking in a frame) occurs only once. It follows from this fact that, as the frame area decreases ($k = 7$–11), the negative binomial distribution transfers into the geometrical distribution, which is apparent in the data presented in Figure 2.8 and in Table 2.1. Thus, the size of an accidental field of view of a remote-sensing instrument plays a key part in forming a statistical model for the breaking field.

To prove the homogeneity of a stochastic process $N$, histograms were constructed in which the data were acquired from the frames taken in various parts of the water area studied. And, all of them yielded results similar to those presented above—so, histograms coincided to an accuracy lower than 10%.

### 2.3.3    Spatial homogeneity and representativeness of a breaking field

To study the spatial features of the statistical characteristics of a gravity wave breaking field and, in particular, of the specific density of breaking centers, the following ranked characteristics were calculated:

(a) the specific density distribution function $F$ depending on the spatial coordinate $x_j$:

$$F^K(x_j) = \sum_{l=1}^{j} N_l^K \left( \sum_{l=1}^{39} N_l^K \right)^{-1} \tag{2.18}$$

where $x_j = \Delta x \cdot j$; in this case the density function (its derivative) of $F^K(x_j)$ determines, in its physical sense, the degree of homogeneity $N^K$ for the field of centers in a space;

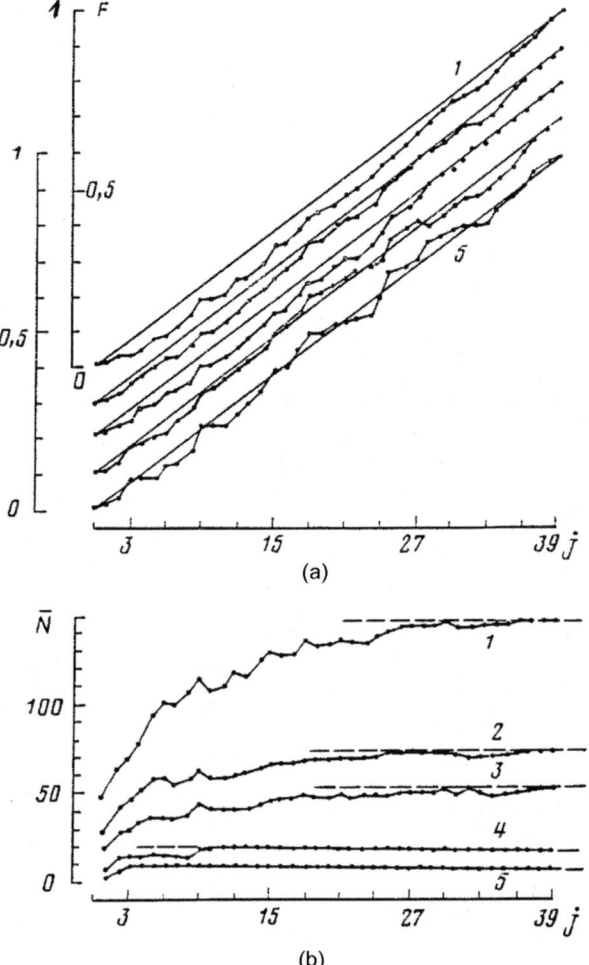

**Figure 2.9.** Experimental spatial dependences of the surface density distribution function (a) $F^K(x_j)$ and of the current value (b) $N^K(x_j)$ as a function of the coordinate $x_j = \Delta x_j$ for zones 1–5. The figures near curves correspond to zone numbers.

(b)  the current value of specific density $N^K(x_j)$ depending on the spatial coordinate $x_j$:

$$N^K(x_j) = (1/\Delta x \cdot j) \sum_{l=1}^{j} N_1^K \tag{2.19}$$

which characterizes, in some sense, the spatial representativeness of a ranked value $N^K(x_j)$ in relation to $N^K$ of the general set. In constructing spatial dependencies the value of $i$ was chosen for zone 1–3 as $i = 1$, and for zones 4 and 5 as $i = 6$ and 11. The results of calculations are presented in Figure 2.9. From analysis of these data we can draw the conclusion—as a result of the high degree of homogeneity of specific density in the physical space—that deviation from the equiprobable law of distribution does not exceed 5–6% for all studied zones. On the other hand, the current value of $N^K(x_j)$ is precise for the spatial radius, on which the ranked value of $N^K(x_j)$ approaches the $N^K$ value characterizing the whole set (the representativeness radius). So, for zones 4 and 5 it equals 4.5 and 1.3 km, respectively, and for zone 13 its value exceeds 12–13 km.

### 2.3.4   Linear non-correlation of breaking fields

To check the hypothesis about the linear spatial non-correlation of the field of specific density of breakings, we calculated the ranked in-pair linear coefficient $R_{XY}$ of correlation of two columns (or two stochastic ranked sets $x$ and $y$) of matrix $N^K(x_{ij})$ for zone 5. Here the $j$-columns $(x, y)$ were arranged to calculate $R_{XY}$ as follows: at first, we took the columns situated on either side ($j = 10, 11$); then, with a separation of $3 \cdot 0.163$ km ($j = 9$ and 12); then, with a separation of $5 \cdot 0.163$ km ($j = 8$ and 13), etc., where subscript $i$ ran through the values of 1–39 in each of the columns. The ranked values of $R_{XY}$ were calculated by well-known relations (Johnson and Leone, 1977):

$$R_{xy} = \frac{\sum_{i=1}^{39}(N_{ix} - N_x)(N_{iy} - N_y)}{\left[\sum_{i=1}^{39}(N_{ix} - \bar{N}_x)^2\right]^{1/2}\left[\sum_{i=1}^{39}(N_{iy} - \bar{N}_y)^2\right]^{1/2}} \tag{2.20}$$

The results of calculations are presented in Table 2.2; the same table gives the confidence boundaries (with a confidence probability of 0.99) of a ranked value of $R_{xy}$

**Table 2.2.** The sampling values of correlation coefficient $R_{xy}$ and their confidence bounds (confidence probability is 0.99) for the five zones.

| Column numbers | | Spatial distance between columns (km) | Sampling value $R_{xy}$ | Confidence bounds | |
|---|---|---|---|---|---|
| $x$ | $y$ | | | Lower bounds | Upper bounds |
| 10 | 11 | 0.163 | 0.31 | −0.018 | 0.6 |
| 9 | 12 | 0.489 | 0.39 | −0.05 | 0.65 |
| 8 | 13 | 0.815 | 0.23 | −0.2 | 0.55 |
| 7 | 14 | 1.141 | 0.39 | −0.05 | 0.65 |
| 6 | 15 | 1.467 | 0.26 | −0.2 | 0.6 |
| 5 | 16 | 1.793 | 0.35 | −0.1 | 0.67 |
| 4 | 17 | 2.119 | 0.21 | −0.15 | 0.5 |
| 3 | 18 | 2.445 | 0.42 | −0.05 | 0.7 |
| 2 | 19 | 2.771 | 0.15 | −0.3 | 0.5 |
| 1 | 20 | 3.097 | 0.09 | −0.35 | 0.44 |

(in accordance with the rules stated in Johnson and Leone, 1977). It follows from analysis of Table 2.2 that, because the confidence boundaries of ranked values of $R_{xy}$ include a zero value in all considered cases, the zero hypothesis $H_0[R_{xy} = 0]$ can be accepted with a probability of 0.99 for all the spatial separations under study of random ranked sets of specific density (from 0.163 to 3 km). A similar operation was also performed for spatially separated lines of matrix $N_{ij}$ (in other words, the ranked sets were formed across the general direction of sea waves). And the possibility of accepting the zero hypothesis $H_0[R_{xy} = 0]$ for random ranked sets over the lines of matrix $N_{ij}$ was demonstrated. Thus, irrespective of the spatial orientation of ranked sets (and the distance between them), the linear non-correlation of ranked sets of the specific density field is observed in the field of specific density of foam structures with respect to the general direction of sea waves (at least, under the studied meteorological conditions). Moreover, since the stochastic processes of the appearance of breaking centers are close to Gaussian ones (for large frames of observations), the spatial ranked sets in the specific density field can be considered to represent independent stochastic processes. This conclusion is important in one experimental respect: it eliminates the restriction on spatial orientation of sea surface surveying when studying the statistical characteristics of foam activity.

### 2.3.5   Azimuth homogeneity of breaking fields

In Section 2.3.4 the linear non-correlation of the breaking field, presented as a two-dimensional matrix, was demonstrated. However, the question as to the azimuth characteristics of stochastic point fields of breaking centers remains open because of the fact that the techniques of whole-number field processing, used in Sections 2.3.1–2.3.4, were not suitable for revealing azimuth features.

On the other hand, the more or less ordered structure of energy-carrying components of the spatial spectrum of elevations—that is, the presence of a prominent maximum of a spectrum and the anisotropy of an angular part of a spectrum (see relation (2.14))—suggests the possibility of the existence of structurally defined features of a spatial breaking field (e.g., on a spatial-azimuth group in the form of a "quasi-grating", or even "a chessboard"-type structure). Below, on the basis of the statistical procedure of processing stochastic whole-number fields offered by Pokrovskaya and Sharkov (1994), it will be shown that the breaking field of developed seas possesses a high degree of azimuth isotropy, despite the existence of a sharply anisotropic field of elevations of basic (carrying) sea waves.

Under hydrometeorological conditions (see Sections 2.3.1–2.3.3), a set of large-scale optical surveys of the disturbed sea surface of the Caspian Sea water area was carried out in the presence of foam formations, at a surveying scale of 1 : 40800, on a rectilinear flight from south to north along the general sea wave direction. The set of optical images obtained of a disturbed sea surface was used, simultaneously, as the input material both for analyzing the breaking field and for restoring the spatial spectral structure of the basic wave field carrying the breaking field traces in the form of foam systems. The aerial photographs taken in field experiments, were grouped as a separate ranked set, where—to ensure statistical independence—only

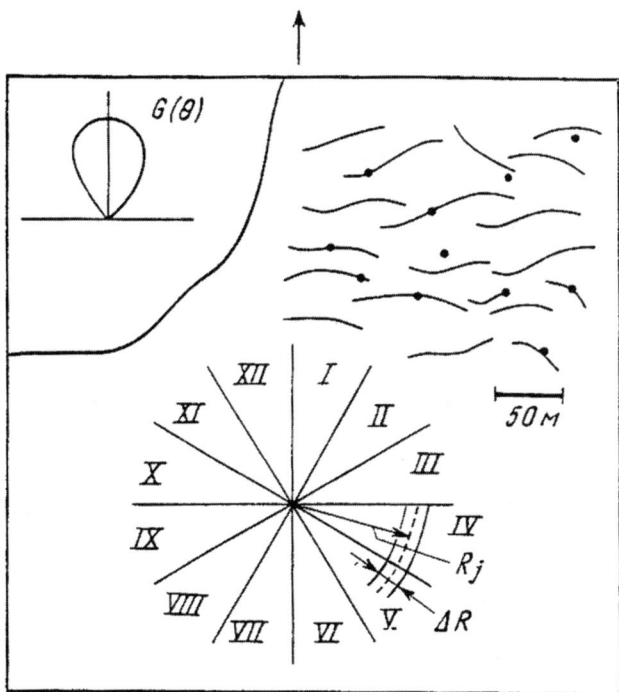

**Figure 2.10.** Aerial photography frame #4318 of instrument MKF-6 on October 31, 1981 involves the schematic sketch of the wave system and the random field of breaking centers (points) in the upper inset of the figure. The geometry is made up of 12 sections that are subdivided into spatial zones with current radius $R_j$ and spacing $\Delta R$. The angle spectra of wave–sea surface displacements are shown as the schematic sketch in the inset.

non-overlapping frames were analyzed. For purposes of efficient analysis, the negatives obtained from the MKF6 instrument were five-fold magnified, and then each individual foam structure was transferred to a paper transparency as a single point. Thus, the point field of breaking centers was made to correspond with the field of foam structures accompanying gravity wave breaking (the random point field approximation).

Those parts of the transparencies that were inside the working zones of the nine independent MKF-6 frames in the developed sea zone were subjected to analysis. The technique used to choose the working zones (beyond the field of solar patches of light) is presented in Section 2.3.1. The spatial size of the analyzed parts of the transparencies correspond to a water area surface size of $1.28 \times 1.28 \, \text{km}^2$.

The analyzed part of each transparency was separated into 12 sectors (I–XII), $30°$ each, with a randomly chosen center ($k$ is the number of a sector) (see Figure 2.10). In its turn, each sector was subdivided into 10 spatial zones according to radius $R_j$ ($j$ is the number of a zone): $R_j = (j - 1/2) \cdot 80$, from $R_1 = 40 \, \text{m}$ to $R_{10} = 720 \, \text{m}$ in steps of $\Delta R = 80$ m. Further, for all chosen zones the quantity of individual breakings that fell in a corresponding zone was calculated, and for all nine MKF-6 frames the ranked sets of specific density $N_j^K(R_j)$ were formed. The volumes of ranked sets of elements (the quantity of breakings) varied from 90 to 212 for each sector of all nine frames. An example of a transparency of one of the MKF-6 frames with a random breaking field and geometry of subdivision into spatial elements is presented in Figure 2.11.

Let us suppose the $N_j^K(R_j)$ set to be the whole-number stochastic flux of indiscernible and independent breaking centers that appeared in the $j$th zone of the $k$th sectors of nine frames. Such an approach to studying the spatial properties of breaking fields can be reliably substantiated experimentally (see Section 2.3).

Proceeding from these statements, we shall consider the spatial-azimuth features of the breaking field.

To study the spatial-azimuth features of breaking fields the ranked character-istic—that is, the distribution function of specific density of centers $F^K$ versus the value of current radius $R_j$—was calculated:

$$F^K(R_j) = \sum_{i=1}^{j} N_i^K(R_j) \left( \sum_{i=1}^{10} N_i^K(R_j) \right)^{-1} \tag{2.21}$$

Thus, the derivative distribution function (the probability density) determines, in its physical sense, the degree of homogeneity in filling the space of azimuth sectors with breaking centers, or, in other words, the degree of azimuth homogeneity.

The results of the calculations are presented in Figure 2.11a, b. Analysis of these data allows us to draw a conclusion as to the high degree of homogeneity of specific density of breaking centers over various azimuths: deviations from the equiprobable law of physical space filling with breaking centers are insignificant in all 12 azimuth directions investigated. This confirms, in particular, the analysis of a confidence corridor, constructed for sector I, in accordance with the Kolmogorov–Smirnov criterion, with a confidence probability of 80% (the significance level is 0.20). The hypothesis on the linear character of the distribution function can be accepted and, consequently, the conclusion can be drawn about the constant value of the probability density of specific filling the sea surface with breaking centers over a space of separately considered sectors (azimuth directions). Moreover, the probabil-ity density has identical values for the whole space (Figure 2.11c)—that is, for the values of azimuth angles from 0 to 360°, which is quite natural, by the way.

Thus, it can be shown experimentally that in the mode of developed and anisotropic seas the field of gravity wave breaking centers represents an azimuth-homogeneous field within spatial scales of 1–2 km.

### 2.3.6   Markov's property of the breaking field

In addition to revealing the large-scale features of the breaking field, the spatial-correlation links of individual breakings of gravity waves among themselves were analyzed on the basis of the same experimental material (Pokrovskaya and Sharkov, 1987a). Analysis was based on studying the law of distribution of the probability

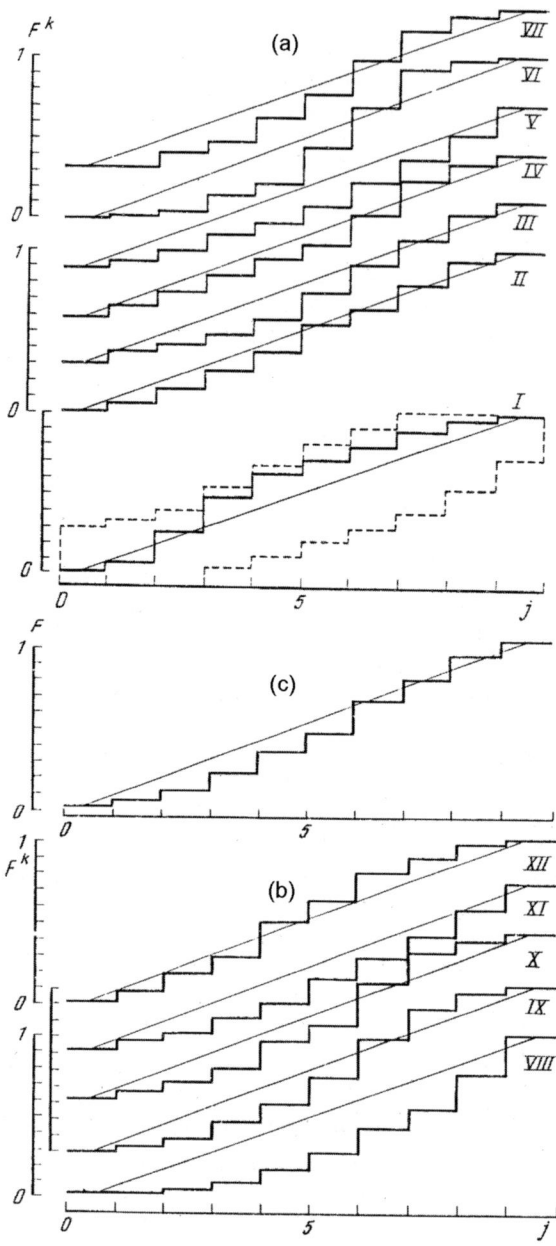

**Figure 2.11.** Experimental density distributions of breaking centers as a function of observation sections: (a) for sections I–VII; (b) for sections VIII–XII; (c) for the density distribution averaged over all azimuths. The solid steps are experimental values of density distribution. The dotted lines show an 80% confidence window using the Kolmogorov–Smirnov criterion. The straight lines are theoretical density distributions.

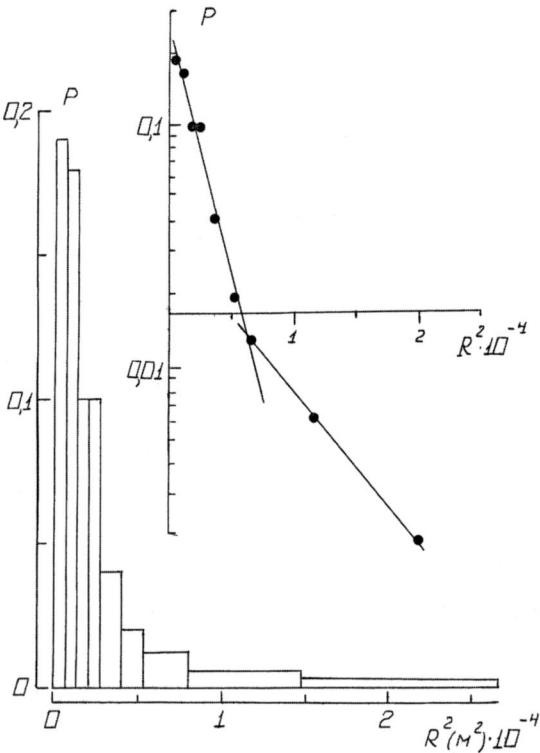

**Figure 2.12.** Experimental histograms of the distribution in frequency of "waiting area" $S_W$ in linear and semi-logarithmic (on inset) scales. The sampling volume is 320. Solid curves show the exponential approximation $p = A \exp(-\alpha S_W)$, $\alpha = 1.44 \cdot 10^{-4}\,\mathrm{m}^{-2}$.

density of "the waiting area" ("the space for life")—that is, the area around the selected center that does not contain any center from the analyzed set (Feller, 1971). To check the hypothesis that the spatial field of gravity wave breakings belongs to after-effect processes (with independent increments), we constructed the histograms of squares of distances from an arbitrary, randomly chosen "driving" center (the breaking center) to its nearest neighbor (the "driven" center) in a spatial window of the basic frame $S_0$ (with a total area of $1.49\,\mathrm{km}^2$). The histogram shown in Figure 2.12 represents a statistical analog of a ranked probability density of "the waiting area" ("the space for life")—that is, the area around the selected center that does not contain any center from the set of centers. Analyzing Figure 2.10, we can easily conclude that the probability density of "the waiting area" has a prominent exponential character, which proves experimentally, by virtue of the well-known theorem (Feller, 1971), the absence of after-effects (Markov's property) for a stochastic, whole-number field of breaking centers. It is interesting to note that the waiting area consists of two sub-areas—the near zone and the far zone—in which the distribution density, being a strictly exponential function (the exponent in the given case), has different numerical parameters. $S_0$, the probability density in the near zone, can be approximated by the following expression for an exponential distribution:

$$P\{R^2\} = A \exp(-\alpha R^2) \qquad (2.22)$$

where the value of $\alpha$ equals $2 \cdot 10^{-4}\,\mathrm{m}^{-2}$; and $A$ is the normalizing factor. It follows from this relation that the mean value of a waiting radius for the near zone will be 70 m. For the far zone the value of $\alpha$ equals $1.4 \cdot 10^{-4}\,\mathrm{m}^{-2}$ and, accordingly, the mean value of a waiting radius will be 85 m. The waiting radius, on which the change of zones takes place, equals 77 m. Note that all the numerical results obtained strictly relate to those hydrometeorological conditions and to those sea wave states (the sea wave spectrum) under which the experiments were carried out (see Sections 2.2.3 and 2.2.5).

## 2.4  SPATIAL–STATISTICAL PROPERTIES OF A DEVELOPING SEA BREAKING FIELD

Unlike Section 2.3, in this section we shall consider the basic statistical characteristics of the point breaking field and their evolution as sea waves develop beginning with the early stages of disturbed surface interaction with wind flux (Pokrovskaya and Sharkov, 1987b).

### 2.4.1  The technique of forming and processing the random point field

The films exposed in field experiments (from the IL-14 carrier) were grouped as separate ranked sets according to various stages of the wind flux interaction with the surface (see Table 2.3). Analysis has shown that the foam structures with a linear size lower than 40 cm show up on negatives unreliably and, for this reason, they were not transferred to a transparency. The analyzed part of a transparency was subdivided into two equal fields, which just became initial elements of a ranked set with the spatial window size $L = 260 \times 280\,\mathrm{m}^2$ (see Figure 2.13). With such a formation of ranked set elements, we managed to completely avoid the overlapping of frames, and, hence, the latter could be considered as independent ranked sets. For all four series in these spatial windows the quantity of individual foam centers ($N$) was calculated, and the ranked sets $N_{ij}^K$ were generated in the form of a matrix point field for the sea areas under study. Here subscript $i$ (the line, conventionally) corresponded to the south–north direction along the general sea wave direction ($i = 1$–18); subscript $j$ (the column, conventionally) corresponded to the west–east direction across the general sea wave direction ($j = 1, 2$); $k$ is the number of a surveying series ($k = 1$–4). As in the case of developed seas (see Section 2.3.1), we shall consider $N_{ij}^K$ to be the whole-number stochastic flux—that is, the random number of indiscernible centers that appear in a spatial window of an element of a ranked set from a corresponding series. Proceeding from this circumstance, we shall consider the spatial characteristics of the matrix field $N_{ij}^K$.

**Table 2.3.** The sea wave fetch conditions and distribution parameters for the waiting area.

| Picture number | Fetch conditions | | | | Sampling volume | Distribution parameter $\alpha \times 10^{-4}$ (m$^{-2}$) | Mean waiting area $\bar S \times 10^{-4}$ (m$^2$) | Mean waiting radius $\bar R$ (m) |
|---|---|---|---|---|---|---|---|---|
| | Fetch $X$ (km) | Nondimensional fetch ($\tilde\chi \cdot 10^{-4}$) | Maximum spectra wavelength ($\Lambda$) (m) | Nondimensional maximum spectra frequency ($\tilde f_m$) | | | | |
| I | 24 | 0.55 | 7 | 0.31 | 47 | 0.28 | 3.57 | 109 |
| II | 57 | 1.32 | 14 | 0.22 | 90 | 1.2 | 0.833 | 51 |
| III | 95 | 2.2 | 18 | 0.19 | 200 | 1.91 | 0.523 | 41 |
| IV | 156 | 3.6 | 30 | 0.15 | 202 | 7.66 | 0.131 | 20.5 |

*Comments*: Nondimensional fetch $\tilde\chi$ and nondimensional maximum spectra frequency $\tilde f_m$ are defined as the following expressions: $\tilde\chi = Xg U_{10}^{-2}$; $\tilde f_m = f_m U_{10} g^{-1}$; where $f_m$ is the maximum spectra frequency (Hz); $U_{10}$ is wind velocity at the 10-m level; and $g$ is free-fall acceleration. Mean waiting radius is $R = (S/\pi)^{1/2}$.

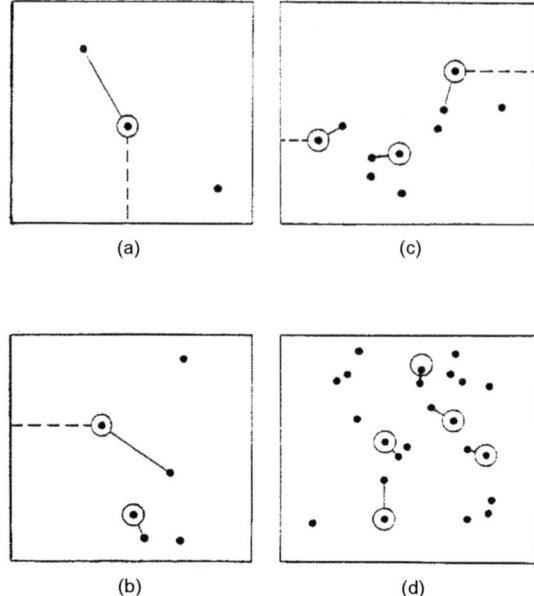

**Figure 2.13.** Aerial photography frames with a random field of breaking centers over a $260 \times 280$-m spatial window: (a) run I with fetch $X = 24$ km; (b) run II with fetch $X = 57$ km; (c) run III with fetch $X = 95$ km; (d) run IV with fetch $X = 156$ km. The leading center is circled.

### 2.4.2  Markov property of the breaking field

To check the hypothesis that the breaking field belongs to processes where an after-effect is absent, calculations were carried out and then the histograms of squares of distances from an arbitrary, randomly chosen "driving" center (the foam structure) to its nearest neighbor (the "driven" center) were constructed in a spatial window of a frame. In so doing the circumstance was taken into account that "the driving" and "the driven" centers should be chosen in such a way that the distance between them be shorter than the distance from the driving center to the nearest boundary of a frame. Thereby, such a processing technique eliminates the effect of the influence of spatial limitation of a studied zone on a ranked set. The examples of transparencies with random fields of centers and with selection of "driving" and "driven" centers for each of four survey series are presented in Figure 2.13. The processing results are presented in Table 2.3 and in Figure 2.14. By "ranked set volume" here we mean the size of distances between the centers used for constructing the histograms. The histograms presented in Figure 2.14 represent a statistical analog of a ranked probability density of "the waiting area" ("the space for life")—that is, the area around the selected center that does not contain any center from the set of centers. Analysis of Figure 2.14 allows us to easily conclude that the probability density of the "waiting area" $p(S)$ has a prominent exponential character (the semi-logarithmic scale of a plot) for all four surveying series. In normalized form the distributions can be

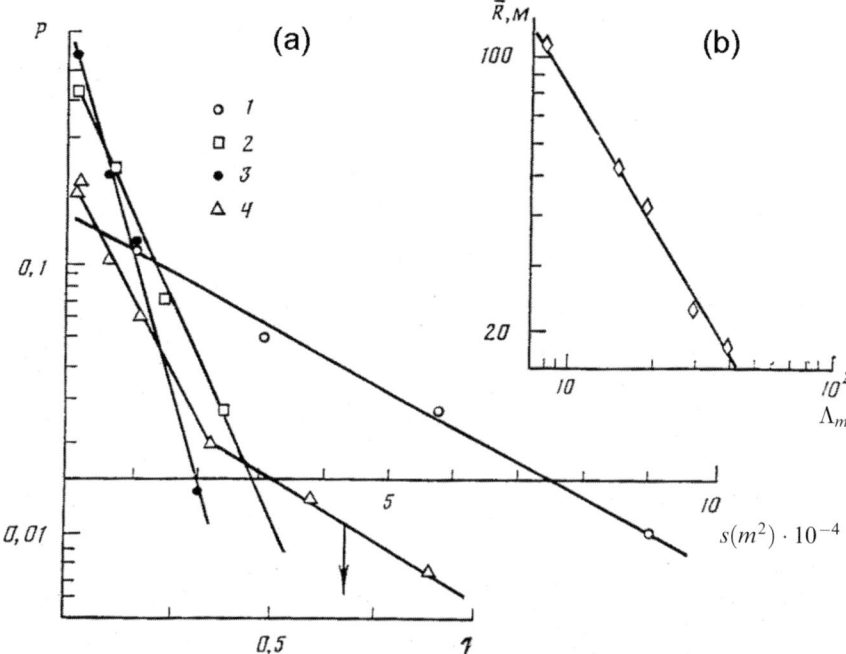

**Figure 2.14.** Experimental histograms of the distribution in frequency of waiting areas with fetch: 1—$X = 24$, 2—$X = 57$, 3—$X = 95$, 4—$X = 156$ km. (a) Solid curves show theoretical distributions using the parameters presented in Table 2.3. (b) Experimental function of the average radius of waiting area as a function of spectra maximum wavelength $\Lambda_m$. The solid line is the theoretical approximation $R = B\Lambda_m^{-1.1}$ ($B = $ const.).

presented as $p(S) = A\alpha \exp(-\alpha S)$, where $\alpha$ is the distribution parameter (see Table 2.3), and $A$ is the normalizing factor. We emphasize that this feature is kept at all studied stages of the wind flux interaction with the sea surface beginning with the earliest stages ($X = 24$ km). A similar property of breaking fields was also revealed (see Section 2.3.6) for the fully developed sea mode ($X > 250$ km; $\tilde{f}_m = 0.13$) under the same hydrometeorological conditions. The exponential character of the probability density of "the waiting area" is, by virtue of the well-known theorem (Feller, 1971), experimental proof of the fact that the after-effect is absent for a random whole-number field of breaking waves (Markov property). In other words, the intervals between the neighboring points of the process represent independent random quantities with identical exponential probability density. And, in this sense, neighboring wave breakings represent spatially independent processes at all stages of wave formation. When analyzing the relation for the wave formation state—which can be characterized in our case by the wavelength of the frequency spectrum maximum $\Lambda_m$ and by the spatial random field structure, specified by the mean radius of the waiting area $R = (S/\pi)^{-1/2}$ (see Table 2.3)—the following interesting circumstance was found: the close correlation between the changes of $\Lambda_m$ and $R$

**Table 2.4.** Distribution parameters for the specific density of breaking centers.

| Picture set number | Sampling volume | Numerical parameter distributions | | | | Approximate law type and the main parameters |
| --- | --- | --- | --- | --- | --- | --- |
| | | Mean value | Variance | Asymmetry coefficient | Excess coefficient | |
| I | 36 | 2.94 | 2.28 | 0.23 | 0.56 | BD; $\bar{N} = 3$; $\sigma = 1.5$ |
| II | 38 | 7.6 | 6.9 | 0.62 | 0.12 | ND; $\bar{N} = 7$; $\sigma = 2.7$ |
| III | 38 | 12.6 | 15.1 | 1.0 | 0.13 | ND; $\bar{N} = 13$; $\sigma = 4$ |
| IV | 36 | 30.6 | 73.2 | 0.76 | 0.75 | ND; $\bar{N} = 30$; $\sigma = 8$ |

*Comments*: Sampling volume is the number of pictures with a spatial window of $260 \times 280$ m. ND and BD are the normal distribution and the binomial distribution, respectively.

(Figure 2.14b). The latter can be expressed by the following functional relation: $R\Lambda_m^n = B$, where $n = 1.1$, and $B$ is a constant that does not depend on fetch conditions (under given hydrometeorological conditions).

### 2.4.3 Laws of specific density distribution

To reveal a particular form of the probabilistic model of the foam field (in the scale of the spatial frame studied) for each of four series (and, accordingly, of the fetch value) the histograms of density of (breaking) centers $N$, being statistical analogs of ranked probability densities, were constructed according to well-known rules (see Section 2.3.2). The processing results are presented in Table 2.4 and in Figure 2.15a–d. In accordance with Pierson's criterion $\chi^2$, to estimate the divergence between theoretical and experimental distributions we obtain: for series I the values $X^2 = 0.08$, whereas the 5th percentile critical value of a unilateral criterion $\chi^2(1; 0.95) = 3.84$ in the given case; for series II and III the $X^2$ values are equal to 2.83 and 0.57, respectively, at $\chi^2(3; 0.95) = 7.8$; for series IV $X^2 = 8.05$ and $\chi^2(5; 0.95) = 11.07$.

Thus, in all cases the observed quantity is insignificant, and hypotheses about the proximity of experimental distributions to binomial (series I) and normal (series II–IV) laws can be accepted.

To a satisfactory degree of accuracy, the mean value of specific density can be approximated as a function of the fetch $\chi$ in the following form:

$$\bar{N} = \begin{cases} AX & X \le 90 \text{ km} \\ BX^2 & X > 90 \text{ km} \end{cases} \tag{2.23}$$

where $A = 0.11 \text{ km}^{-1}$; and $B = 1.2 \cdot 10^{-3} \text{ km}^{-2}$.

### 2.4.4 Linear non-correlation of breaking center specific density

To check the hypothesis about the linear spatial non-correlation of the field of specific density of breaking centers, the ranked in-pair linear correlation coefficient $R_{xy}$ of two adjacent lines of increment field elements $N_{ij}$ was calculated. The ranked $R_{xy}$

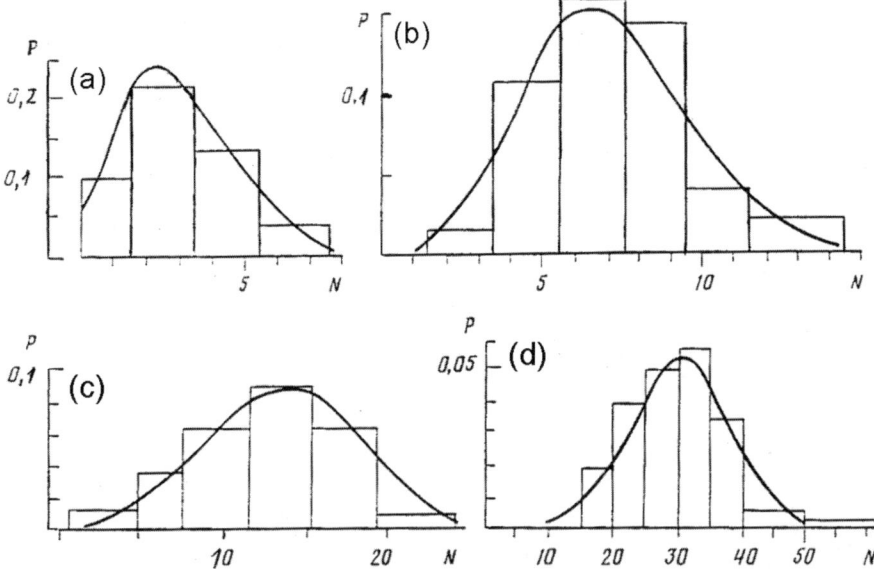

**Figure 2.15.** Experimental density distributions of breaking centers as a function of fetch $X$: (a) 24, (b) 57, (c) 95, (d) 156 km. The solid lines are theoretical distributions using parameters obtained while processing corresponding histograms (see Table 2.3).

values were calculated by well-known relations (2.20) (Johnson and Leone, 1977). The results of calculations are presented in Table 2.5; the same table presents the confidence boundaries (at a confidence probability of 0.95) of the ranked $R_{xy}$ value according to the rules (Johnson and Leone, 1977). Analysis of the data of Table 2.5 indicates that, since the confidence boundaries of ranked values included a zero value in all considered cases, the zero hypothesis $H_0[R_{xy} = 0]$ can be accepted (with a probability of 0.95) at all sea wave fetch stages.

Thus, irrespective of the degree of fetch, the linear non-correlation of breaking field increments is observed. Recalling the experimentally found property of the absence of after-effect between neighboring points of the process, we can easily see

**Table 2.5.** The sampling values of correlation coefficient $R_{xy}$ and their confidence bounds (confidence probability is 0.95).

| Pictures set number | Sampling values $R_{xy}$ | Confidence bounds | |
|---|---|---|---|
| | | Lower bound | Upper bound |
| 1 | −0.48 | −0.75 | 0.05 |
| 2 | −0.38 | −0.70 | 0.15 |
| 3 | 0.364 | −0.20 | 0.68 |
| 4 | 0.279 | −0.25 | 0.62 |

that the spatial non-correlation obtained represents, as a matter of fact, a corollary from the properties of processes with independent increments—that is, of processes in which the quantity of points in non-overlapping areas is independent (Feller, 1971). Moreover, since the stochastic processes of the appearance of breaking centers in the frame are close to Gaussian ones, the experimental spatial field of breaking center increments can be considered to strictly possess the property of independence in increments.

### 2.4.5    Spatial homogeneity of breaking center specific density

To study the spatial features of the statistical characteristics of the specific density of foam structures, the following ranked characteristic (i.e., the specific density distribution function) was calculated depending on the spatial coordinate $x_j$ (along the flight crossing):

$$F^K(x_j) = \sum_{l=1}^{j} N_l^K \left( \sum_{l=1}^{18} N_l^K \right)^{-1} \tag{2.24}$$

where $(x_j) = \Delta x \cdot j = 280$ m; in this case the probability density $N_{ij}^K$ (the derivative of $F^K(x_j)$) determines in its physical sense the degree of homogeneity of the mean value of $N^K$ for the field of centers in the space. Analysis of the confidence corridor constructed indicates, in accordance with Kolmogorov–Smirnov's criterion (Johnson and Leone, 1977) at a confidence probability of 0.80 (the significance level of 0.20), that the hypothesis about the linear character of distribution function $F^K(x_j)$ is valid and, thus, the probability density is constant over the space. Note that deviations of the empirical distribution function from the law of uniform distribution of specific density essentially differ for various stages of fetch. So, for $X = 156$ km the deviations do not exceed 2–3%, whereas for the initial stage ($X = 24$ km) they reach 15%. The latter circumstance is most likely related to spatial limitation of the size of ranked set element $L_i$ and to the proximity of $L_i^1 \approx 200$ m to the values of a waiting zone radius, which is equal, for $X = 24$ km, to $R^1 \sim 109$ m.

## 2.5    FRACTAL PROPERTIES OF WAVE-BREAKING ZONES IN STATIONARY AND DEVELOPING SEAS

Gravity waves breaking on the water surface represent some peculiar example of turbulent motion, in which the resulting picture is formed as a result of the complex processes of the interaction of wind and wave motions (see, e.g., Phillips, 1977; Kitaigorodskii, 1997; Zacharov and Zaslavskii, 1982). There are many conventional modes of such surface wave turbulence. These modes depend on wind speed and, to a certain degree, they can be characterized by a relative fraction of the distribution of breaking waves. The places and instants of breaking are certainly stochastic parameters. However, without any doubt, the spatial and temporal positions of breakings possess a certain degree of spatiotemporal correlation which, however, cannot be revealed by standard statistical techniques (see Sections 2.3, 2.4). Note that there also exist some other possibilities of describing spatial-stochastic structures formed by breaking wave positions. In particular, one such possibility is related to determining the scale-invariant (fractal) properties of the surface formed by the breaking wave region.

Ideas about the fractality of structures have already found some applications in describing various phenomena in nature (Mandelbrot, 1982) and, in particular, in hydrodynamical turbulence theory (Frisch, 1995).

Below we shall present the experimental results of a work by Zaslavskii and Sharkov (1987), which show that the spatial structure of breaking gravity wave centers on a disturbed sea surface possesses fractal properties as well. Processing was carried out based on the same experimental material as the investigation of spatial-statistical characteristics (see Sections 2.3, 2.4).

### 2.5.1 Techniques of formation and processing of the random-point field for fractal processing

Generally speaking, an initial stochastic field can be generated for subsequent fractal processing by various techniques. In the given case we shall make use of the methodology of constructing the initial field as a random-point field. Disregarding only the detailed structure of wave breaking, we shall depict the breaking wave as a point on the plane. As a result, the restricted section of the surface (the range) can be presented by a set of points on it, which represent the state of breaking waves (i.e., of individual breaking centers) at some time instant. We shall use the definition of Hausdorff's dimension. Let the set of points be covered by densely packed multidimensional cubes with side $\varepsilon$. We shall designate by $N(\varepsilon)$ the number of cubes into which at least one point of a set falls. Then Hausdorf's dimension $d$ of this set is presented by the quantity:

$$d = \lim_{\varepsilon \to 0} \frac{\ln N(\varepsilon)}{\ln(1/\varepsilon)} \tag{2.25}$$

In this case the set of points depicting breaking waves is disposed on the plane, and the entire problem of covering it by squares (or rectangles) and performing the limit (2.25) is reduced to the technique of processing optical images of the range.

### 2.5.2 Fractal properties of breaking fields under developed sea conditions

For developed sea conditions ($X = 250\,\text{km}$, $\Lambda_m = 42\,\text{m}$, $\tilde{f}_m = 0.13$) the basic elements of ranked sets were represented by 39 maps, each of which corresponded to the surface of a sea range having a size of $3264 \times 457\,\text{m}^2$. Thus, the total investigated water area of the sea was equal to $39 \times 1.5 = 58.5\,\text{km}^2$ (see Section 2.3.1).

Further processing of the maps obtained was as follows. Each map was subdivided into $n_K = 5 \cdot 2^K$ ($K = 1, \ldots, 8$) equal parts. These parts will be called cells. A cell relates to the breaking zone, if at least one breaking wave falls into it. Otherwise, cells relate to "the silence area". Obviously, at initial values of $K$ all cells contain several breakings, since a vast sea surface area corresponds to the size of a

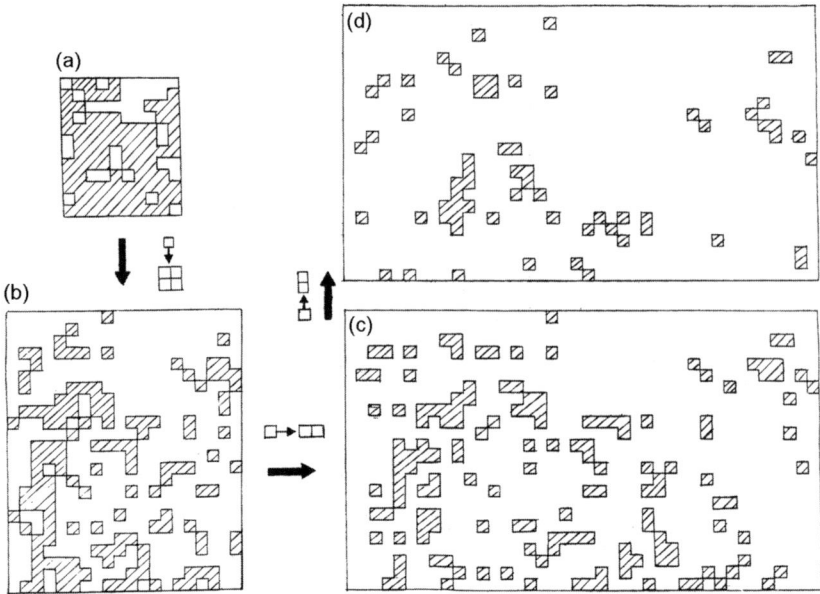

**Figure 2.16.** Cell structure of sea breaking fields for the same basin in fully developed state ($X = 250$ km and $\Lambda_m = 42$ m) with the following division scales: $k = 4$ (a), 6 (b), 7 (c), 8 (d). The shaded cells contain no less than one breaking $N_K > 1$. The empty cell contains $N_K = 0$. The character of cell area division is shown in the insets.

cell. However, as $K$ grows, the resolution of breaking map construction increases and silence areas begin to appear on it. The critical value $K_0 = 2$ corresponds to this transition. This value can be conferred a simple physical meaning. At $K_0 = 2$ a surface of area $0.075 \, \text{km}^2$ falls into a cell. This area approximately corresponds to the characteristic maximum silence area size $l_0 = 273$ m for developed seas and for given conditions of heat- and mass-exchange in the sea–atmosphere system (see Section 2.2.3).

Figure 2.16 presents the same fragment of maps for four various $K$ values: $K = 4, 6, 7, 8$. For analysis convenience, all cells are reduced to the same size (a pixel), since as $K$ grows the size of a map and of its fragment sharply increases. Breaking zones correspond to dashed cells. As $K$ increases, the picture of breaking zones and silence areas sharply changes. Such instability is typical of fractals.

Analysis of Hausdorff's dimension for breaking zones was carried out according to formula (2.25). Let $N_B = N(\varepsilon_K)$ be the number of breaking zones on all 39 maps, and $\varepsilon_K = (S_K)^{1/2}$ be the linear size of a cell on a map at the $K$th step of its subdivision. Figure 2.17 presents the data for the dependence of $\ln N_B = N(\varepsilon_K)$ on $\ln(1/\varepsilon_K)$. This dependence consists of two sections, which can be well approximated by straight segments (zones 1 and 2). The section of curve 1 corresponds to the dimension of 2. This is natural because—if the cells of map subdivision are very large—they always have at least one breaking. Hence, the dimension $d_1$ for rather

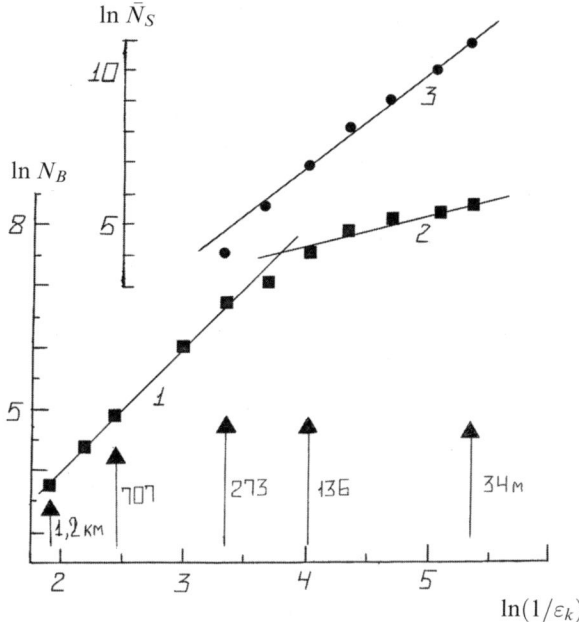

**Figure 2.17.** Dependence of the number of shaded $N_B$ and empty $N_S$ cells as a function of the linear dimension of the $\varepsilon_K$ pixel in fully developed seas ($X = 250$ km and $\Lambda_m = 42$ m). The first three boxes represent a pixel area with 1.5, 0.75, and 0.5 km$^2$, and the areas of later boxes are $1.5/n_K$, where $n_K = 5 \cdot 2^K$ ($K = 1$–8). Arrows and numbers show the values of the linear dimension for division cells.

high $\varepsilon_K$ values should coincide with the dimension of a plane. However, for $K_0 = 3$, which corresponds to $\varepsilon_0 = 190$ m, the curve for $\ln N_B(\varepsilon_K)$ enters section 2, which corresponds to a quite different dimension ($d_2 = 0.5$). The transition region lies in the interval of values $K_T = (2$–3) and corresponds to the interval of linear dimensions of subdivision cells $\varepsilon_T = 270$–190 m. We shall interpret this interval below.

The small dimension $d_2$ of breaking zones indicates the specific role of silence areas in the mode of surface wave turbulence under consideration. Since the silence area in this case always has a positive measure, it should be attributed to "fat fractals". For these fractals one can also introduce some similarity parameter characterizing their fractal properties. We shall designate by $N_K$ the number of all subdivision cells at the $K$th step. Then $N_S = (N_K - N_B)$ is the number of cells falling on the "silence" area. The dependence of $\ln(N_K - N_B)$ versus $\ln(1/\varepsilon_K)$ is presented in Figure 2.17 (curve 3). Beginning with $K = 3$ it corresponds to the law

$$N_S = (N_K - N_B) = (1/\varepsilon_K)^d = (1/S_K)^{d/2} \qquad (2.26)$$

where $d = 2.0$.

### 2.5.3   Fractal properties of breaking fields under developing sea conditions

Under developing sea conditions, optical sensing of the sea surface was performed from onboard an IL-14 airplane laboratory using the optical instrument AFA TE-100 (see Section 2.2) for the following fetch values $X$: $X_1 = 24$ km and $\Lambda_m = 7$ m (set 1); $X_2 = 57$ km and $\Lambda_m = 14$ m (set 2); $X_3 = 95$ km and $\Lambda_m = 18$ m (set 3); $X_4 = 156$ km and $\Lambda_m = 30$ m (set 4); $X_5 = 250$ km and $\Lambda_m = 42$ m (set 5) (full fetch) (see Table 2.3). The technique of optical image processing was reduced to representing the limited section of the surface with breaking waves in the form of the field of a random set of indiscernible centers with subsequent field fragmentation into $n_K$ cells with side $a_K$ ($a_K = 240, 120, 80, 40, 20, 14, 10$ m). The cell was attributed to the breaking zone, if at least one breaking fell into it. Otherwise, the cells were attributed to "the silence area". Analysis of Hausdorff's dimension for breaking zones $N_B(a_K)$ and silence areas $N_S(a_K)$ showed (Figure 2.18) that these zones and areas possess fractal properties. Here the dimension of breaking zones $N_B$ coincides with the dimension of a plane (equal to 2) up to some transition zone, and then the dimension

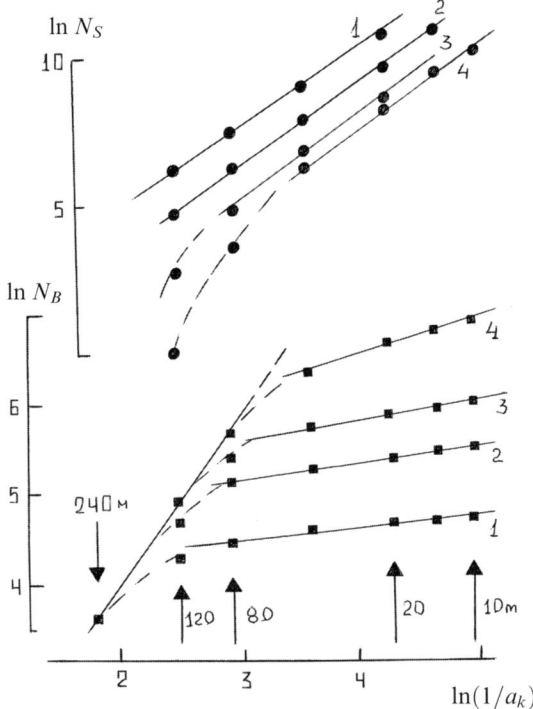

**Figure 2.18.** Dependence of the number of shaded $N_B$ and empty $N_S$ cells as a function of the linear dimension of the $a_K$ ($a_K = 240$ m, $120, 80, 40, 20, 14, 10$) pixel in developing seas: 1—fetch $X = 24$ km and $\Lambda_m = 7$ m; 2—$X = 57$ km and $\Lambda_m = 14$ m; 3—$X = 95$ km and $\Lambda_m = 18$ m; 4—$X = 156$ km and $\Lambda_m = 30$ m. Arrows and numbers show the values of the linear dimension for division cells.

sharply drops. So, the transition zone is $a_K = 240{-}120$ m for a low fetch ($X = 24$ km), and at $a_K < 120$ m the dimension sharply drops down to the value $d = 0.06$. This indicates that the number of breaking zones very weakly depends on the value of a subdivision zone element. The number of breaking zones is virtually constant. Then, as sea wave fetch increases, the transition zone decreases down to 60–40 m, and the dimension $d$ of a breaking field increases up to 0.3 (for $X = 156$ km). In turn, for $a_K < 120$ m we can introduce the proper similarity parameter for silence areas, which characterizes their fractal properties. Namely, the number of cells falling on silence areas $N_S$ at the $K$th step of subdivision corresponds to the following scaling (Figure 2.18):

$$N_S = (N_{NK} - N_B) = (1/a_K)^d = (1/S_K)^{d/2} \qquad (2.27)$$

where $S_K = a_K^2$, $d = 2.1$.

A surprising result is the stability of the index of fractality of silence areas at all stages of wind flux interaction with waves, which is probably related to the homogeneity of a picture of the stochastic fetch of sea waves at all stages of their development.

The obtained result can be understood as follows. The interaction of near-surface waves with each other and with the wind exceeds some threshold, under certain conditions, and results in the appearance of chaotic dynamics in the form of break-ings. As a result of wave–wind interaction, wave fetch begins. The fetch terminates when the wave reaches a critical velocity and then breaks. The obtained fractal characteristic actually relates to wave fetch zones, which in fact determine the silence area. The mean size of silence areas falls to 110 m under the same conditions in which the data used were obtained. The interval of silence areas in which steady scaling is revealed lies in the region of 34–275 m, and the characteristic length of a gravity wave equals $\Lambda = 40$ m. These data show that the sharp change in fractal properties of breaking zones can be correlated with scales characterizing the size of silence areas (i.e., the distance between successive breakings). Thus, a distance of about five wave-lengths corresponds to the maximum size of a silence area. This quantity is certainly a function of wind speed and other parameters of near-surface processes. The above considerations led Zaslavskii and Sharkov (1987) to a non-trivial conclusion on a possible stochastic mechanism of wave fetch. It is similar to Fermi's stochastic mech-anism of particle acceleration and exists for waves as well, though in the latter case it looks a little bit more complicated (Zaslavskii and Sagdeev, 1988). The qualitative picture of the stochastic fetch of waves looks like this. The spectral structure of waves near the overturning boundary is very wide. In the near-surface layer, wind-generated disturbance is periodic and falls in resonance with any harmonic of a wave. The energy pumping into a wave and its acceleration occur most intensively as a result of these resonant interactions.

Another consequence of the obtained result relates to the possibility of diagnos-ing sea waves on water based on the fractality factor $d$ of "silence" areas. Its change should obviously characterize the process of origin and development of a picture of wave breakings in the field of external effects (internal waves, for instance).

## 2.6   CONCLUSION

**1**   Large-scale optical sensing of the sea surface in sea wave fetch mode and in a developed sea zone, the results of which were analyzed in this chapter, was carried out in the water area of the Caspian Sea within the framework of the "Kaspii-81" complex of works undertaken by the USSR Academy of Sciences. Experimental remote techniques used to study laws of the statistical spatial structure and scale-invariant properties of chaotic fields of gravity wave breaking in the ocean–atmosphere system are proposed and developed.

**2**   On the basis of the techniques developed, it is experimentally shown that under conditions of net spatial fetch (at all stages of development) the probabilistic model of the gravity sea wave breaking field can be presented in the form of a whole-number stochastic spatial field with homogeneous and independent increments (a Markov-type field) with uniformly distributed and independent centers of gravity wave breakings. The density of distribution of the specific density of breaking centers (foam structures) in developed sea mode evolves from the normal (Gaussian) law (a spatial frame of $1.5\,\text{km}^2$) up to the geometrical distribution (a spatial frame of $10^3\,\text{km}^2$). The linear non-correlation of ranked sets in the spatial field of specific density is found, which in turn possesses a high degree of spatial and azimuth homogeneity.

**3**   The spatial field of wave-breaking centers possesses scale-invariant properties; in particular, the parameter for similarity of silence areas is revealed. The stability of the parameter for similarity of silence areas is revealed at all stages of wind flux interaction with a disturbed surface.

# 3

# Linear and two-dimensional geometry of whitecapping and foam structures

This chapter describes the results of remote investigation of the geometrical characteristics (linear and two-dimensional) of the process of individual breaking of gravitational waves that result in whitecapping and foam fields of various types. The investigation technique is optical surveying at high spatial resolution using low-speed platforms (helicopters). The statistical models of breaking processes are constructed on the basis of experimental data. Existing theoretical views on the possibility of describing breaking waves on the basis of the threshold mechanism for a Gaussian three-dimensional stochastic field are critically analyzed.

## 3.1 THE PROBLEM OF STUDYING THE SPATIAL–STOCHASTIC STRUCTURE OF INDIVIDUAL BREAKING WAVES

As already noted (Chapter 1), investigation of the spatiotemporal characteristics of foam activity at the breaking of individual gravitational waves is a topical means of solving numerous scientific and applied problems of oceanology, such as studying the ocean–atmosphere interaction (Monahan, 2001), wind–wave dynamics (Phillips, 1977), the development of techniques for remote sensing of the ocean surface (Phillips, 1988; Glazman, 1991a; Sharkov, 1998, 2003), the development of reliable quantitative measures for determining the state of a disturbed sea surface, instead of descriptive estimations of the Beaufort Scale that still have not satisfied practical demands (Alcock and Morgan, 1978; Bortkovskiy, 1983). The most important quantitative characteristics of foam activity include the individual dimensions of various types of formations (crests and strip structures), as well as their fractional coverage areas, depending on surface wind velocity, sea surface state, size of sea waves, season of observation, biological activity at the given World Ocean area, etc. Here, since the wave-breaking and subsequent whitecapping process depend on many factors of a physical, chemical, and biological nature, it should be considered as a stochastic process, and its quantitative characteristics should be described in appropriate, correct statistical language. In a series of earlier papers devoted to studying foam structures (Monahan, 1971; Ross and Cordon, 1974; Samoilenko *et al.*, 1974; Bortkovskiy and Kuznetsov, 1977; Bortkovskiy, 1983, 1987), no correct statistical processing of experimental data was performed—for example, the character and specific type of distributions of measured parameters was not clarified; no data were presented on volumes of ranked sets of elements, on confidence intervals of measurement errors, etc. In their turn, experimental data obtained as a result of shipborne investigations have some serious limitations. First, by virtue of their natural specificity, these measurements (Samoilenko *et al.*, 1974; Bortkovskiy and Kuznetsov, 1977; Bortkovskiy, 1983, 1987) are local to some extent and cannot provide the necessary information over considerable areas for short observation times. Second, they do not provide the full statistical independence of ranked sets when surveying the surface, since the lifetime of individual foam structures can reach a few minutes or longer (see Chapter 4). Third, perspective photographing causes serious distortions in shipborne observations (Bortkovskiy and Kuznetsov, 1977; Bortkovskiy, 1983). On the other hand, the application of aerial photography from high-altitude carriers or surveying from low altitudes (100–300 m) by means of high-speed (more than $70 \, \text{m s}^{-1}$) carriers without using complicated compensation devices in the instrument does not allow the high spatial resolution required for detailed analysis of foam structures. In the present chapter, based on the assumption of the quasi-stationary and quasi-ergodic character of a random whitecapping process, statistical processing of the data of aerial photography, carried out at high spatial resolution onboard a low-speed carrier ($15–20 \, \text{m s}^{-1}$) from 100 m in altitude, is performed (Bondur and Sharkov, 1982, 1990).

## 3.2    REMOTE INVESTIGATION OF INDIVIDUAL FOAM STRUCTURES IN THE WAVE-BREAKING PROCESS

To perform small-scale surveying from onboard a research helicopter, an optical camera was used that had a view angle of $44°$ and an objective focal length of 20 cm; it was placed in a special mechanical system for eliminating "blurring" from the optical image. The spatial resolution, accounting for the resolution capability of the optical instruments on a vibrating carrier, was about 2 cm at the flight altitude of 100 m. Rectilinear flights with tack lengths of 30–60 km were performed during the investigation. The experiments were carried out in water areas of the Black and Barents Seas during the summer seasons of 1976–1980, at a distance from the coast of 20–40 km, within the framework of the "Barents-76"–"Barents-80" complex of programs of the USSR Academy of Sciences on studying the structure of sea waves by remote-sensing techniques (Bondur and Sharkov, 1982, 1990). The data on hydrometeorological conditions were taken from measurements on accompanying shipborne and coastal meteorological stations. The experiments were carried out at a sea surface state of 1, 2–3, 3–4, 4–5 using Beaufort notation and, accordingly, at surface wind velocities (at the level of 19.5 m) of 2, 3.2, 5.7, 9.5, and 10.5 m s$^{-1}$. The values of wind velocities were determined by averaging the results of measurements at several places in the sea area studied and during the term of a cycle of measurements. The root-mean-square error of wind velocity measurements was about 0.4 m s$^{-1}$. In this case the wave–wind conditions were close to the mode of steady (developed) sea waves, and the near-surface layer stratification was close to neutral.

These experiments were devoted to investigation of the spatiotemporal charac-teristics of individual foam systems of various structure and, principally, of the two most optically contrasting types—crests and a widespread, spotty foam. Despite the fact that foam activity passes the whole gamut of intermediate states—from a boiling breaker up to an emulsion monolayer that is breaking up—the analysis of brightness, density, and specific form of foam systems on optical images allows us to confidently identify at least two classes (types) of foam formations: (1) "wave crest foam" (i.e., "whitecaps"), the so-called "short-living form" (i.e., "dynamic foam") of foam activity with a lifetime of units of seconds, and spotty structures (or "foam streaks"); (2) "static foam" (or "residual foam") with a lifetime of about 10 seconds to several minutes. It is important to note that these types of foam structures exhibit not only obvious optical distinction, but possess essentially different electromagnetic properties (see Chapters 6 and 8), which is quite important in constructing radiation models and models of the ocean–atmosphere system dispersion (Sharkov, 1998, 2003). At wind velocities higher than 15 m s$^{-1}$ there arises a special class of stable foam systems: the thread-like systems caused by capture of air bubbles by Langmuir vortices (i.e., Langmuir circulation) (Thorpe, 1982).

As an example, Figure 3.1 presents optical pictures of a disturbed sea surface (the Barents Sea), obtained during scheduled aerial photography in the 1 : 500 scale under optimum solar illumination conditions. At the center of the frame in Figure 3.1a the bright image of a breaking crest (the whitecap) is clearly distinguished, and Figure 3.1b shows, at the center and in the right lower corner of the frame, the optical images of three fields of spotty foam with subsequent formation (in the windward direction) of thread-like fields of foam systems.

The optical classification of foam systems under consideration on a disturbed oceanic surface was proposed for the first time by the author of this book while performing airborne remote sensing in the Barents Sea (1976) within the framework of the "Barents-76" program of the USSR Academy of Sciences (Bondur and Sharkov, 1982, 1990). Later on, such a classification was repeatedly used in per-forming similar experimental investigations both by Russian and Western researchers (Bortkovskiy, 1983, 1987; Glazman and Weichman, 1990; Raizer and Novikov, 1990; Raizer *et al.*, 1994; Kokhanovsky, 2004; Marmorino and Smith, 2005). And now this classification has got "rights of citizenship" (Monahan, 2001).

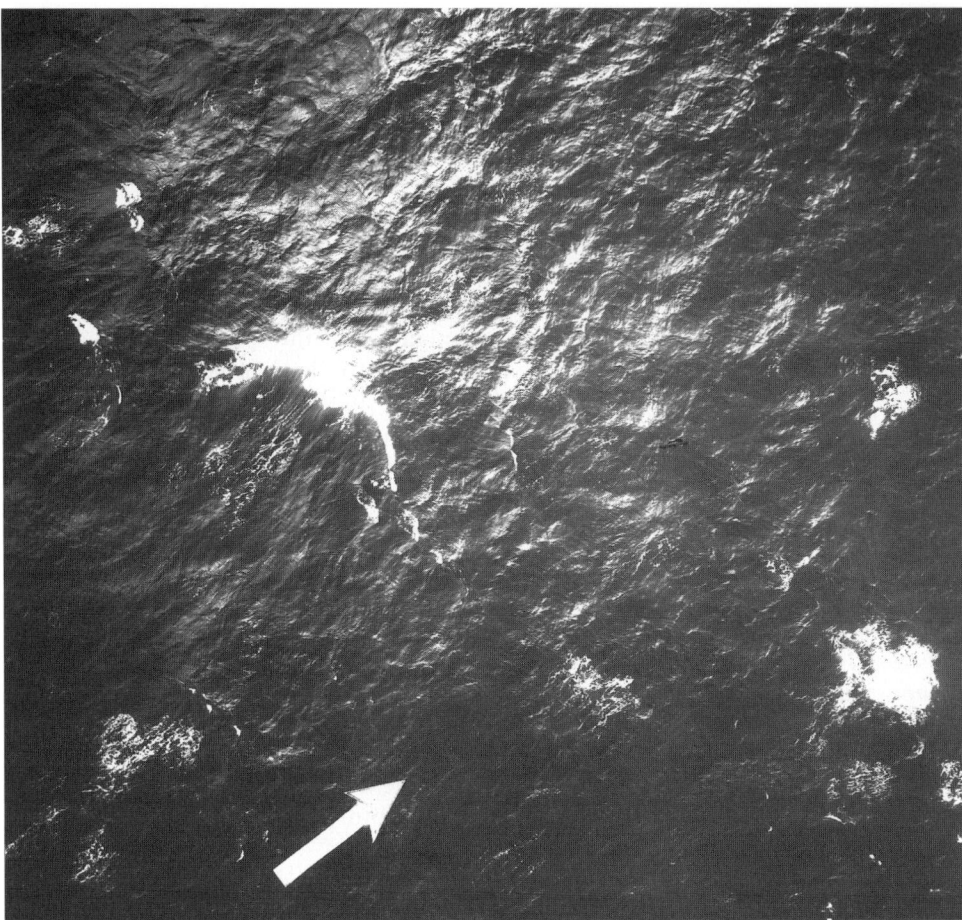

**Figure 3.1.** Aerial images of a disturbed sea surface taken with the AFA optical instrument over the Barents Sea at 200 m altitude. The white arrows show the near-surface wind direction. (a) A whitecap and a foam streak are shown.

## 3.3   PROCESSING THE DATA FROM REMOTE SENSING OF INDIVIDUAL FOAM STRUCTURES IN THE WAVE-BREAKING PROCESS

Films taken during remote-sensing experiments were grouped into separate ranked sets corresponding to similar hydrometeorological conditions. To provide the statistical independence of ranked sets, only non-overlapping frames were analyzed. The total number of processed frames was 205 for $V = 5.7\,\mathrm{m\,s^{-1}}$, 66 for $9.5\,\mathrm{m\,s^{-1}}$, and 63 for $10.5\,\mathrm{m\,s^{-1}}$. In each working frame the images of whitecap and strip foam were contoured, the geometrical shape of each individual formation was estimated, their characteristic (maximum and minimum) linear and two-dimensional sizes were

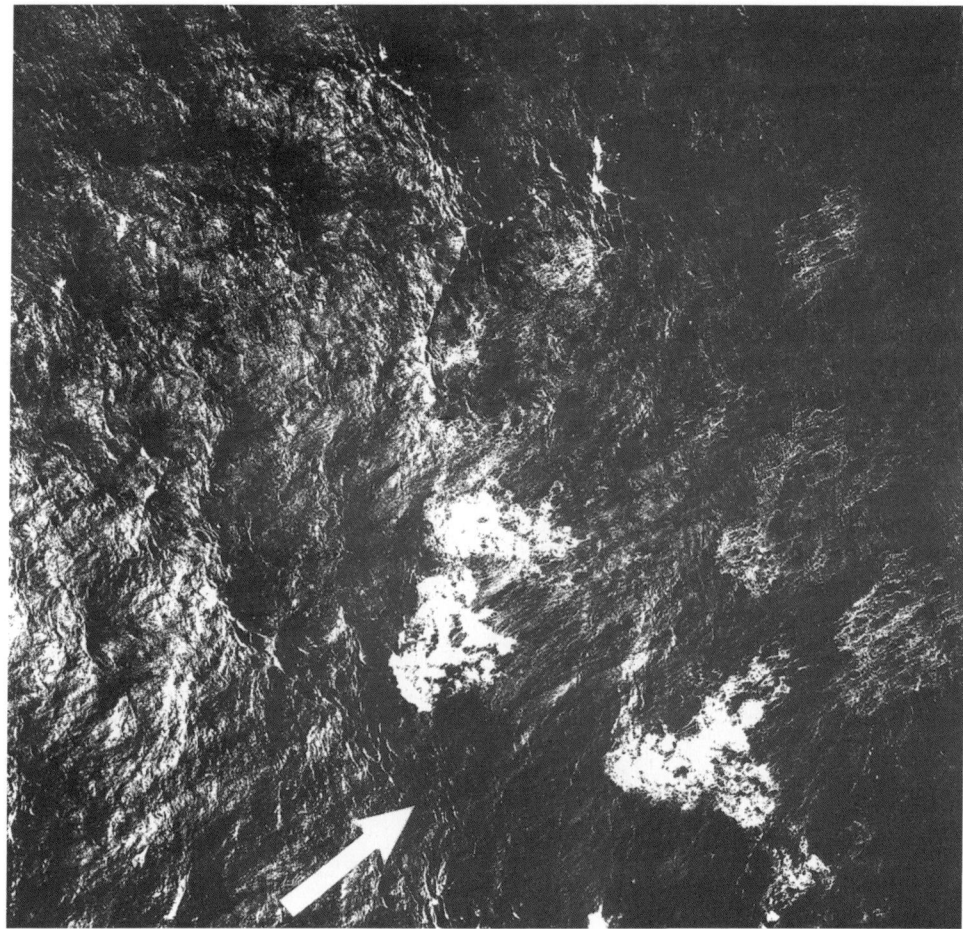

**Figure 3.1** (*cont.*).  (b) Three fields of residual foam with thread foam fields extending along the wind direction.

calculated, and the position of characteristic dimensions relative to each other and with respect to the general direction of sea waves was determined. Then, realizations of the ranked sets and quantitative characteristics obtained for various hydrometeorological conditions were constructed. The total volume of a ranked set of elements of foam system images for the studied hydrometeorological conditions varied from 165 to 311 elements (see Tables 3.1, 3.2, and Figures 3.2, 3.5 for more details). Issues concerning the contribution of subjective estimation by operators in recognizing optical images of various foam systems have been repeatedly discussed (Monahan, 1982; Bortkovskiy, 1983, 1987; Glazman and Weichman, 1990). Special checking in the course of processing these data has shown that the individual features of the operators who recognized the various foam structures did not introduce substantial

**Table 3.1.** The parameters and confidence bounds for the distributions of foam structure geometrical sizes.

| Wind velocity V (m/s) | Foam structure type (sampling volume) | | Parameters of distributions | | | | The confidence bounds | | | | Approximate law type |
|---|---|---|---|---|---|---|---|---|---|---|---|
| | | | $M$ | $M/\sigma$ | $\gamma_1$ | $\gamma_2$ | $P=0.6$ | | $P=0.9$ | | |
| | | | | | | | LB | UB | LB | UB | |
| 1 | 2 | | 3 | 4 | 5 | 6 | 7 | 8 | 9 | 10 | 11 |
| 5.7 | Foam (165) | $L_{min}$ | 3.65 | 0.34 | 1.68 | 4.13 | 2.7 | 4.9 | 2.2 | 6.3 | G |
| | | $L_{max}$ | 6.6 | 0.41 | 3.18 | 19.01 | 4.8 | 8 | 3.7 | 11.5 | G |
| | Whitecappings (212) | $L_{min}$ | 0.39 | 0.71 | 1.51 | 3.24 | 0.16 | 0.52 | 0.08 | 1.0 | LN |
| | | $L_{max}$ | 2.09 | 0.46 | 1.42 | 4.21 | 1.25 | 2.8 | 0.8 | 3.05 | G |
| | Orientation | | −45.9 | 0.78 | 1.12 | −0.72 | | | | | N |
| 9.5 | Foam (216) | $L_{min}$ | 3.86 | 0.40 | 2.33 | 9.70 | 2.7 | 4.9 | 2.2 | 6.3 | G |
| | | $L_{max}$ | 7.5 | 0.37 | 1.05 | 1.21 | 5.4 | 10.1 | 3.9 | 13.9 | G |
| | Whitecappings (221) | $L_{min}$ | 0.55 | 0.37 | 2.41 | 10.6 | 0.41 | 0.69 | 0.3 | 0.9 | LN |
| | | $L_{max}$ | 2.71 | 0.18 | 1.29 | 1.94 | 1.7 | 3.25 | 1.4 | 4.8 | G |
| | Orientation | | −12.9 | 3.1 | 0.46 | −0.26 | | | | | N |
| 10.5 | Foam (308) | $L_{min}$ | 4.20 | 0.49 | 2.27 | 11.56 | 2.3 | 5.9 | 1.1 | 8.3 | G |
| | | $L_{max}$ | 7.45 | 0.38 | 1.02 | 1.27 | 5.4 | 10 | 3.8 | 13.4 | G |
| | Whitecappings (311) | $L_{min}$ | 0.56 | 0.52 | 2.11 | 7.19 | 0.37 | 0.67 | 0.15 | 1.1 | G |
| | | $L_{max}$ | 2.91 | 1.07 | −0.81 | −1.94 | 1.8 | 3.75 | 1.35 | 5.5 | G |
| | Orientation | | −9.4 | 4.5 | 0.25 | −0.56 | | | | | N |

*Comments*: G is the gamma distribution; LN is the lognormal distribution; N is the normal distribution; and UB and LB are the upper and lower bands.

**Table 3.2.** The parameters and confidence bounds for the distributions of foam area sizes.

| Wind velocity $V$ (m/s) | The foam structure type (sampling volume) | Symbol | Parameters distributions | | | | Confidence bounds with probability $P$ | | | |
| --- | --- | --- | --- | --- | --- | --- | --- | --- | --- | --- |
| | | | | | | | $P = 0.6$ | | $P = 0.9$ | |
| | | | $M$ | $M/\sigma$ | $\gamma_1$ | $\gamma_2$ | Lower bound | Upper bound | Lower bound | Upper bound |
| $V = 5.7$ | Foam (165) | $S^S$ | 17,7 | 1.02 | 3.09 | 10.52 | 8.8 | 31,2 | 2.4 | 72,9 |
| | | $S^S_O$ | 0.712 | 1.26 | 3.2 | 12.6 | 0.17 | 1.14 | 0.04 | 2.8 |
| | Whitecappings | $S^W$ | 0.52 | 1.11 | 3.73 | 22.19 | 0.16 | 0.98 | 0.04 | 1.95 |
| | (212) | $S^W_O$ | 0.0146 | 1.24 | 2.31 | 6.97 | 0.0031 | 0.022 | 0.002 | 0.045 |
| $V = 9.5$ | Foam (216) | $S^S$ | 23,2 | 0,67 | 2.46 | 10,77 | 10,4 | 38 | 3,8 | 72,9 |
| | | $S^S_O$ | 1.13 | 0.93 | 2.36 | 7.10 | 0.4 | 2.2 | 0.1 | 3.0 |
| | Whitecappings | $S^W$ | 0.93 | 0.75 | 3.28 | 15.2 | 0.34 | 1.21 | 0.16 | 2.08 |
| | (221) | $S^W_O$ | 0.04 | 0.71 | 1.67 | 3.22 | 0.016 | 0.067 | 0.004 | 0.12 |
| $V = 10.5$ | Foam (308) | $S^S$ | 26.3 | 1.02 | 3.64 | 25.01 | 10.08 | 51 | 4.2 | 80 |
| | | $S^S_O$ | 1.62 | 0.80 | 1.22 | 0.92 | 0.6 | 3.2 | 0.16 | 4.8 |
| | Whitecappings | $S^W$ | 1.2 | 2.1 | 10.87 | 6.17 | 0.36 | 1.39 | 0.21 | 2.24 |
| | (311) | $S^W_O$ | 0.07 | 1.08 | 4.36 | 16.6 | 0.015 | 0.11 | 0.003 | 0.21 |

*Comments:* Unit symbols for foam area quantities $S^S$–$S^W$ are presented in m². Foam fractional coverage quantities $S^S_O$ and $S^W_O$ are given as percentages.

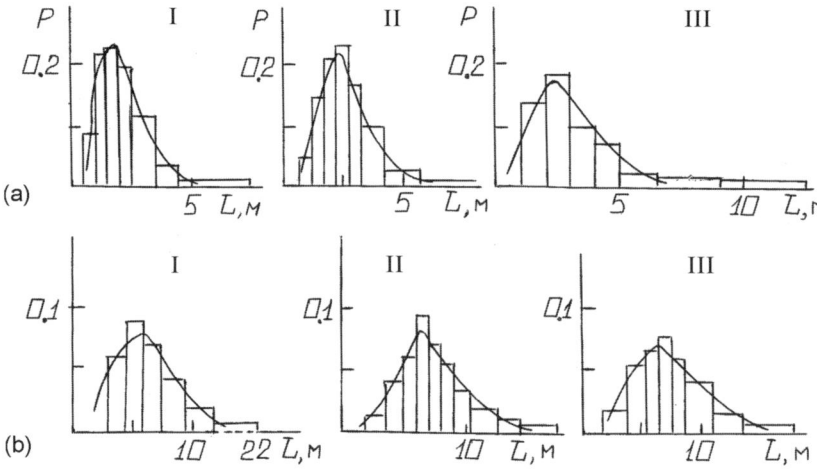

**Figure 3.2.** Experimental histograms of the distribution in frequency of maximum sizes of individual foam structures for two types within a $100 \times 100\,\mathrm{m}^2$ spatial window and their approximated theoretical distributions: (a) whitecap foam at wind velocity $V = 5.7\,\mathrm{m\,s}^{-1}$ and sampling volume $n = 212$ elements (I); $V = 9.5\,\mathrm{m\,s}^{-1}$, $n = 221$ (II); $V = 10.5\,\mathrm{m\,s}^{-1}$, $n = 311$ (III); (b) residual foam at $V = 5.7\,\mathrm{m\,s}^{-1}$, $n = 165$ (I); $V = 9.5\,\mathrm{m\,s}^{-1}$, $n = 216$ (II); $V = 10.5\,\mathrm{m\,s}^{-1}$, $n = 308$ (III). The solid lines show the gamma distributions using the parameters obtained while processing the corresponding histograms (see Table 3.1).

error into the statistics of the parameters studied. As a result of digitizing, we obtained:

- samples of data files of individual formation areas $S_j^K$ for strip foam $S^S$ and whitecap foam $S^W$;
- samples of data files of coverage areas $S_{j0}^K$ of strip foam $S_0^S$ and whitecap foam $S_0^W$ (in magnitudes of coverage areas) $(\mathrm{m}^2)$ for recalculation of total sea surface area and fractional area values (as percentages);
- samples $L_j^K$ of data files of characteristic minimum $(L_{\min})$ and maximum $(L_{\max})$ linear dimensions of individual formations of whitecap $(L^W)$ and strip $(L^S)$ foam;
- samples of data files of fractional gravity wave breaking event number $N_0$ in the spatial window of $100 \times 100\,\mathrm{m}$ at wind velocities of 5.7, 9.5, and $10.5\,\mathrm{m\,s}^{-1}$ and at a sea surface state of 1, 2–3, and 3–4 Beaufort, respectively.

No whitecap and strip foam structures were found on the film shot at wind velocities of 2 and $3.2\,\mathrm{m\,s}^{-1}$. In addition to studying geometrical size statistics, a special procedure was performed to study the statistical characteristics of azimuth orientation $(\Theta_j, \deg.)$ of the axis of minimum extension of crests $(L_{\min}^W)$ relative to the direction of general wave propagation $(\Theta = 0°)$, which was determined from the data of optical processing of the optical image fragments studied (with a spatial window of $100 \times 100\,\mathrm{m}$).

For each group of data files ($X_j$) of the parameters studied ($X_j = S_j, S_{j0}, L_j, \Theta_j$) histograms (plots of sampling probability) $p(X_j)$ were constructed, these histograms being statistical analogs of the density function of sampling probability[1] of random values under particular experimental conditions (Korn and Korn, 1961):

$$p(X_j) = \frac{1}{n}\sum_{i=1}^{K}\frac{n_i(X_j)}{h_i} \tag{3.1}$$

where

$$n_i(X_j) = \begin{cases} n_i(X_j), & (X_j)_i \le X_j \le (X_j)_{i+1} \\ 0, & X_j < (X_j)_i; \quad X_j > (X_j)_{i+1} \end{cases} \tag{3.2}$$

Here $n_i(X_j)$ is the group frequency of a random ranked set $X_j$ hitting the $i$th class interval with an interval width of $h_i = (X_j)_{i+1} - (X_j)_i$; $n$ is the total volume of a random ranked set of the given element (see above); and $K$ is the number of intervals (categories).

For each group of data files, ranked mean values $M$ and dispersions $D$ were calculated (with regard to Sheppard corrections on grouping) (Korn and Korn, 1961) as:

$$M[X_j] = \bar{X}_j = \frac{1}{n}\sum_{i=1}^{K}n_i(\bar{X}_j)_i \tag{3.3}$$

$$D[X_j] = \sigma^2[X_j] = \frac{1}{n-1}\sum_{i=1}^{K}n_i\{(\bar{X}_j)_i - M[X_j]\}^2 - \frac{1}{12}h^2 \tag{3.4}$$

where $(X_j)_i$ is the mean value of the parameter studied for each category; and $n_i$ is the quantity of items in each category. Statistics of type (3.3) and (3.4) are unbiased estimates of the mean value and dispersion and satisfy efficiency and consistency conditions. Moreover, to select the form of the distribution function of ranked sets $X_j$ we calculated the values of coefficients of asymmetry $\gamma_1$ and excess $\gamma_2$ (with regard to Sheppard corrections on grouping) (Korn and Korn, 1961) as:

$$\gamma_1 = \frac{1}{D^{3/2}[X_j]}\left\{\frac{1}{n}\sum_{i=1}^{K}n_i\{(\bar{X}_j)_i - M[X_j]\}^3 - \frac{1}{4}M[X_j]h^2\right\} \tag{3.5}$$

$$\gamma_2 = \frac{1}{D^2[X_j]}$$

$$\times\left\{\frac{1}{n}\sum_{i=1}^{K}n_i\{(\bar{X}_j)_i - M[X_j]\}^4 - \frac{1}{2}\frac{1}{n-1}\sum_{i=1}^{K}n_i\{(\bar{X}_j)_i - M[X_j]\}^2h^2 + \frac{7}{400}h^4\right\} - 3 \tag{3.6}$$

The results of processing for all the parameters studied are presented in Tables 3.1 and 3.2 and in Figures 3.2–3.8.

---

[1] Note that, in accordance with the definition of formula (3.1), this function is just the named quantity.

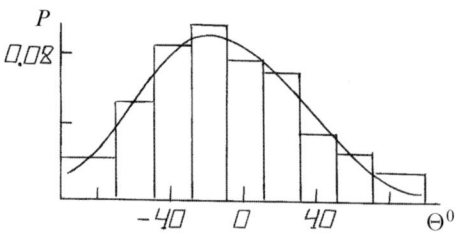

**Figure 3.3.** Experimental histogram of the distribution in frequency of minimum axis for whitecap crests which relates to the general sea direction within the spatial window $100 \times 100\,\text{m}^2$ at wind velocity $V = 10.5\,\text{m s}^{-1}$. The solid line shows the normal distribution using the parameters obtained while processing the corresponding histograms (see Table 3.1).

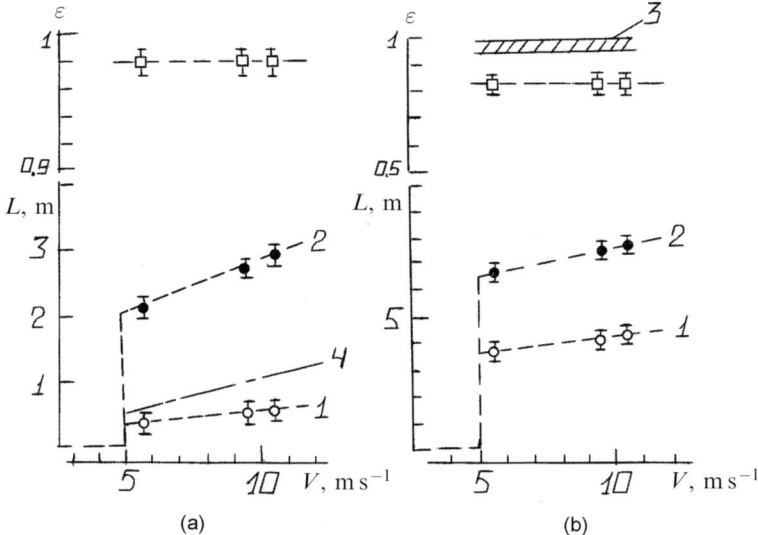

**Figure 3.4.** Dependence of the average values for minimum (1) and maximum (2) dimensions and the eccentricity of whitecap (a) and residual (b) individual foam structures as a function of wind velocity. The vertical line segments show 60% confidence intervals with regard to sampling volume. Zone (3) is the 60% confidence interval of the eccentricity of whitecaps; (4) is the size approximation of crests in the wind direction using data by Bortkovskiy (1983).

## 3.4  STATISTICS ON THE ELEMENTS IN THE LINEAR GEOMETRY OF INDIVIDUAL FOAM STRUCTURES IN THE WAVE-BREAKING PROCESS

Results of detailed statistics on the elements involved in the linear geometry of individual foam structures in the wave-breaking process—namely, characteristic minimum and maximum linear dimensions of individual formations of crest and strip foam fields and spatial orientation of crests—are presented in Figures 3.2 and 3.3 and in Table 3.1 as functions of wind velocity.

The distribution densities of the linear dimensions of foam structures of two classes, obtained from the experimental data, strongly differ from the normal distribution (positive asymmetry and excess), as expected. This is clearly seen from analysis of Figure 3.2 and the data of Table 3.1. Experimental ranked sets have been equalized in a wide class of distributions according to well-known rules (Johnson and Leone, 1977). Analysis of the probabilities of concordance between theoretical and experimental distributions has shown that the linear geometrical dimensions of foam structures obey the gamma distribution with the parameters indicated in Table 3.1. The solid curves in Figure 3.2 illustrate theoretical gamma distributions:

$$p(X_j) = \frac{1}{\Gamma(\eta)} \lambda(\lambda X_j)^{\eta-1} \exp(-\lambda X_j) \tag{3.7}$$

where $\eta = 6/\gamma_2 = 4/\gamma_1^2$; $\lambda = (\eta/\sigma^2)^{1/2}$; and $\Gamma(\eta)$ is the gamma-function.

The experimental distributions obtained are stable and only weakly deform depending on wind force variation in limits from 5 to $10.5\,\mathrm{m\,s^{-1}}$. So, we can observe small variations in the asymmetry coefficient within the limits of 1.5–3 and in the excess coefficient within the limits of 2–10. Since the experimental distributions possess strong tails ($\gamma_1 > 1$), the confidence intervals indicate a certain asymmetry between upper and lower limits with respect to a ranked mean value. We should note here that, in such a case, ranked dispersion cannot be assumed to be an acceptable measure of the reliability of results; and, so, when studying the foam field statistics it is necessary to perform special estimations of confidence intervals with corresponding levels of confidence probability (see Table 3.1 and Figure 3.2).[2]

The ranked mean values of $L_j^K$ have a weak, but clear tendency toward linear growth with increasing wind force (Figure 3.4). The dependencies of mean $L_j^K$ values clearly exhibit "thresholds" and, so, corresponding stepwise approximations can be offered for linear dimensions of crests (the "width" and "length" of a crest) and spotty structures:

$$L_j^K = \begin{cases} 0, & V < 5\,\mathrm{m\,s^{-1}} \\ a[1 + b(V - 5)], & V \geq 5\,\mathrm{m\,s^{-1}} \end{cases} \tag{3.8}$$

where for $L_{\min}$ of crests $a = 0.39$, $b = 0.034$; for $L_{\max}$ $a = 2.09$, $b = 0.16$; for $L_{\min}$ of spotty foam $a = 3.65$, $b = 0.11$; and for $L_{\max}$ $a = 6.6$, $b = 0.17$.

Figure 3.4 also presents experimental data on the linear dimensions of crests known from the literature (Bortkovskiy, 1983, 1987). With regard to distinguishing

---

[2] Related attempts at discovering the statistics of the linear dimensions of foam systems of gamma-distribution type were obtained much later, after analysis (using a similar technique) of aerial photography of the oceanic surface in the northwestern part of the Pacific Ocean, near the Kamchatka Peninsula (Raizer *et al.*, 1994), as well as in the tropics and in Southern Ocean waters from onboard a research ship (Bortkovskiy, 1987).

the technique of surveying and data processing, the conformity of our experiments with literature data seems quite satisfactory.

Analysis of the whole data file of samples of foam whitecap structures, localized at the breaking wave front, showed that in the overwhelming majority of cases the maximum dimension $L_{max}$ of crests is oriented along the local front of a breaking wave and can be (by Bortkovskiy's terminology, 1983) called the "width" of a crest, while the minimum extension $L_{min}$ of a crest is oriented perpendicular to the local front of a breaking wave—by the terminology of Bortkovskiy (1983) the length of a whitecap. We emphasize that—in analyzing the orientation of linear dimensions—the question is about local fronts of breaking waves and, so, this orientation should not be associated with wind vector orientations (as done, for example, in Bortkovskiy's, 1983, paper), because essential azimuthal "blurring" of the orientation of axes of crest formations (in the spatial frame of $100 \times 100$ m) with respect to some chosen direction is observed (see below).

In the practice of remote observations of a disturbed sea surface, the orientation of a breaking wave is usually made to correspond with the direction of a vector of the velocity of near-surface wind or with the general direction of sea waves. However, analysis of histograms of the distribution of azimuth orientations of the axis of minimum dimension of a crest with respect to the direction of a dominating energy-carrying component in the spatial window of the $100 \times 100$ m frame (Figure 3.3 and Table 3.1) shows that histograms can be approximated by a distribution that is close to a Gaussian one, and with a wide angular dispersion sector (of about $40°$). No obvious dependence of dispersion on wind velocity is seen; only small variations of $\sigma(\theta)$ are observed within the limits of $36$–$42°$ (see Table 3.1). Thus, estimation of the direction of the wind velocity (or sea wave) vector based on individual observations in the spatial window of $100 \times 100$ m can result in considerable errors in the values of these parameters.

The high spatial resolution provided in the experiments under consideration (Bondur and Sharkov, 1982, 1990) made it possible to obtain information on the geometrical shape of individual foam structures of the two types studied. Here, an interesting, from a physical viewpoint, and practically important, from a remote sensing viewpoint, is the following experimental fact: foam systems possess a steady geometrical ellipsoidal shape with a stable (on average) eccentricity $\varepsilon = [1 - (L_{min}/L_{max})^2]^{1/2}$, which is equal to $0.98 \pm 0.007$ for crest structures and to $0.82 \pm 0.08$ for spotty foam. Note that (with a reliability of 0.5) foam systems can be distinguished by the value of their eccentricity (see Figure 3.4), and this circumstance can serve as one of the criteria for automatic recognition and classification of foam systems by their optical images against the background of a lightening sea surface. On the other hand, however, the geometrical shape does not depend and linear dimensions only weakly depend on wind conditions (i.e., strength of the wind). For this reason, neither the shape nor linear dimensions can serve as an effective criterion for estimating the force of near-surface wind. This, however, is not true in relation to the dependencies of fractional foam coverage and fractional density of foam structures (the dissipation centers of wave energy) on wind conditions (see Sections 3.5 and 3.6).

The statistical characteristics of the linear dimensions of foam systems that form as a wave breaks, presented in Section 3.4, have been used as a fundamental basis in constructing a model of radar scattering from the sea surface at small grazing angles (Malinovskii, 1991).

## 3.5   STATISTICS OF ELEMENTS OF THE TWO-DIMENSIONAL GEOMETRY OF INDIVIDUAL FOAM STRUCTURES IN THE WAVE-BREAKING PROCESS

The results of investigation of the statistics of the two-dimensional geometry elements of individual foam structures in the wave-breaking process—namely, characteristic areas of individual formations of crest and strip foam fields—as well as of the statistics of fractional foam coverage areas of crest and strip foam fields in the spatial frame of $100 \times 100$ m are presented in Figures 3.5–3.8 and in Table 3.2 as functions of wind velocity.

The densities of the distribution of fractional foam coverage areas for two classes of foam structures, obtained from the experimental data, strongly differ from the Gaussian distribution (i.e., positive asymmetry and positive excess). This is clearly seen from analysis of the data of Table 3.2 and Figure 3.5.

The experimental distributions that were obtained as a whole (Figure 5.15) are stable and only slightly deform depending on a change in wind strength within the limits of 5 to $10.5\,\mathrm{m\,s^{-1}}$ (e.g., the asymmetry coefficient changes within limits of

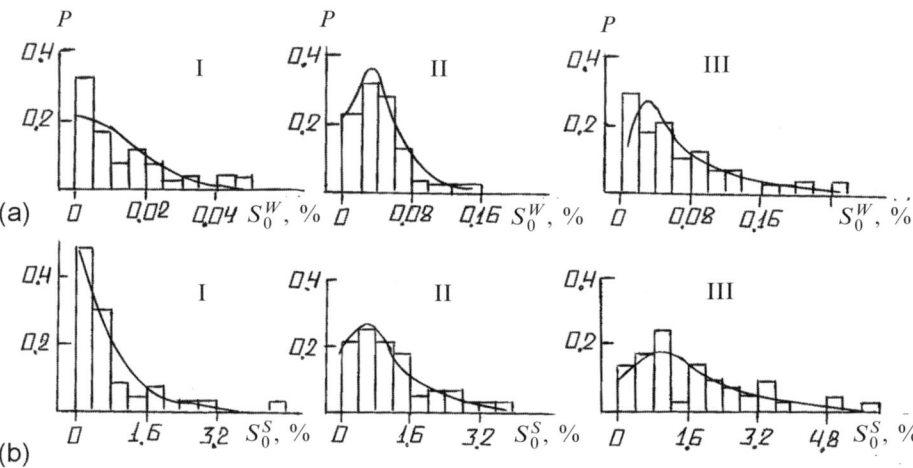

**Figure 3.5.** Experimental histogram of the distribution in frequency of relative foam coverage within a spatial window $100 \times 100\,\mathrm{m^2}$; (a) whitecap foam; (b) residual foam at various wind velocities (I at $V = 5.7\,\mathrm{m\,s^{-1}}$ ; II at $V = 9.5\,\mathrm{m\,s^{-1}}$; III at $V = 10.5\,\mathrm{m\,s^{-1}}$). The solid line shows the gamma distribution using the parameters obtained while processing the corresponding histograms (see Table 3.2).

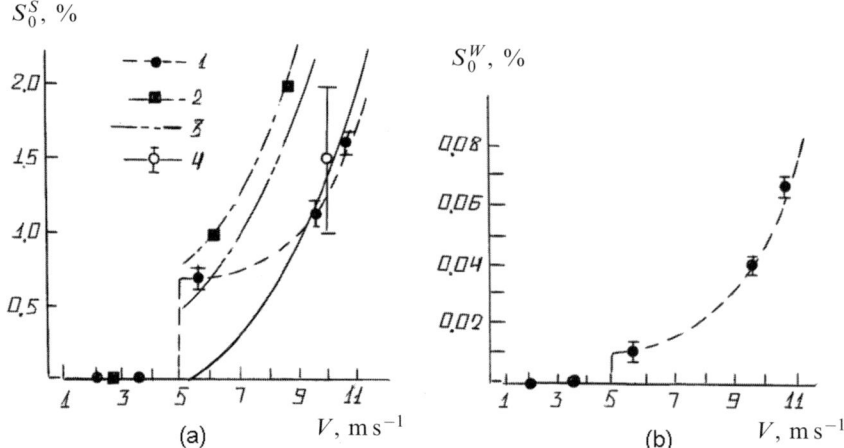

**Figure 3.6.** Dependence of the average values for relative foam coverage within a spatial window $100 \times 100 \, m^2$ for residual (a) and whitecap (b) foam structures as a function of wind velocity (at $H = 19.5 \, m$): 1—average values (points) with confidence intervals of variance estimation for sampling element volumes of two foam types (see data in Table 3.1 and Figure 3.2) (the dotted curves are approximations (3.9) and (3.10)); 2—experimental data (Samoilenko *et al.*, 1974); 3—approximation of data by Bortkovsky and Kuznetsov (1977 ); 4—the experimental point and approximation of Ross and Cordon (1974) .

2 to 4). The probability of correspondence of experimental histograms (3.7) to gamma distributions was estimated according to Pearson's criterion of consent and constituted not less than 0.5–0.6 for all cases. The ranked mean value of distributions has an obvious tendency to grow with increasing wind strength (these dependencies will be discussed in more detail below); such a feature was also shown for a sampling dispersion. By virtue of the presence of a strong "tail" (high asymmetry coefficients) in experimental distributions, the sampling dispersion cannot be considered as a measure of distribution-type characterization, though it can be defined as the degree of reliability of measurements (as done, for example, in Ross and Cordon, 1974). Special calculations of confidence intervals with corresponding levels of confidential probability were carried out (presented in Table 3.2). The results of calculations of confidence intervals, carried out with reliability levels of 0.6 and 0.9, exhibit the sharp asymmetry of the upper and lowest limits with respect to a ranked mean value (see Table 3.2 and Figure 3.5).

Let us now consider a practically important (in oceanology and remote sensing) question about the dependence of the mean values of fractional foam coverage areas for two classes of foam structures in the spatial frame of $100 \times 100 \, m$ on wind velocity. Analysis of the data obtained makes it possible, first, to reveal the presence of a rigid threshold region of wind velocity values in forming foam structures, which was also noticed earlier (Monahan, 1971; Ross and Cordon, 1974; Samoilenko *et al.*, 1974). The threshold region varies for the water areas studied (the Black Sea with a water temperature of $12°C$) from 4.5 to $5.5 \, m \, s^{-1}$ (Figure 3.6). The second, more important, point is the fact of finding—while analyzing our results—the threshold in the mean values of fractional foam coverage areas (Figure 3.6). This result can be characterized as a concept of the threshold genesis of foam systems, which was offered for the first time by Bondur and Sharkov (1982). The mean values of fractional foam coverage areas (as a percentage) can be approximated—as well as the wind dependencies of the characteristic dimensions of foam structures of the two types (3.8), based on experimental data (Bondur and Sharkov, 1982)[3]—by stepwise functions (the dotted line in Figure 3.6). These are for strip structures

$$S_0^S = \begin{cases} 0, & V < 5\,\mathrm{m\,s}^{-1} \\ 0.65[1 + 4.76 \cdot 10^{-2}(V - 5)^2], & V \geq 5\,\mathrm{m\,s}^{-1} \end{cases} \tag{3.9}$$

and for crests

$$S_0^W = \begin{cases} 0, & V < 5\,\mathrm{m\,s}^{-1} \\ 0.015[1 + 2.2 \cdot 10^{-2}(V - 5)^2], & V \geq 5\,\mathrm{m\,s}^{-1} \end{cases} \tag{3.10}$$

It is interesting to note that approximations of the fractional foam coverage areas obtained by other authors virtually coincide with our value ($S_0^S = 0.65\%$) for wind velocity values close to $5\,\mathrm{m\,s}^{-1}$. So, according to the data of Bortkovskiy and Kuznetsov (1977), $S_0^S = 0.5\%$ by approximating these data to a wind velocity of $0.5\,\mathrm{m\,s}^{-1}$ (curve 3 in Figure 3.6a). By approximating the data by Samoilenko *et al.* (1974) to a wind velocity of $0.5\,\mathrm{m\,s}^{-1}$ (curve 2 in Figure 3.6a) the fractional area of coverage by strip foam equals 0.75%. For better clarity the approximation curves of these authors are interrupted at a wind velocity of $5\,\mathrm{m\,s}^{-1}$. Figure 3.6a presents an experimental point and the approximation by Ross and Cordon (1974) (curve 4), which has frequently been used for application purposes. Analysis of the data of Figure 3.6a indicates that for wind velocities exceeding $9\,\mathrm{m\,s}^{-1}$ this approximation agrees quite well with other authors' data, and for wind velocities lower than $9\,\mathrm{m\,s}^{-1}$ there is no obvious correspondence of the approximation with the concept of threshold genesis of foam systems.

When discussing the data of Figure 3.6 we should pay attention to the fact that in their original paper Bondur and Sharkov (1982) chose values of confidence intervals at a probability of 0.6, as an estimate of actual scattering around three points corresponding to the mean foam coverage values for three values of wind velocity, because—by virtue of the presence of a strong "tail" (high asymmetry coefficients) in experimental distributions—sampling dispersion cannot be considered as a measure of distribution-type characterization, though it can be defined as the degree of reliability of measurements.

It was this circumstance, which led to vigorous discussion (Glazman and Weichman, 1989, 1990; Monahan, 1990) regarding the interpretation of the data in Figure 3.6. Glazman and Weichman (1990) correctly interpreted the original data of Bondur and Sharkov (1982). Since the volume of ranked sets (*n*) of foam structure elements, from which experimental histograms were constructed and the mean values

---

[3] This concept was confirmed much later by wave experiments (Liu, 1993).

of distribution parameters were determined, constituted 165 to 311 elements (see Table 3.1), Figure 3.6 presents, as an estimate of the degree of reliability of experimental measurements of the mean value, the values of ranked standard deviations at estimating the ranked mean value with a ranked set volume $n$; namely, $[\mathbf{D}(S_{S0})]^{1/2} \approx (\sigma^2/n)^{1/2}$, where $\sigma^2$ is the sampling dispersion of a corresponding distribution—see formula (3.4). Analysis of the data in Figure 3.6, which presents estimates of ranked standard deviations at estimating the ranked mean value (the gravity center) for three experimental sets (files) of data around gravity centers, convincingly demonstrates the reliability of the data obtained by Bondur and Sharkov (1982) and the groundlessness of the claims advanced by Monahan (1990).

Third, contrary to the statement of Ross and Cordon (1974)—widespread in the literature—that strip foam only appears at wind velocity values higher than $9\,\mathrm{m\,s^{-1}}$,[4] Bondur and Sharkov (1982) stated that the "threshold" genesis of both crests (whitecappings) and strip structures occurs virtually simultaneously, which is a natural continuation of the time evolution of "dynamic" foam formations.

It was found in Bondur and Sharkov (1982) that—under the hydrometeorological conditions studied—strip foam accompanies the crests at all wind strength values and the contribution of strip structures to total foam coverage (in the relative area) is overwhelming (see Figure 3.6b). The ratio of the area of these structures to the area of crests $R = S_0^S/S_0^W$ is determined by the following approximate formula:

$$R = R_0 - a(V - 5) \quad \text{for } V > 5\,\mathrm{m\,s^{-1}} \tag{3.11}$$

where $R_0 = 40$; and $a = 3.4\,\mathrm{m\,s^{-1}}$.

It follows from formula (3.11) that as wind strength grows the ratio $R$ decreases from 40 down to 23 according to a linear law. Such a nonzero value of coefficient $R$ for threshold values of wind velocity $V \approx 5\,\mathrm{m\,s^{-1}}$ is naturally and most likely associated with the considerable lifetime of "strip" foam. Because the area of generated crests grows as the wind strengthens, the ratio $R$ will decrease. However, as follows from the results of Bortkovskiy and Kuznetsov (1977) and Ross and Cordon (1974), the size of the relative area occupied by crests probably tends to saturation ($S_0^W \rightarrow 6$–7%) at wind velocities $V = 15$–$20\,\mathrm{m\,s^{-1}}$. So, the ratio $R$ will most likely begin to increase with further strengthening of the wind ($V > 20\,\mathrm{m\,s^{-1}}$).

Analysis of confidence intervals $S_0^S$ and $S_0^W$ with confidential probabilities of 0.6 and 0.9 (Table 3.2) indicates that wind velocity estimations within the range of $5$–$11\,\mathrm{m\,s^{-1}}$, based on individual measurements of fractional foam coverage on spatial sections of water areas of size $100 \times 100\,\mathrm{m}$, can result in considerable errors. A similar conclusion can also be drawn indirectly from analysis of the data of many authors (Ross and Cordon, 1974; Bortkovskiy and Kuznetsov, 1977; Bortkovskiy, 1983, 1987), though these authors did not present estimates of the reliability of their results. Some unambiguity in the solution of this practically important question can be obtained by using larger two-dimensional realizations of the spatial frame of observation. To determine the size of a spatial frame $S_Z$ that provides the required statistical

---

[4] In Ross and Cordon (1974) no experiments were carried out at wind velocity values lower than $10\,\mathrm{m\,s^{-1}}$, so the statement of the authors cannot be substantiated experimentally.

accuracy of estimation of mathematical expectation ($M$), we shall make use of the relation (Bendat and Piersol, 1966)

$$S_Z = (2\beta\varepsilon^2)^{-1}(\sigma/M)^2 \qquad (3.12)$$

where $\varepsilon^2$ is the square of a normalized root-mean-square error; and $\beta$ is the width of a band of spatial frequencies.

Calculations have shown that the 5% accuracy of estimation of a mathematical expectation of fractional foam coverage areas for the wind velocity range of 5–11 m s$^{-1}$ can be obtained at a spatial frame size greater than $3 \times 3$ km$^2$, which can be accomplished both in aircraft and in ship conditions of observation of a disturbed sea surface.

The high spatial resolution provided in the experiments by Bondur and Sharkov (1982) allowed us to obtain—unlike earlier papers by Monahan (1971); Ross and Cordon (1974); Samoilenko *et al.* (1974); Bortkovskiy and Kuznetsov (1977); Bortkovskiy (1983, 1987)—information about areas of individual formations such as crests and strip structures. The character of the distributions of individual foam coverage areas of the two classes is virtually the same as for distributions of total areas. They are close, in their class, to the gamma distribution; and they have high positive values in their excess and asymmetry coefficients (see Table 3.2 and Figure 3.7). The parameters of distributions remain stable as wind velocity changes. It is interesting to note that, as in the case of fractional foam coverage area, the dependencies of the mean values of absolute areas of individual formations on wind velocity (Figure 3.8) exhibit clear "thresholds", and corresponding stepwise

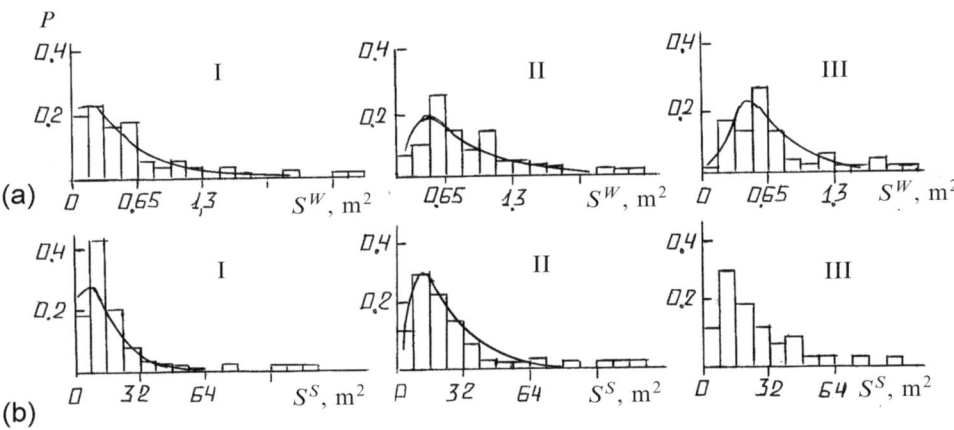

**Figure 3.7.** Experimental histogram of the distribution in frequency of coverage areas of individual structures within a spatial window $100 \times 100$ m$^2$; (a) whitecap foam; (b) residual foam at various wind velocities (I at $V = 5.7$ m s$^{-1}$; II at $V = 9.5$ m s$^{-1}$; III at $V = 10.5$ m s$^{-1}$). The solid line shows the gamma distribution using the parameters obtained while processing the corresponding histograms (see Table 3.2).

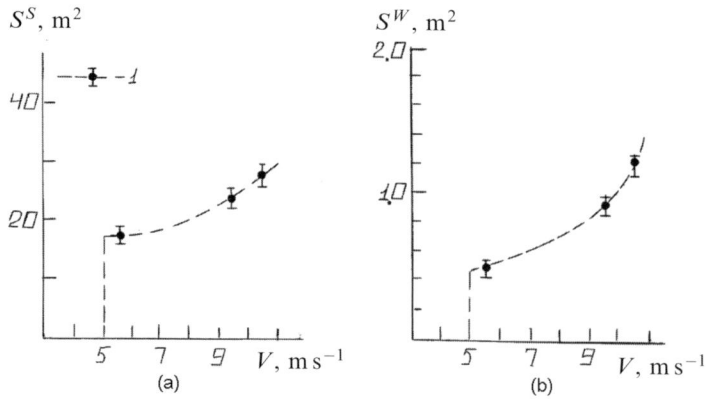

**Figure 3.8.** Dependence of average values for coverage areas of individual structures within a spatial window $100 \times 100\,\text{m}^2$ for residual (a) and whitecap (b) foam structures as a function of wind velocity (at $H = 19.5\,\text{m}$): 1—average values (points) with confidence intervals of variance estimations for sampling element volumes of two foam types (see data in Table 3.1 and Figure 3.7) (the dotted curves are approximations (3.13) and (3.14)).

approximations can be offered. These are for the absolute area ($\text{m}^2$) of individual strip formations in the form:

$$S^S = \begin{cases} 0, & V < 5\,\text{m s}^{-1} \\ 17 + 3.07 \cdot 10^{-2}(V - 5)^2, & V \geq 5\,\text{m s}^{-1} \end{cases} \qquad (3.13)$$

and for crests

$$S_0^W = \begin{cases} 0, & V < 5\,\text{m s}^{-1} \\ 0.4 + 3.8 \cdot 10^{-2}(V - 5)^2, & V \geq 5\,\text{m s}^{-1} \end{cases} \qquad (3.14)$$

Analysis of Figure 3.8 indicates that there exists a tendency (being quadratic with respect to wind velocity) for the areas of individual formations both for strip and for crest foams to grow, though this increase is not so highly expressed in magnitude. So, as the wind velocity changes from $5.7\,\text{m s}^{-1}$ to $10.5\,\text{m s}^{-1}$, the mean area of crest foam changes from $0.5\,\text{m}^2$ to $1.6\,\text{m}^2$ and the mean area of strip foam increases from 17 to $26\,\text{m}^2$.

We should pay attention to the considerable values of confidence intervals in statistical data files $S^S$ and $S^W$, especially for $p = 0.9$ (see Table 3.2). This circumstance does not allow us to be sufficiently confident about the possibility of estimating the wind velocity from measurements in the spatial frame of $100 \times 100\,\text{m}$ of areas of individual formations, though such a question could arise under appropriate statistical reliability.

An important conclusion of the qualitative theory of the evolution of energy spectra in developed sea waves is the dependence between the dissipated energy of a breaking wave and the foam coverage area (the crests). Based on this conclusion and on the experimental data by Bondur and Sharkov (1982), we can suppose the existence: (1) of a "threshold" of dissipated energy (per unit of sea surface area) at wave breaking; and (2) of a weak cubic dependence of this energy on growing wind strength (at least in the range of wind velocities of 5–10.5 m s$^{-1}$). In addition, simultaneous reception of information on relative and absolute areas of coverage by whitecap foam in our experiments (only one type of foam systems is considered here) allows us to introduce the following coefficient:

$$Q = \frac{S_0^W}{S^W} \qquad (3.15)$$

which represents, in its physical sense, the specific (per unit sea surface area) density of energy dissipation centers. Using the data of Table 3.2, the coefficient can be approximated by the following stepwise function:

$$Q(\text{km}^{-2}) = \begin{cases} 0, & V < 5 \text{ m s}^{-1} \\ 281[1 + 6.37 \cdot 10^{-3}(V-5)^3], & V \geq 5 \text{ m s}^{-1} \end{cases} \qquad (3.16)$$

Here we have chosen a more natural—for remote-sensing practice—normalization of the density of energy dissipation centers with respect to a sea water area that is equal to 1 km$^2$. Note that, unlike the aforementioned quadratic approximations of wind dependencies of the areas and linear dimensions of foam systems, the present dependence (3.16) is cubic, which was confirmed subsequently from theoretical considerations (Phillips, 1988). It follows from (3.16) that the threshold density of wave energy dissipation centers, arising at $V = 5$ m s$^{-1}$, equals about 280 centers per 1 km$^2$ and, further, this value sharply grows[5] (according to the cubic law) and reaches about 580 centers per 1 km$^2$ at $V = 10.5$ m s$^{-1}$.

As already noted (see Chapter 1 and Section 2.5), the geometrical shape of foam systems formed on a disturbed sea surface has no smooth outlines, and the boundaries of foam fields possess a clearly rough structure, most likely of fractal type (see Figure 3.1). Proceeding from these considerations, Raizer and Novikov (1990) and Raizer et al. (1994) undertook attempts to establish the possible fractal dependencies for objects of complicated shape by constructing the so-called $P^2$–$A$ diagram, where $P$ is the perimeter and $A$ is the area of a complicated object with rough boundaries. Should analysis of the geometrical characteristics of a complicated object reveal the following dependence:

$$P \sim (A^{1/2})^{d_F}$$

then $d_F$ is the fractal dimension of a line of contour of a planar figure. Such an approach was applied in studying the structure of space images of various types of cloudiness (Lovejoy, 1982; Baryshnikova et al., 1989). To study the fractal properties of a random two-dimensional point field some other approaches are used (see Section 2.5). The fractal statistics of foam systems was revealed (Raizer and Novikov, 1990; Raizer et al., 1994) from analysis of aerial photography of the oceanic surface in the

[5] Note that Bortkovskiy (1987), after rather complicated approximation procedures, came to the conclusion that $Q$ does not depend on wind velocity, which, of course, contradicts numerous observational data, including those presented in the present book.

northwestern section of the Pacific Ocean, near the Kamchatka Peninsula, under three conditions of sea disturbance (according to the Beaufort Scale): 3–4, 4–5, and 5–6. The authors present files of the processed data in the form of clusters for all foam fields in the optical image frame for three wave states on the $\log P$–$\log(A)^{1/2}$ diagram (fig. 3 in Raizer and Novikov, 1990). According to their data, the coefficient of regression between $\log P$ and $\log(A)^{1/2}$ for a set of points in each cluster gives an estimate of quantity $d_F$ (the authors do not present here the technique used for determination of $d_F$) that equals 1.3, 1.48, and 1.57 for the three gradations in sea surface state. On the other hand, making use of the data presented in fig. 3 of Raizer and Novikov (1990) in the bi-logarithmic coordinates, we can easily see that the average slope of each data file gives a quantity $d_F$ close to unity. In its turn, this implies that the considered structures represent objects with smooth boundaries. This circumstance is most likely related to the insufficient spatial resolution obtained by the authors at discretization (by 256 elements) of an initial optical frame. Namely, the real size of the processed frame was 1.6 km and, thus, the spatial pixel of resolution of the discretized frame was 6.25 m, which is certainly insufficient to reveal the fractal properties of natural foam fields. And, so, the hopes of Raizer and Novikov (1990) to use fractal properties as new characteristics determining the character and extent of sea waves cannot be justified with such spatial resolutions. In a subsequent paper (Raizer *et al.*, 1994) the authors essentially improved the spatial resolution of a discretized frame, having increased its size up to 1.6 m. In this case they managed to perform fractal analysis of individual foam structures and to demonstrate certain distinctions in the values of fractal dimension for crest and strip foam structures: for crest foams it was equal to 1.39, and for strip foam 1.23. Analysis of the data presented results in the following principal conclusion: by studying such delicate characteristics as the fractal properties of complex sets, all geometrical characteristics of both an observation system and a studied physical object should be closely analyzed.

## 3.6  STATISTICS OF THE SPECIFIC DENSITY OF BREAKING CENTERS

The unique spatial resolution (better than 2 cm) over a significant spatial frame of $100 \times 100$ m, achieved in optical remote-sensing experiments (Bondur and Sharkov, 1982, 1990), allows us to obtain a file of statistical data that can be used to show the detailed statistics of individual foam structures of two types, considered as point dissipation centers, as was done in the experiments described in Chapter 2 (Figure 2.6).

The densities of specific density distribution (the quantity of individual formations) of foam structures in the spatial frame of $100 \times 100$ m (and, accordingly, having an area $A = 10^4\,\mathrm{m}^2$), obtained from the experimental data, highly differ from the Gaussian distribution, while possessing positive asymmetry and positive excess (see Table 3.3 and Figure 3.9). Here by "ranked set volume" we mean the number of optical surveying frames of spatial size of $100 \times 100$ m, in which the quantity of foam

**Table 3.3.** Distribution parameters for the specific density of breaking centers ($N_0$).

| Wind velocity $V$ (m/s) | Sampling volume | Distribution parameters | | | |
|---|---|---|---|---|---|
| | | $N_0$ | $\sigma/N_0$ | $\gamma_1$ | $\gamma_2$ |
| 5.7 | 205 | 2.09 | 4.11 | 9.03 | 1.92 |
| 9.5 | 66 | 3.95 | 6.75 | 3.48 | 0.68 |
| 10.5 | 63 | 6.94 | 16.88 | 2.26 | 0.42 |

*Comments*: $N_0$ is the mean value of specific density of breaking centers.

structures of whitecap type, considered as individual point formations, is determined. The solid curves in the same figure illustrate the theoretical Poisson distributions:

$$p(N, A) = \frac{(N_0 A)^N}{N!} \exp(-N_0 A) \qquad (3.17)$$

where $N$ is the specific (per studied area of the frame) density of foam structures; $N_0$ is the mathematical expectation mean of the process; and $A$ is the area of a studied frame. It is known from the theory of random processes (Cramer and Leadbetter, 1967; Karlin, 1968) that $N_0$ can be interpreted as the intensity of the Poisson process (or, what is the same for the Poisson process, its mean value) considered on the area of the frame under study. In the case under consideration this interpretation is obvious: this is the intensity in the appearance of gravitational oceanic wave breaking centers on the water area under study.

The experimental distributions obtained for specific density have a stable character carry, as in the case of linear dimensions; quantities $\gamma_1$ and $\gamma_2$ are weakly deformed depending on the change in wind velocity (Table 3.3). Of most importance

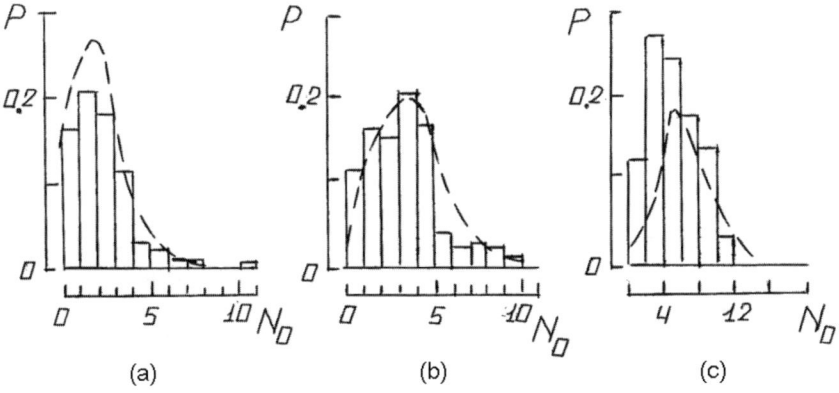

**Figure 3.9.** Experimental histograms of fractional breaking event number within a spatial window $100 \times 100$ m: (a) at $V = 5.7\,\mathrm{m\,s^{-1}}$ and sampling volume $n = 205$ (frames of optical images ); (b) at $V = 9.5\,\mathrm{m\,s^{-1}}$, $n = 66$; (c) at $V = 10.5\,\mathrm{m\,s^{-1}}$, $n = 63$. The dashed lines are Poisson distributions at $N_0 = 2$ (a), 4 (b), and 7 (c) .

for us is the degree of proximity between experimental ranked probabilities and theoretical Poisson distributions. The estimates of sum $X^2$, obtained in accordance with the rules of Pierson's criterion $\chi^2$ (Johnson and Leone, 1977), give for $V = 5.7$ m s$^{-1}$ the value $X^2 = 9$, for $V = 9.5$ m s$^{-1}$, $X^2 = 8.5$, and for $V = 10.5$ m s$^{-1}$, $X^2 = 10$. On the other hand, the 5% critical value of a single-sided criterion $\chi^2(9; 0.35)$ in the case of $V = 5.7$ and $9.5$ m s$^{-1}$ equals 16.9, and for $V = 10.5$ m s$^{-1}$ $\chi^2(5; 0.95) = 11$. That is, the hypothesis as to the Poisson character of specific density distribution can be accepted. In order to avoid misunderstandings, we emphasize once again that the features of foam field statistics found by us relate to the studied spatial window of $100 \times 100$ m. So, with certain confidence we can believe that, as the area of a studied frame (under assumption of spatial steadiness of the whitecapping process) expands, the specific density distributions will tend to a Gaussian one (as actually follows from the known property of Poisson's distribution for $N \to \infty$). The mean value of the specific density of foam structures $N_0$ has an obvious and strong (cubic) tendency to grow as wind velocity strengthens and can be approximated by a stepwise function

$$N_0(\text{km}^{-2}) = \begin{cases} 0, & V < 5\,\text{m s}^{-1} \\ 210[1 + 2 \cdot 10^{-2}(V - 5)^3], & V \geq 5\,\text{m s}^{-1} \end{cases} \tag{3.18}$$

Here we chose a more natural—for remote-sensing practice—normalization of foam structure density per area of $1\,\text{km}^2$. We should have in mind here that the technique of obtaining the approximation dependence (3.18) essentially differs from the technique of forming the dependence (3.16). This fact gives rise to some quantitative distinctions in the parameters of approximation dependencies (3.16) and (3.18), though their qualitative character (the cubic dependence on wind velocity) is rigidly kept.

The Poisson character of the distribution density $P(N)$ gives some reasons to consider the spatial field of wave breakings to be a set of spatially distributed independent events without any consequence. In other words, instances of the appearance of overshoots (i.e., crests) are independent in such a spatial field when considering the field both in a spatial and in a temporal presentation. This is a quite important experimental result concerning wave energy dissipation problems; namely, the breaking of an individual wave occurs irrespective of the breaking of neighboring waves.

It is interesting to note that Glazman and Weichman (1990) and Glazman (1991) made attempts at characterizing (interpreting) the Poisson distribution (3.17) for a completely different breaking field parameter—namely, for the fractional coverage areas (as percentages) for whitecap foams by using the experimental data presented in Section 3.5. If we suppose the Poisson distribution for a number of breaking crests encountered simultaneously in the observed region of the oceanic surface with area $A$ to be valid, then we can assume the statistics of the given whitecap's area to be independent of whitecap number distribution. In such a case the mean number of crests $N_0$ encountered in the given region with area $A$ can be supposed to be linearly related to the region's area (i.e., $N_0 = \rho A$, where $\rho$ is the mean surface density of breakings per unit of surface). In this case the mean fractional whitecap coverage $S_0^W$ can be calculated as $S_0^W = \rho S(W)$, where $S(W)$ is the mean area of an individual whitecap. $\rho$ is the direct analog of the mean number density of particles in the classical ideal gas theory. It follows from this fact that the mean number of crests observed in the region with area $A$ equals

$$N_0 = S_0^W \frac{A}{S^W} \qquad (3.19)$$

Let us estimate the validity of such an approach using the experimental data presented above. So, according to the experimental data of Figure 3.6b, at wind velocity $V = 9.5\,\mathrm{m\,s^{-1}}$ mean fractional crest coverage equals $S_0^W = 4 \cdot 10^{-4}$ and the mean value of an individual crest area (from the data of Figure 3.7b) equals $S^W = 1\,\mathrm{m^2}$. Thus, the mean number of crests observed in the region with area $A = 10^4\,\mathrm{m^2}$ equals 4, which exactly corresponds to the value of the mean number of crests obtained using another technique—namely, by the method of directly counting the number of crests in region $A$ (see Table 3.3 and Figure 3.9). Similar results were also obtained under other meteorological conditions. So, at a wind velocity of $5.7\,\mathrm{m\,s^{-1}}$ the mean number of crests observed in a region with area $A = 10^4\,\mathrm{m^2}$ and calculated by formula (3.19) equals 2, and at a wind velocity of $10.5\,\mathrm{m\,s^{-1}}$ $N_0 = 6.8$. These values almost exactly correspond to figures obtained by the direct counting method (see Table 3.3 and Figure 3.9). Thus, relation (3.19) can be supposed to be valid in considering and analyzing Poisson (and, probably, close to Poisson) processes not only for the mean values of the parameters mentioned, but for their instantaneous values as well. In fact, the same supposition was implicitly used by Glazman and Weichman (1990). The distribution density for fractional whitecap coverage in the spatial frame of $100 \times 100\,\mathrm{m^2}$ can be determined from (3.17), having made the transformation $N = S_0^W (A/S^W)$. Figure 3.10 demonstrates the Poisson distribution (3.17) for a case in which wind velocity was $9.5\,\mathrm{m\,s^{-1}}$, and for $S_0^W = 4 \cdot 10^{-4}$. Figure 3.10 also shows an experimental histogram for fractional whitecap coverage in the spatial frame of $100 \times 100\,\mathrm{m^2}$—borrowed from Figure 3.5b—which demonstrates satisfactory agreement with Poisson's model.

We have already shown in this section that the hypothesis as to the purely Poisson character of the specific density distribution of breaking centers can be accepted. Nevertheless, a more detailed analysis of breaking center statistics that accounts for frame size, carried out in Section 2.3, has shown that as frame size decreases the number density distribution of breaking centers tends to a negative binomial distribution (Pascal's distribution or to a mixed Poisson distribution) and, then, as the frame area further decreases—in the special case of Pascal's distribution—to the geometrical distribution. There is no contradiction in these analyses (Sections 2.3 and 3.6). The fact is that the following characterization is known from the theory of random processes for a negative binomial distribution (Johnson and Leone, 1977). The negative binomial distribution can be presented as a mixed Poisson distribution if the mean value (and, accordingly, the intensity) of the initial primary Poisson process is attributed by the gamma distribution. From the viewpoint of wave interpretation this implies that the intensity of the gravitational wave breaking

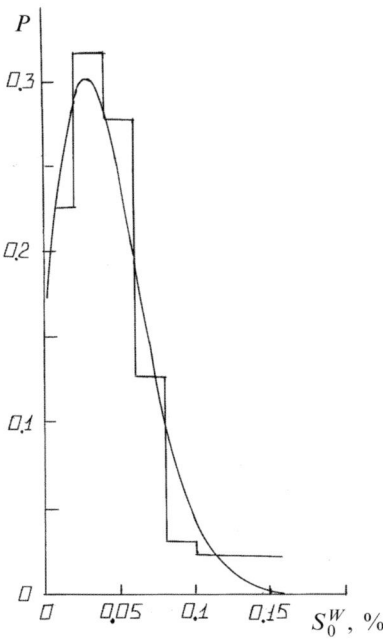

**Figure 3.10.** Comparison of the experimental histogram (Bondur and Sharkov, 1982) in Figure 3.5(a)—II with a theoretical prediction based on the Poisson distribution (3.17), as explained in the text.

process is non-uniform in spatial and temporal presentations and can be presented as a random value with its own gamma distribution. By further decreasing the frame area the negative binomial distribution degenerates into the geometrical distribution—namely, when the event (breaking) can take place exactly once in a studied frame. The importance of geometrical distribution is explained by a property called the "absence of an after-effect"—that is, by the independence of events (breakings) in non-crossed regions (see Section 3.7). As the frame area increases, the experimental distribution tends to a Gaussian one (Section 2.3), which actually follows from well-known properties of the Poisson distribution. In this case the intrinsic features of a process (such as, for example, the inhomogeneity of breaking process intensity in space and time) are not shown in the explicit form.

### 3.7  A SPATIAL FIELD OF WAVE BREAKINGS AND THE OVERSHOOT THEORY FOR A RANDOM GAUSSIAN FIELD

A full description of the spatial features of the field of breaking waves of a moving sea surface remains a complicated and unsolved problem. One possible approach to quantitative description of a gravitational wave breaking field consists in attracting the results of the theory of overshoots (crossings) of the random Gaussian field over the fixed level (Davenport and Root, 1958; Cramer and Leadbetter, 1967; Karlin, 1968; Bunkin and Gochelashvili, 1968; Tikhonov, 1970; Khusu et al., 1975). Below we shall discuss the issues surrounding applicability of the overshoot theory to the description of experimental results obtained on the real sea surface. These results include, in particular, general breaking field statistics and the dependencies of specific density on the degree of fetch and on the statistics of both the linear and aerial geometry of structures. In addition, the experimental results obtained above (Chapter 2 and Sections 3.2–3.6) are compared with qualitative and some quantitative conclusions of the overshoot theory for random fields, and the possibility of applying the obtained results to the physical foundations of sea wave dynamic theory is also discussed.

Early attempts at describing theoretically the quantitative characteristics of wave breaking and whitecapping processes mainly leaned on the natural attraction of the results of the overshoot theory for a random Gaussian two-dimensional field (a disturbed sea surface represents such a field) over the fixed level (Cramer and Leadbetter, 1967; Karlin, 1968; Bunkin and Gochelashvili, 1968; Tikhonov, 1970; Khusu et al., 1975). Here we should note that detailed investigations of the statistical characteristics of overshoots were performed for the one-dimensional case (Cramer and Leadbetter, 1967; Karlin, 1968; Tikhonov, 1970), whereas the results for the two-dimensional case are more qualitative in character (Bunkin and Gochelashvili, 1968) and were performed for the fields using Gaussian statistics.

### 3.7.1 Properties of a flux of fixed level crossings by the Gaussian field

According to the fundamental results of Cramer and Leadbetter (1967), the flux **a** of a fixed level crossing converges at $\mathbf{a} \to \infty$ according to the distribution of Poisson's flux, provided that the covariance function $R(t)$ of the Gaussian field satisfies the following two conditions:

$$R(t) = 1 - \frac{\lambda_2}{2} t^2 + \frac{\lambda_4}{4} t^4 + O(t^4) \tag{3.20}$$

at $t \to 0$; $\lambda_2$ and $\lambda_4$ are finite quantities, and

$$R(t) = O(t^{-\gamma}) \tag{3.21}$$

at $t \to \infty$ for the some $\lambda > 0$.

These conditions correspond to the double-differentiability of a random field, at least in the average square law. More general conditions for the Poisson character of a flux of fixed level crossings were obtained later by Cramer and Leadbetter (1967).

When investigating a disturbed sea surface we can easily see that the necessary and sufficient condition (see Tikhonov, 1970; Khusu et al., 1975) of double-differentiability for realization of the studied process is satisfied, since the one-dimensional spatial spectrum of surface elevations decreases faster than $K^{-4}$ for $k > k_1$—see expression (2.14)—and, accordingly, the one-dimensional frequency spectrum drops faster than $\omega^{-5}$. The latter conclusion corresponds to limiting the fourth derivative of the covariance function $R(t)$ at zero, as well as to meeting the condition (3.21). Conditions (3.20) can be looked upon from more general positions as well. Analysis of versatile experimental material allowed us to establish, for the covariance function of a sea wave displacements field, a typical form of attenuating cosinusoid (or of a sum of cosinusoids) (Roshkov, 1979):

$$R(t) = \exp\{-\alpha(t)t\}\cos\{\beta(t)t\} \tag{3.22}$$

where the inequality $\beta \gg \alpha$ is characteristic of the typical values of parameters $\alpha$ and $\beta$. For example, $\alpha \approx 0.07\,\text{s}^{-1}$, $\beta \approx 0.6\,\text{s}^{-1}$ (Roshkov, 1979). It can easily be seen that under these conditions relations (3.20) are satisfied: so, at $t \to 0\,\exp(-\alpha t) \approx 1$ the cosine decomposition gives the form of (3.20); and at $t \to \infty$ the limit is known (for $\gamma > 0$) to be

$$\lim\{\exp(-t)/t^{-\gamma}\} = 0 \tag{3.23}$$

which implies satisfaction of condition (3.21).

Since the random Gaussian surface under study satisfies conditions (3.20) and (3.21), then, according to the theory of crossings, the statistical properties of a breaking field of this surface should be close to a Poisson process. And, the experimental material presented above really does show that a satisfactory probabilistic model of a gravitational wave breaking field at all fetch stages can be a spatial field of Poissonian type. So, according to the data of Chapter 2 and Section 3.6, for a breaking field the conditions characterizing the Poisson properties of a flux (Cramer and Leadbetter, 1967) are satisfied—namely, (a) the independence of events (breakings) in non-crossed regions (frames) or, in other words, the property of the absence of an after-effect between the events; (b) ordinariness as a consequence of the geometrical law of the breaking number distribution density as the observed area of a frame decreases in size (Karlin, 1968); (c) stationarity or, on other words, the probability of events in the region $A^M$ (the specific density distribution law) depending on $N^M$ and $A^M$ and not depending on the position of the region in the water area studied.

The experimental density of distribution $p(N^M)$ being binomial in character (Section 2.3) can be interpreted as the conventional probability of the presence of a given number of events $N^M = n$ in a limited spatial frame $A^M$ under the condition of the Poissonian character of the process as a whole—that is, over the macro-area $A^O$ of the water area with the total number of events $N$. And, then, by virtue of the well-known property of Poisson distributions (Karlin, 1968), we can write:

$$P\{N(A^M) = n | N(A^O) = N\} = \frac{P\{N(A^M) = n;\ N(A^O) - N(A^M) = N - n\}}{P\{N(A^O) = N\}}$$

$$= C_N^n p_0^n (1 - p_0)^{N-n} \tag{3.24}$$

where $p_0 = A^M / A^O$.

Another characterization of experimental distribution can also be supposed: this is associated with the appearance of $N^M$ events in the studied spatial frame in performing $N$ independent, homogeneous Bernoulli tests, each test having a probability $p_0$ of individual event appearance. The Bernoulli scheme reduces the general statistics of a system to the binomial distribution (see, e.g., Karlin, 1968). As the frame area or the sea disturbance degree (and, accordingly, $N^M$) increases, the experimental distribution tends to a Gaussian one (Section 2.3), which actually follows from well-known properties of a binomial distribution. As the spatial frame $A^M \to 0$ decreases the experimental distribution density $p(N^M)$ tends to the geometrical distribution by virtue of the ordinariness properties of the field of events (the breaking field) and the spatial independence of events (individual breakings). The latter feature of distributions $p(N^M)$ represents also a well-known property of Poisson processes (e.g., characterization can be performed according to the Bernoulli scheme).

### 3.7.2   The intensity of a flux of fixed level crossings by the Gaussian field and experimental observations

Using the results of the overshoot theory, we shall consider the dependence of the mean value of the intensity of a flux of fixed level crossings by the Gaussian field $N$ on the degree of state of the rough surface, using the following expression for the density of overshoots of the anisotropic Gaussian field (Karlin, 1968):

$$N = (m_{20}m_{02} - m_{11}^2)(2\pi m_{00})^{-3/2}\exp\left\{-\frac{a^2}{2m_{00}}\right\} \tag{3.25}$$

where $m_{ij} = \int_{-\infty}^{\infty}\int_{-\infty}^{\infty} S(k_x, k_y)k_x^i k_y^j\, dk_x\, dk_y$ are the moments of a spatial spectrum of rough surface heights; and $k_x$, $k_y$ are wavevector projections. By accounting for the approximation of the experimental spectrum of disturbed sea surface elevations (2.14) and after some transformations that take approximation $k_m > k_1 \gg k_2$ into account, which is natural under field conditions, we have:

$$m_{00} = \frac{3\pi}{8}(B_0 + X)k_m^{-1.5} + B_1\pi k^{-3}$$

$$m_{02} = \frac{\pi}{2}\left(\frac{1}{4}(B_0 + X)k_1^{1/2} + B_1 k^{-1}\right)$$

$$m_{20} = \frac{\pi}{2}\left[\frac{5}{4}(B_0 + X)k_1^{1/2} + B_1 k^{-1}\right] \tag{3.26}$$

$$m_{11} = 0$$

It follows from relation (2.14) that $B_1 = 0.1k_1^3 = B_0 k_1^{1.5}$. Substituting the expressions found for the spectrum moments into (3.25), we obtain:

$$N = \frac{64}{\pi^4\sqrt{135}}\frac{(ak_m)k_m^{5/4}}{(B_0 + X)^{5/2}k_1^{1/2}}\exp\left[-\frac{4}{3}\frac{(ak_m)^4}{(B_0 + X)k_1^{1/2}}\right] \tag{3.27}$$

Of great importance for further analysis are the physical prerequisites applied when choosing the fixed level "a". So, following Longuet-Higgins (1969), Snyder and Kennedy (1983) assumed level "a" to be the point of a field at which "instantaneous" negative acceleration exceeds the value of $(0.4–0.5)g$. In some works (e.g., Avanesova *et al.*,1984) the authors do not define level "a" at all. If we assume that level "a" for a narrow-band noise spectrum (according to Longuet-Higgins, 1969) is the quantity $a = (g/2)(2\omega_m^2)^{-1}$, then we have $ak_m = 0.25$. And this expression does not depend on hydrometeorological conditions or on sea wave fetch conditions. If, however, we accept "a" to be the height of an essential wave of the Pierson–Moskowitz spectrum, then we shall also have for $ak_m$ a constant quantity equal to 0.18. With regard to this circumstance and remembering our experimental dependence of $\tilde{k}_m$ on the fetch $\tilde{k}_m = 100\pi^2 X^{-0.66} g U_{10}^{-2}$, we shall obtain from relation (3.27) that the mean value of the specific density of breakings $N$ decreases with increases in fetch as $N \sim \chi^{-3.3}\exp(-X^{-0.67})$. It is interesting to note that in the "threshold" model (Snyder and Kennedy, 1983) similar results were obtained using a spatial spectrum of JONSWAP type. So, for the breaking density $N$, Snyder and Kennedy (1983) used another parameter: the probability of breakings per unit of time and per unit of surface also rapidly drops with increases in fetch as $N \sim X^{-n}$, where $n$ varies from 2 to 4 depending on the stage of fetch (see fig. 5 in Snyder and Kennedy, 1983). Certainly, similar conclusions radically contradict both the experimental results obtained in the present book and the numerous qualitative observations of other authors. So, according to the data presented in Chapter 2—see relation (2.22)—$N \sim X^N$, where $n = 1-2$ at various phases of the fetch stage. However, the authors of the "threshold" model, as well as the authors of subsequent papers (e.g., Bordugov *et al.*, 1986), in which the same point of view is developed, try to explain the obvious contradiction with experimental data by references to the fact that "steeper waves break more intensively" and that "the quantity of breakings should decrease in a developing wave field". A similar point of view was given earlier by Bortkovskiy (1983). Moreover, the "overshoot" theory also demonstrates its inconsistency while interpreting the statistics of a linear geometry of individual breakings (see below), where it indicates the sharply decreasing dependence of the mean density of crests on wind velocity, which also contradicts experimental data (see Section 3.6).

The obvious contradiction of the "threshold mechanism" concept in the overshoot theory for a Gaussian field (and its modifications) with experimental data testifies to essential losses in the physical notions underlying these concepts. At least two corrections are required here—the nonlinearity of a Gaussian field needs to be taken into account and inclusion of the effect of atmospheric pressure fluctuations in the model. The importance of these factors can be demonstrated by the results of numerical experiments (Krasilnikov *et al.*, 1986; Krasilnikov, 1987). The results of such numerical modeling indicate that harmonious waves with small distortions of a profile (caused, for example, by wind pressure fluctuations) very rapidly (for the time of a wave period) evolve to nonlinear forms with saturation by the highest harmonics and to subsequent breaking (the ambiguity of form). This process takes place due to inclusion of the "intrinsic feedback" via the dynamic cauchy–Lagrange boundary conditions. The presence of large waves results to a certain extent in "pushing" ("acceleration of") the breaking mechanism. Proceeding from this hypothesis, it becomes clear that both the absence of "close" correlations between neighboring breakings ("the absence of an after-effect"), and the impossibility of formation— under the conditions of ordinariness of a flux of events (breakings)—of a "grouping" in the breaking wave field, as follows from the "threshold" mechanism theory (Snyder and Kennedy, 1983). Certainly, in this case the question is about free gravitational waves, without any "external" effects of internal wave type or any influence of the lower boundary on the water medium (the "shallow" water concept). In the latter case the spatial statistics of breakings will certainly sharply change. However, the latter reasoning should not be taken to mean that both the "distant" and weak correlations between spatial breaking zones, related to the stochastic dynamics of the fetch process, should be completely absent in the breaking field of developing waves. So, for example, Section 2.5 presents the results of the experimentally found fractal character of a sea wave breaking field (discovered for the first time by Zaslavskii and Sharkov, 1987) both in the mode of its complete fetch and under unsteady disturbance.

### 3.7.3 Regions of overshoots of an isotropic Gaussian field and experimental observations

According to the results of Bunkin and Gochelashvili (1968), the regions of overshoots of a two-dimensional isotropic Gaussian field represent the single-connected regions of elliptic shape with randomly fluctuating parameters (eccentricity and area, in particular) and randomly, uniformly distributed (over azimuth angles) orientations of major axes. Quantitative estimations can be performed following the results of their paper for mean values of specific density $N_0$ and the mean area of a single crest $S^W$, which can be presented, for a simplified exponential spatial spectrum $F(k) = (\alpha/2)k^{-4}$ for $k_0 < k < k_m$ and $F(k) = 0$ for $k_m < k < k_0$, as follows:

$$S^W \sim \alpha\pi^2 k_0^{-2}\left(\ln\frac{k_m}{k_0}\right)(k_0 a)^{-2} \tag{3.28}$$

$$N_0 \sim (2\pi)^{-3/2} k_0^2 \frac{k_0 a}{2\pi}\ln\frac{k_m}{k_0}\exp\left(-\frac{k_0^2 a^2}{\alpha\pi}\right) \tag{3.29}$$

where $k_0$ is the wavevector of a random disturbance of the sea surface corresponding to the energy-carrying component (for the Pierson–Moskowitz spectrum $k_0 = 0.77g U_{10}^{-2}$, $U_{10}$ is wind velocity at an altitude of 10 m and $g$ is free-falling acceleration); and $k_m$ is the boundary of high-frequency cutoff of a spectrum, $\alpha = 0.8 \cdot 10^{-2}$ (Zacharov and Zaslavskii, 1982).

Analysis of relations (3.28) and (3.29) reveals the essential dependence of mean values of $N_0$ and $S^W$ (in the considered approximation) on the "mean" steepness of breaking waves $(ak_0)$, on hydrometeorological conditions, and on sea wave fetch conditions (the value of $k_0$). If we accept the value "a" to be the height of an essential wave for the Pierson–Moskowitz spectrum $h_S = 2.38\alpha^{1/2}g^{-1}U_{10}^2 \approx 2.46 \cdot 10^{-2}U_{10}^2$, then we can easily see that $(ak_0)$ does not depend on hydrometeorological conditions (and, in particular, on $U_{10}^2$), being equal to the value of 0.18. Therefore, the mean value of the area of an individual crest grows and that of specific density drops with increasing wind velocity as the fourth degree (i.e., $S^W \sim U_{10}^4$ and $N_0 \sim U_{10}^{-4}$).

Turning to the investigations of Bondur and Sharkov (1982, 1990)—see relation (3.14)—we see that the mean area of crests (3.28) qualitatively correctly describes the growth of the crest area with increasing wind velocity (in the experiment the dependence is really quadratic: $S^W \sim U_{10}^2$). However, as to the mean value of specific density, a qualitative distinction is observed here: the experimental data demonstrate a cubic growth in specific density with increasing wind strength—see (3.18)—whereas theoretical considerations give a sharp drop in specific density with wind velocity, which, certainly, radically contradicts observed data.

Theoretical analysis of the distribution densities $N_0$ and $S^W$ of a two-dimensional field was not carried out in detail in Bunkin and Gochelashvili (1968), except for the general-physical note on the possibility of using a gamma-distribution type $p(x) \sim x^\nu \exp(-\beta x)$ for the $S^W$ class, where $x = S^W$. The latter circumstance fully complies with the results of experiments by Bondur and Sharkov (1982) on the statistics on $S^W$ of individual foam structures (see Section 3.4).

Looking at more detailed results for the one-dimensional case (Tikhonov, 1970; Khusu et al., 1975), we find that expressions for the densities of distribution of overshoot duration (in our approach they are linear dimensions) and specific density can be written down in an explicit form by taking the following limitations into account: the Gaussianicity of an initial field; excess in the fixed level "a" of the root-mean-square deviation of heights $\sigma_H$—that is, $(a/\sigma_H) > 1$; and the small breaking time $(\tau_0)$ compared with the time of correlation of the basic random process $(\tau_1)\tau_0 \ll \tau_1$. It is known that, in the first approximation, the Gaussianicity of the field of an elevated sea surface is observed with a high degree of accuracy, and in this case $\tau_1$ equals a few hundredths of a second (Roshkov, 1979), whereas the time of over-shooting (breaking) is of the order of a few seconds (see Chapter 4). Thus, both the first and last conditions are observed.

The second limitation can be substantiated by proceeding from the semi-empirical relations between the height of a breaking wave $(H = a)$ and its period $(T) - H \geq 0.02gT^2$ (Ochi and Tsai, 1983). Having accepted the power-law spatial spectrum for estimating $\sigma_H^2$ (see above), we have:

$$\sigma_H^2 = 2\pi \int F(k)\, dk \approx \frac{\pi\alpha}{2} k_0^{-2} \tag{3.30}$$

and, using the relation between the length and period for a gravitational wave $\Lambda = (g/2\pi)T^2$, we can easily see $(a/\sigma_H) > 1$–2. Thus, in the approximation considered the second condition for overshoot theory applicability to a breaking wave is also satisfied. So, theoretical relations for the distribution densities of linear dimensions of overshoots $(x)$ for the Gaussian field (Tikhonov, 1970) can be presented for the power-law spectrum $F(k)$ as the Rayleigh distribution:

$$p(x) = Ax \exp\left(-\frac{A}{2}x^2\right) \tag{3.31}$$

where $A = k_0^2 \ln(k_m/k_0)\pi\alpha(k_0 a)^2$; and for the specific density as the Poisson distribution.

Additional estimations of the correspondence between experimental histograms and the theoretical Rayleigh distribution (3.31) using Pierson's criterion $\chi^2$ showed that there are no reasons to discard (at the significance level of 5%) the hypothesis on the Rayleigh character of distributions of linear dimensions of foam structures, the more so as our approximation by gamma-distributions (3.7) is close to those predicted by the overshoot theory ($\eta - 1 = 1$–2). However, the essential dependence of the coefficients $A$ of distributions (3.31) on wind velocity (the fourth power) is not confirmed experimentally—coefficients do not depend on hydrometeorological conditions, and no regular dependence is found in any case.

In conclusion to this section, where theoretical concepts and experiments were discussed, we note that in such delicate experiments as the study of breaking sea waves a single-parameter description of the sea surface in the form of specifying wind velocity (and, accordingly, the state of the sea surface—for example, according to the Beaufort Scale[6]) is clearly not satisfactory. It is necessary to attract data on the state of near-surface layer turbulence and on the spectral characteristics of sea disturbance within a wide range of wavenumbers, because there are certain experimental indications (Ochi and Tsai, 1983) to the fact that wave-breaking efficiency is mainly influenced by higher (than the first and second) moments of the sea wave spectrum, as well as by the fluctuation of a turbulent near-surface wind.

### 3.7.4  On the relationship between the dissipation and transparency intervals in spectra of sea wave heights

The experimental data given in this chapter allow us to draw some important, in our opinion, conclusions concerning estimations of the relationship between the dissipation and transparency intervals in the spectra of sea wave heights.

The notion of the need for spatial separation of wavenumbers at given intervals underlies, as a physical basis, the applicability of Kolmogorov's weak-turbulent spectra ideology to describe the energy characteristics of real sea waves (Zacharov and Zaslavskii, 1982). It is known that, beginning with a wind velocity higher than $5$–$6\,\mathrm{m\,s^{-1}}$, the main dissipative factor for wind waves is the breaking of crests. And energy dumping is localized in crests and turbulent spots in the near-surface layer of water, where the foam actually gathers and is kept without destruction at the emulsion state for a considerable time (some tens of seconds) due to the effect of "turbulent" calming of waves over the turbulized water layer directly.

The presence of optically contrasting foam formations in a turbulent spot is a confident attribute for identifying optical systems that can record spatial scales of the order of wave energy dissipation regions. We can easily see from the data presented in Table 3.1 and in Figure 3.2 that characteristic scales of crest foam are concentrated (at a probability of 90%) in the range of wavenumber scales of $1$–$50\,\mathrm{m^{-1}}$, and those of spot-like foam (which characterizes dissipation regions) in the range of $0.5$–$2\,\mathrm{m^{-1}}$. To

---

[6] Note that the question of the non-universal character of the Beaufort Scale in performing delicate wave experiments has been repeatedly discussed in the science literature (e.g., Alcock and Morgan, 1978; Bortkovskiy, 1983).

compare the theoretical estimates of wave–wind interaction, the mentioned scales of regions are expediently presented in dimensionless form (following, for example, Zacharov and Zaslavskii, 1982) as $\tilde{k} = kU_{10}^2 g^{-1}$, where $U_{10}$ is wind velocity at an altitude of 10 m and $g$ is free-fall acceleration. Having performed this operation, we find that at low wind velocities (5.7 m s$^{-1}$) the region of dimensionless wavenumbers of crests is concentrated in the range of 4.5–50, and at high velocities (10.5 m s$^{-1}$) in the range of 15–500. The ranges $\tilde{k}$ of turbulent spots are localized in a more large-scale region: for $V = 5.7$ m s$^{-1}$ in the range of 1.5–6, and for $V = 10.5$ m s$^{-1}$ in the range of 5–30.

So, for developed sea waves the dimensionless spectral maximum $\tilde{k}_m$ occupies the region (according to Zacharov and Zaslavskii, 1982) $\tilde{k}_m \lesssim 2$–4. It can easily be seen that at low wind velocities there is considerable overlapping ("intermixing") of dissipation interval limits in the $k$-space, $\tilde{k} = 1.5$–50 (at the 90% probability), and of the interval that energy is carried surrounding a spectral maximum and an inertial interval.

At moderate winds (10–11 m s$^{-1}$), however, there occurs a certain separation between dissipation and transparency intervals: the dimensionless maximum of a spectrum tends to unity (for the Pierson–Moskowitz spectrum of developed sea waves $\tilde{k}_m = 0.77$); the lower boundary of the dissipation region, however, establishes in the region of $\tilde{k}^- \sim 5$. In this case the dissipation interval sharply extends into the high-frequency region, reaching the values of $\tilde{k}^- \sim 500$, which corresponds to the characteristic scales of gravitation-capillary waves.

Thus, notions about the considerable separation of the dissipation interval and the wave spectrum maximum in the $k$-space, both at early wave formation stages and at stages close to full fetch ($\tilde{k}_m \sim 1$), can hardly be considered as experimentally substantiated, contrary to the opinion of Zacharov and Zaslavskii (1982).

With these estimations, however, we should keep in mind that the total relative foam coverage area on a disturbed sea surface (the physical space) is rather small (0.5–2%) even at moderate wind velocities (see Figure 3.6).

In addition, because of the threshold mechanism of the origin of wave energy dissipation centers (Section 3.5), we should arguably take the "rigid" inclusion of the dissipation mechanism in wave field energetics into account in theoretical models.

## 3.8   CONCLUSIONS

1    Large-scale optical sensing of the sea surface in a developed sea wave zone, whose results were analyzed in the present chapter, were performed in the basins of the Barents and Black Seas within the framework of the "Barents-76"–"Barents-80" series of programs of the USSR Academy of Sciences. Experimental remote-sensing techniques for studying the regularities of the statistical spatial structure of chaotic disperse media, which appear as a result of breaking gravity waves in the ocean–atmosphere system, are presented and developed.

**2**   Experimental distribution densities of linear geometry elements of foam systems
of two types on a disturbed sea surface, according to optical surveying data, are
close to having gamma distributions and have stable parameters. In this case the
mean values of the linear geometry of individual elements linearly grow with
increasing wind velocity, while possessing a threshold character. The geometrical
shapes of foam structures possess a steady ellipsoidal configuration with parameters statistically differing for crest and spotty foam.

**3**   The experimental distribution densities of the values of the area of individual
foam formations and fractional coverage area of crests and strip structures, using
a surveying frame size of $100 \times 100$ m, are close to having gamma distributions
with stable parameters (for wind velocity variations from 5.7 up to $10.5\,\mathrm{m\,s}^{-1}$).

**4**   Dependencies of the mean values of the fractional foam coverage area and of
areas of individual formations of two classes on wind velocity have a "threshold"
character. These dependencies are quadratic in character, and the density of
dissipation centers is a cubic function of wind velocity.

# 4

# The lifetime dynamics of sea wave breakings

This chapter presents the results of experimental remote optical investigations of regularities in the temporal evolution of the natural disperse media that form in the ocean–atmosphere system at moderate sea and breaking gravity sea wave conditions. The results were obtained using a specially proposed and developed technique of processing the data of large-scale photography and filming of the sea surface, carried out in the Pacific Ocean and Caspian Sea basins at wind velocities of 7.7 to 13.5 m s$^{-1}$ from a ship. The purpose of these investigations was to reveal the detailed temporal structure of the evolution of the gravity wave breaking process directly and of the further evolution of strip foam fields that form in this process. The time dependence of foam formation evolution was experimentally established, and a specific group of gravity wave breaking—mesobreaking—was found.

## 4.1  THE PROBLEM OF STUDYING THE LIFETIME DYNAMICS OF THE SEA WAVE BREAKING PROCESS

As already noted (Chapter 1), one of the most complicated wave dynamics tasks is solving the problem of wave energy dissipation at gravity wave breaking. Investigations have taken various directions—namely, spatiotemporal estimations of breaking parameters using dimensional considerations (Phillips, 1977); simplified numerical models (Vinje and Brevig, 1981; Krasilnikov, 1987); wave breaking, carried out under laboratory conditions, which certainly cannot be correlated with field measurements of breaking gravity waves in natural situations (in "deep" water, in particular). The most important structural element of gravity wave breaking is the disperse phase (foam and drop-spray structures and an aerated layer) that appears in the breaking process, which to a considerable extent (according to the conclusion of some models, such as Krasilnikov, 1987) "incorporates" in itself the dissipated energy of a wave. This circumstance places on the agenda the necessity for a statement of theory and execution of successive field experiments, under open sea conditions, to study the time characteristics of the evolution of the disperse phase that forms as individual gravity waves break.

On the other hand, a topical problem in aerospace oceanography is description of the mechanisms of the interaction between electromagnetic radiation and a disturbed sea surface. Investigations of recent years (Phillips, 1988; Vorsin et al., 1984; Cherny and Sharkov, 1988; Glazman, 1991a; Sharkov, 2003; Bulatov et al., 2003) indicate the necessity for an in-depth study into the detailed structure (both temporal and spatial) of individual sea wave breaking zones, because demonstration of purely statistical links between the scattered signal intensity and geometry elements of circulating waves (as, for example, done in works by Malinovskii, 1991, Sletten et al., 2003, Mouche et al., 2006 and Hu et al., 2006) does not provide a final solution to the question on the mechanism of scattering and radio emission of electromagnetic waves.

As noted in Chapter 1, it is expedient at this point to give the following classification of optical methods for studying gravity wave breaking processes:

- Type I—aerial photography from onboard a high-altitude air carrier to reveal regularities in the distribution of sea wave dissipation centers (gravity wave breaking) over large sea basins (of several $km^2$) without differentiating the type of foam system. The experimental method was proposed by the author of the present book and was described in detail by Pokrovskaya and Sharkov (1986, 1987a, b) and Sharkov (1993a, b) (see Chapter 2).
- Type II—aerial photography from onboard a low-speed air carrier and perspective filming from onboard the research vessel to study the spatial–statistical characteristics of the area and the linear geometry of foam structures of various types, but without analyzing the temporal dynamics (the method was offered and advanced in Bondur and Sharkov, 1982; see Chapter 3). Recently, high-sensitivity panoramic systems in the infrared range (in the mid-wave infrared and thermal sub-ranges) began to be used in this type of observation for detailed identification of disturbances in the thermal regime of a surface skin layer in the wave-breaking and foam field formation processes (Marmorino and Smith, 2005).
- Type III—investigation of the temporal dynamics of the process that takes place when individual gravity sea waves break; and the study of temporal evolution of geometrical properties of individual disperse formations on the sea surface by means of fast-acting photography from onboard the research vessel (filming and photography with long-focal lenses). This method was offered and further developed in Cherny and Sharkov (1988), Timofeev and Sharkov (1992), and Sharkov (1994, 1995) (see Chapters 4 and 8).
- Type IV—investigation into the disperse structure of the aeration layer and surface disperse systems by means of macro-photography of foam mass samples taken from the surface (Raizer and Sharkov, 1980), or directly from the sea surface layer by means of sealed boxes (Timofeev and Sharkov, 1992; Bortkovskiy and Timanovsky, 1982) (see Chapter 4).

The carriers of optical instruments introduce a certain specificity in the observational process, as a result of which each of these types of remote observations of short-life disperse systems possesses certain limitations. Principally, we should note here that observations according to types I and II do not reveal the temporal dynamics of foam structures, since in these types of observations the individual foam structure is recorded at a certain moment in its own "life" that is unknown to the researcher. And, so, statistics of the time of "life" of foam structures can be considered as "embedded" in the statistics of the general geometry of the foam systems determined by observations of type II (see Section 3.5). The latter feature leads to substantial "blurring" of the statistical densities of distributions of the geometrical parameters of foam systems found in works by Bondur and Sharkov (1982, 1990) (see Section 3.5 for more details). This feature (which is important for understanding the physics behind such "blurring") was a subject of lively discussion on the results of works by Bondur and Sharkov (1982, 1990) (Glazman and Weichman, 1989, 1990; Monahan, 1990).

However, despite general accessibility to visual observation of the sea wave breaking process, we do not possess any reliable literature data on the temporal evolution of a sea wave breaking zone obtained in field experiments.

The purpose of this chapter is to present—on the basis of the observation technique (photography) offered by Timofeev and Sharkov (1992), as well as on the basis of special processing of optical data—experimental results on the temporal evolution of the geometrical parameters of foam systems of two classes ("whitecaps" and strip foam), as well as data on detecting a new class of disperse structures that appear as sea waves break—so-called mesobreakings (Timofeev and Sharkov, 1992; Sharkov, 1994, 1995, 1996a, b).

## 4.2  TECHNIQUE AND CONDITIONS OF FIELD EXPERIMENTS

Field experiments were carried out in three basins that substantially differ in their hydrological and hydrophysical parameters: (a) in the northwest region of the Pacific Ocean at a distance from the coast (Kamchatka Peninsula) greater than 50 miles; (b) in the Frieze Strait (Kuril Islands); and (c) in the Caspian Sea at Makarov's Shoal (at a 20-mile distance from the coast, with a sea depth of about 20 m). The experiments were carried out during the summer season of 1982 within the framework of the "Sakhalin-82" and "Kaspii-82" complex of programs of the USSR Academy of Sciences, which studied sea wave structure using remote-sensing techniques under scientific guidance of the author of the present book. The flight composition of research means included the AN-30 airplane-laboratory (with optical instruments) and the AN-24 airplane-laboratory equipped with the side-looking radar station "TOROS". Participating in the "Sakhalin-82" program was the research hydrographical vessel (RHV) *Abkhazia*, which provided hydrometeorological maintenance and contact measurements. In addition, optical observations and large-scale photography and filming of disturbed sea surfaces were carried out from onboard this ship. Data on hydrometeorological conditions were taken from the results of detailed meteorological measurements carried out on RHV *Abkhazia*.

Near-surface experiments in the Pacific Ocean and Caspian Sea were carried out under wind–wave conditions close to a fully developed sea wave regime, and near-surface layer stratification was close to being neutral (see Table 4.1). The values of near-surface wind velocities were determined by averaging the results of RHV measurement (at a level of 20 m) during 2–3 hours before performing a cycle of optical measurements. Root-mean-square deviations in the state of natural wind velocity variability were less than $0.3$–$0.4 \, \mathrm{m \, s^{-1}}$ during the term of observation.

Experiments in the Frieze Strait were carried out in drift mode from onboard the *Abkhazia* in the strait from the Okhotsk Sea side. Despite the fact that wind disturbance was, in general, quite stationary in character, intensive tidal currents are known to be present in the region studied, and, in this connection, considerable intrusions of Pacific Ocean water masses into Okhotsk Sea near-surface waters were recorded, and, moreover, internal wave passages to the sea surface were recorded (according to data of synchronous radar surveying performed by the AN-24). As an example, Figure 4.1 presents the image of a disturbed sea surface in the backscattering field at that part of the Okhotsk Sea basin directly adjacent to the Frieze Strait, at the moment of Pacific Ocean water mass intrusion into Okhotsk Sea near-surface waters due to tidal flows. The general view of intrusion is characterized by a spiral-wise contour. When water masses collide, intensive breakings of surface wave systems arise, which, in their turn, are reflected into the backscattering field in the form of strips and strings with sharply heightened backscattering intensity.

### 4.2.1   Optical observations of the temporal evolution of the breaking process

Zenit-TTL and Zenit-E 35-mm cameras were used for optical field photography with Industar-50 and Jupiter-37 lenses (their focal length was 50 and 134 mm, respectively). Application of a long-focus lens made it possible to more fully use the descriptive properties of a photo-film (a whitecap or a foam spot "filling the whole frame") in wave-breaking photography at a distance up to 30 m from the ship. All photography was carried out from the main deck of the ship. A camera was installed at a height of 15 m from the sea level. Photography was performed manually, which allowed—despite the ship rolling—foam formation to be kept inside the field of view of the lens. Photography was carried out in a series of 20–30 frames, with equal time intervals between the frames (of 1–2 s usually). This allowed direct imprinting of the wave-breaking moment on 1–2 frames, and studying the temporal dynamics of an accompanying foam spot and of a spot of turbulent calming of capillary disturbance. The total time of photography of one series was strictly monitored, and the period between the frames was calculated from this time.

**Table 4.1.** Conditions of field experiments.

| Site of experiments | Data | Type of HMC | Hydrometeorological conditions (HMC) | | | | | | |
|---|---|---|---|---|---|---|---|---|---|
| | | | Wind velocity ($h = 20$ m) (m/s) | Salinity (‰) | Wind-generated waves (no.) | Swell (m) | Water temperature (°C) | Air temperature (°C) |
| NW Pacific Ocean: | 21.08.82 | Ia | 7.7 | 35 | 3 | 1 | 11.6 | 12.1 |
| 51° 20′ N, | 11.08.82 | Ib | 9.2 | | 4 | 2 | 11 | 12 |
| 162° 00′ W | 16.08.82 | Ic | 10 | | 4 | 2 | 10 | 11 |
| | 11.08.82 | Id | 10.4 | | 5 | 2.5 | 11 | 12 |
| | 12.08.82 | Ie | 13.5 | | 6 | 2.5 | 11 | 12 |
| Frieze Strait, Kuril Islands | 01.08.82 | II | 8.5 | 35 | 3 | 1 | 9.1 | 10.0 |
| Caspian Sea, Makarov Bank | 10.10.82 | III | 9.0 | 14 | 4 | | 11 | 12 |

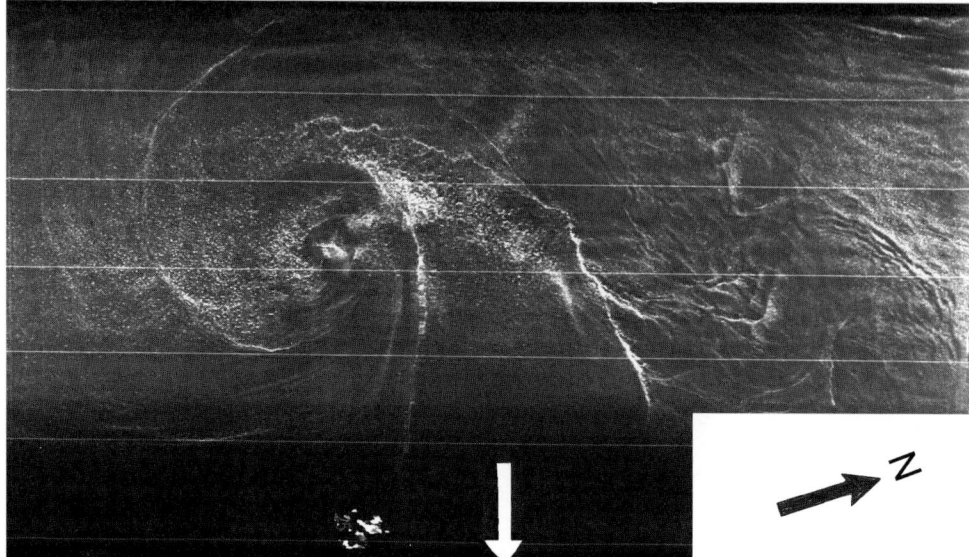

**Figure 4.1.** Side-looking radar image of a disturbed sea surface near the Frieze Strait (Kuril Islands) from the Sea of Okhotsk side during Pacific water intrusion in the form of a helix into Okhotsk Sea water. The radar survey was carried out by TOROS side-looking radar onboard the Russian airplane laboratory AN-24. The position of the Russian hydrographic vessel where the optical photography and filming were performed is in the center of the radar frame. The black arrow shows the northward direction. The white arrow is directed toward Frieze Strait. The space interval between the fine white lines is equal to 5 km.

During the operation the angle between the camera's optical axis and the sea surface was 30 or 45°. Knowledge of these angles was necessary for subsequent deciphering of the geometrical scales of obtained facsimiles. Depending on the observation geometry, the size of the general-view frame varied from 60 to 75 m$^2$ (the area was recalculated to the sea surface). In addition to detailed photography, perspective sea surface photography from onboard the RHV was also carried out. In this case the optical camera was oriented along the horizon—the horizon line was in the upper part of a frame (this technique was described in Bortkovskiy and Kuznetsov, 1977). Photography was performed both in multiple photography mode—at the rate of a frame per second—and in perspective photography mode at a slowed-down rate—a frame per minute—to provide the statistical independence of frames. This technique possesses some peculiarities. On the one hand, the operator easily catches and keeps an object within the field of view of a camera with a long-focus lens, which cannot be achieved by rigidly fastening the camera, even on a stabilized platform. On the other hand, the operator cannot keep pace with the photography rate accurately enough without special training (for photography). However, a sufficiently skilled operator, having undergone special training, provided an error in the photography rate, not exceeding 10%. For this purpose, 10, 20, 30, 40 frames were made at a rate of 1 frame per second, and the total time was fixed by a stopwatch. Another peculiarity relates to photographic perspective. This complicates processing of the images taken, since sea disturbance strongly deforms real areas of foam formations (a foam spot partially disappears, and now opens more fully as a result of sea wave crests). In this case, only frames clearly "identifying" foam systems were analyzed, whenever possible.

In addition to multiple photography—whose technique was just described— filming was also carried out by means of a standard movie camera (at a rate of 24 frames per second). Analysis of the film material obtained has shown, however (Timofeev and Sharkov, 1992), that such a photography mode is not conducive to further processing for the following reasons. First, the photographic rate (in relation to the temporal evolution of disperse systems) is rather high, and, second, the small size of film negatives does not allow effective processing to be fully performed.

During work in the Pacific Ocean (Table 4.1) gravity wave breakings were studied under the surface by means of a floating, leak-tight photobox. The shutter of the Zenit-TTL camera was triggered by a remote signal from the operator, when wave breaking was noticed in the box area. The camera was focused on a cut in the photobox window. Photography was performed at a scale of 1:2 on 35-mm film. The working volume, displayed in a frame, was about 20 cm (it depended on the size of photo-graduated bubbles). The photobox distance from the RHV exceeded 50 m on the weather side of vessel, and the absence of any hydrodynamical effect of the RHV's body on the sea surface area studied was monitored visually.

## 4.2.2  Optical image processing technique

Positive prints at four-fold magnification were made. The photographic material was sub-divided into series (perspective photos, individual breakings, and their remnants). In their turn, the series were collected into groups with similar hydrometeorological conditions at the moment of photography. For the various slope angles of the camera and for various lenses special templates were designed and manufactured, which allowed perspective distortions of linear dimensions and areas on the sea surface to be taken into account. The technique used in designing the templates was, in general, similar to the technique described by Bortkovskiy and Kuznetsov (1977). The template grid was deposited on a transparent plastic foil. By imposing templates on photographic prints, we can determine the areas of foam formations to an accuracy of $0.1 \, \text{m}^2$ and their linear dimensions to an accuracy better than 0.1 m.

The crests (at the boiling whitecap stage) and the spotty (strip) foam (at the spread stage) were counted in each working frame. Then, the geometrical shape of each individual formation was estimated, and its characteristic linear and area dimensions were calculated. In this estimation by "crest length" we mean its characteristic size along the breaking wave front.

Strictly speaking, the two types of foam systems are closely interrelated from the temporal evolution point of view, but, nevertheless, a skilled observer can find distinctions between them, though the spotty foam is rather conventional. So, the distinction can be made precisely enough: by "wave crest" we mean a breaking stage

**Figure 4.2.** Typical ship-borne view of the wave-breaking process at $12\,\mathrm{m\,s}^{-1}$ wind velocity. The whitecap is in the foreground, and the dissipating residual foam field stands in the background. Photo: Bortkovskiy and Kuznetsov (1977).

in which there is a characteristic height of a crest foam mass (the "boiling" wave crest), whereas its accompanying foam mass in "spread" mode relates to strip foam. A typical example is presented in Figure 4.2.

In addition to calculating the area of a foam field as such, the total area of a turbulent spot countered by a foam mass, inside which the foam mass settles down as a "flocculent" (strictly speaking, fractal) field, was also calculated. Optical identification of a turbulent spot is emphasized by the absence of capillary waves (slicks) in a spot.

## 4.3   TEMPORAL EVOLUTION OF THE BREAKING PROCESS

Analysis of the characteristic temporal cycle of an individual wave as it breaks, presented in Figure 4.3, indicates that large breaking (whitecap foam) actually lasts a few seconds (its total lifetime is 1–3 s) and is noticeably shorter than the carrier wave period. The "boiling up" stage as such lasts a few tenths of a second, and recording the evolution of this initial stage is troublesome in field experiments. In our experiments we shall (conventionally) consider this stage as the beginning of breaking, and the whitecap area is maximum in this case.

Two characteristic time areas can be distinguished in the time dependence of a strip foam area (Figure 4.3). First, the growth of the foaming area (some kind of "spread" in wave crest foam mass), its reaching a maximal value, and, second, the

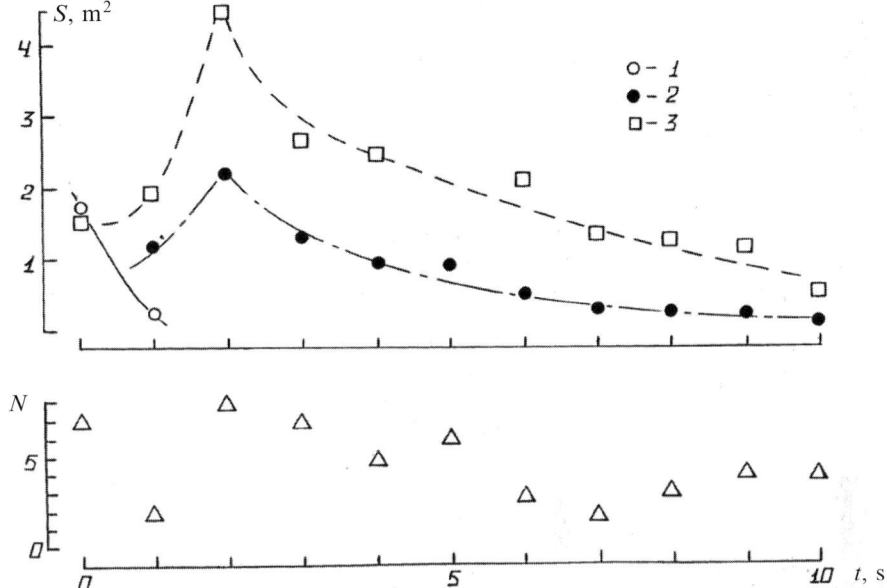

**Figure 4.3.** Time series of the cycle of separate waves as they break: areas ($S$) of a whitecapping foam field (1), a residual foam {2}, and a turbulent spot (3) on the sea surface. $N$ is the "momentary" number of mesobreakings in the 74-m$^2$ area of the spatial image of the sea surface.

time interval of foam mass diffusion. Similar to this is the time dependence of a turbulent spot (slick) area—that is, growth in area after spontaneous breaking, reaching a maximum, and diffusion of the area. We should especially note that the whole process of foam turbulent spot evolution, while starting on a carrier (broken) wave, then "transfers" to neighboring waves.

It is interesting to note that the timescale of evolution of a breaking zone, considered above, corresponds well with the results of filming, radiothermal, and scatterometric observations of an individual large wave breaking (its period exceeded 10 s), carried out in the Indian Ocean (see Cherny and Sharkov, 1988, and Chapter 8). The full lifetime of the crest of such a wave was 1.5–2 s, and the full lifetime of a foam field was 7–10 s.

Let us turn now to more detailed analysis of strip foam area evolution. The behavior of air bubbles in water is known to be caused by their turbulent diffusion into deeper layers from the surface and subsequently floating up under the effect of buoyancy. The dynamics of gas bubble motion in the upper turbulent part of the sea was analyzed in detail by Thorpe (1982). Leaning on the characteristic size of bubbles in the breaking zone (see below), we can conclude that the velocity of bubbles floating up is about 30 cm s$^{-1}$, and that such air inclusions quickly emerge on the surface. For a time after breaking (1–3 s), air bubbles propagating through the water depth replenish the foam spot on the surface, not allowing it to noticeably decrease. Then

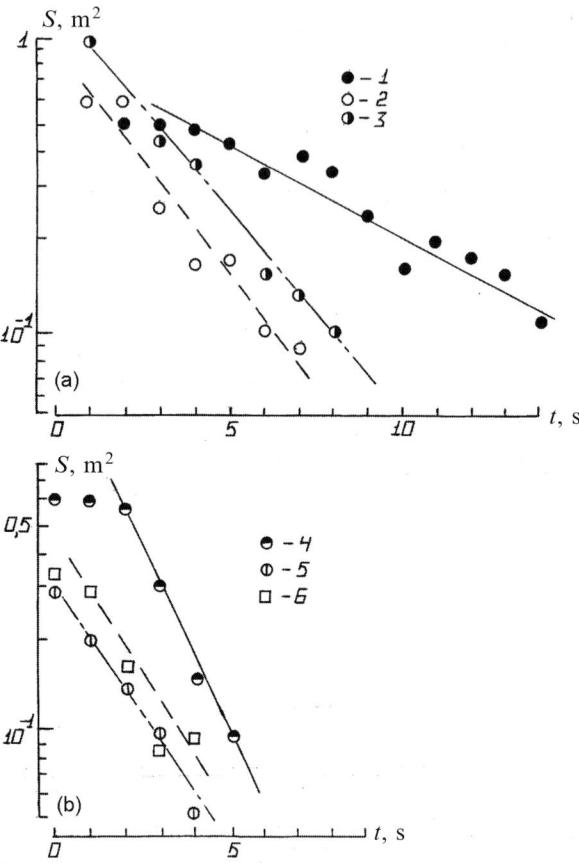

**Figure 4.4.** Time series of dissipation of residual foam fields under the following hydrometeorological conditions (see Table 4.1): (a) I a, (b) I b, (c) II, (d) III. Different labels are used to indicate the results of measurements of individual dissipation of foam fields with the following half-life constants: 1—7.7 s, 2—3.1 s, 3—3.0 s, 4—1.8 s, 5—2.7 s, 6—2.4 s, 7—0.96 s, 8—0.91 s, 9—1.3 s, 10—0.52 s, 11—0.52 s, 12—0.41 s.

there comes the foam spot diffusion phase, where the foam density sharply drops because foam-forming bubbles are destroyed. The lifetime of an individual bubble floating on the surface depends on many parameters (mainly, size, surface-active substances, temperature, salinity). The temporal characteristics of foam are even more uncertain, since foam represents a set of densely packed bubbles of various sizes, shapes, and wall thicknesses (Raizer and Sharkov, 1980; Raizer *et al.*, 1976; Weaire and Hutzler, 2000).

   In order to characterize foam spot dynamics in detail, it is necessary to understand the many quantities characterizing the time dependencies of a structure: disperse composition, the geometry of a turbulent spot on the surface, and the "flocculence" of foam inside this spot. Similar measurements have not yet been

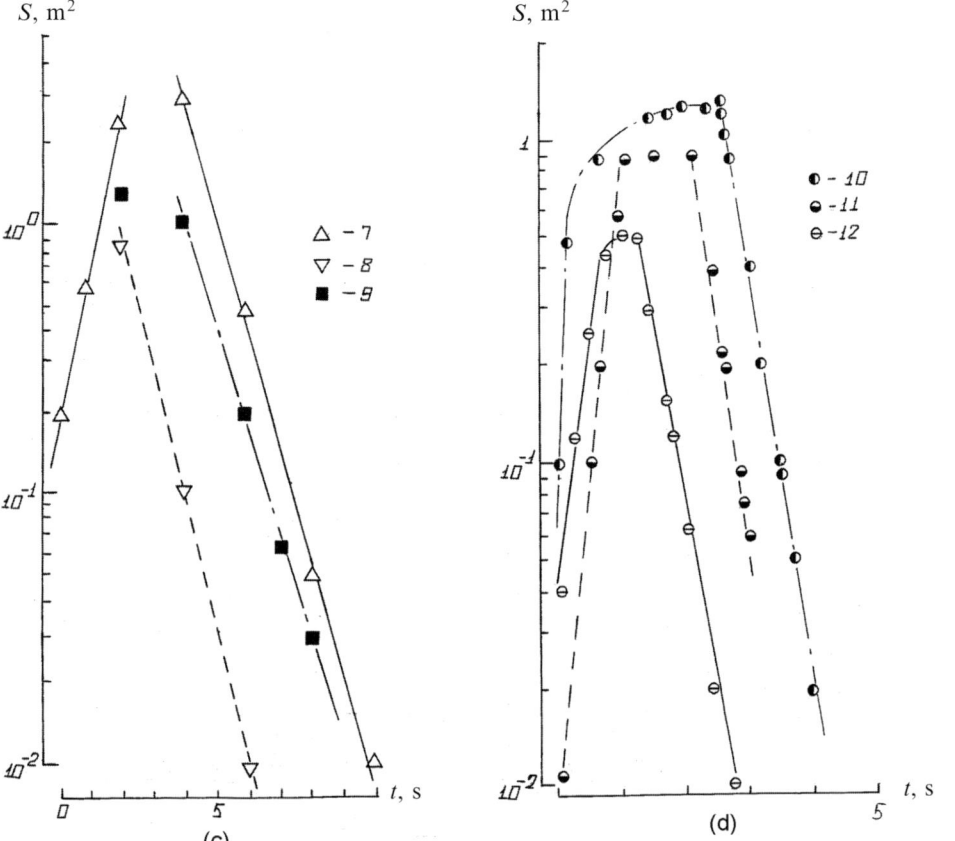

**Figure 4.4** (*cont.*)

performed in detail. In the work of Timofeev and Sharkov (1992) and Sharkov (1994, 1995, 1996a, b) the most general characteristic was studied—namely, the area $S$ of substantial foam in a spot and its time dependence $S(t)$. $S(t)$ was found to decrease almost exponentially with time at spot diffusion stages:

$$S(t) = S_0 \exp\left(-\frac{t}{\tau}\right) \tag{4.1}$$

where $S_0$ is the maximum value of a spot area.

Quantity $\tau$ will be called the foam "lifetime" constant (the "half-time"), and the full lifetime of the foam field will be $(3\text{--}4)\tau$.

The exponential character of foam field diffusion is clearly seen from a series of time pictures of the diffusion of individual foam spots (separate realizations), which are presented in Figure 4.4a–d in the semi-logarithmic scale for various hydrometeorological conditions and various regions: (a) and (b) for the Pacific

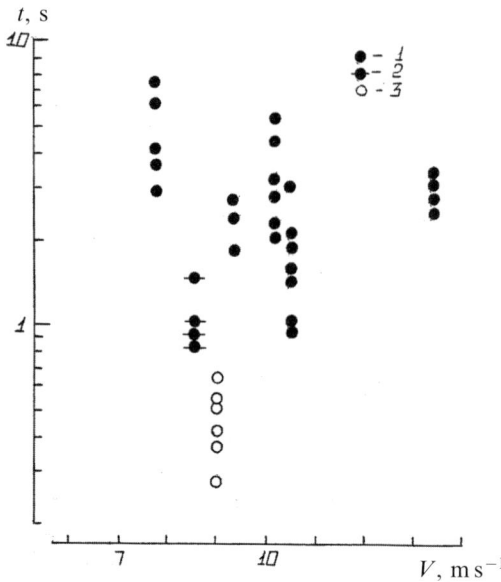

**Figure 4.5.** Dependence of the half-life dissipation time constants of residual foam fields on surface wind velocity (at a height of 20 m): 1—hydrometeorological conditions of type I, 2—of type II, 3—of type III (see Table 4.1).

Ocean; (c) for the Frieze Strait, and (d) for the Caspian Sea. Deviations of points from approximating straight lines are most likely explained by perspective distortions when recording foam spots on a disturbed sea surface.

Specially performed filming of the breaking process (Figure 4.4d) has shown that as the foam field grows the exponential law prevails. It is interesting to note here that the foam field growth constant is close to the diffusion constant, which is clearly seen from filming results (Figure 4.4d).

The half-diffusion time constants that characterize the destruction of separate spots, which were obtained in the experiments, are combined in Figure 4.5 in the form of dependence of $\tau$ on wind velocity at an altitude of 20 m from the sea level. The greatest number of $\tau$ values fall in the interval of 1.5–3.0 s; however, there are considerable deviations (from 0.7 to 8.0 s). The physical explanation of such considerable distinctions in half-diffusion time is most likely related to the type and character of the gravity waves that are breaking. According to Longuet-Higgins and Turner (1974), the breaking process can be sub-divided into two types: "plunging" and "spilling". In these two cases the aerated sub-surface layer will certainly be formed in completely different ways and, accordingly, the spatiotemporal characteristics of spectral composition of an aerated layer and its diffusion-turbulent dynamics will be quite different—see Thorpe (1982) and Bortkovskiy (1983). So, for example, the increase in foam field lifetime due to replenishment of foam spots, floating up as large bubbles for some time (20–30 s), is practically impossible, since relatively large bubbles, comprising the foam structure volume (Raizer and Sharkov,

1980), float up after breaking much faster (just a few seconds) (Thorpe, 1982), whereas smaller bubbles, arising for most of the breaking process and taken to depth (Thorpe, 1982; Deane and Stokes, 2002), cannot give any noticeable foaming on the surface, and do not appear on the surface at all. On the other hand, a wave crest that "dives" deeper at breaking may probably cause outflow of a much higher concentration of surface-active substances and, thereby, increase the temporal stability of a foam structure. Similarly, we can explain the decrease in lifetime at spilling gravity wave breaking.

Analysis of Figure 4.5 shows no prominent dependence of lifetime on wind velocity; though, as the wind velocity increases, the dispersion in scattering of lifetime values slightly decreases. This is probably associated with greater intermixing of the sea surface layer as the wind strengthens and, hence, as the chemical and physical parameters of water decrease in their effect. The values for lifetimes of foam structures, observed in the Frieze Strait region, are lower on average than those obtained in the open ocean. This circumstance is more likely related to the fact that the character of gravity waves that break in the Frieze Strait principally differ from those in the open ocean. The presence of intensive tidal flows and internal wave outflows to the sea surface has a strong effect (Figure 4.1).

The results of processing photographic material obtained in the Caspian Sea—near Baku (at a location called Makarov's Shoal)—are especially emphasized in Figure 4.5. Characteristic lifetimes of whitecaps (breakings) varied, for given hydrometeorological conditions (see Table 4.1), in the range of 0.5–1.7 s, with a major gravity wave period of 2.0–2.5 s. Time constants of foam spot diffusion turned out to be much lower than those in the open ocean (0.26 to 0.65 s instead of 1.0 to 5.0 s).

Figure 4.6 presents a comparison of the areas and linear (maximum in the wave front) dimensions of crest foam (separate realizations correspond to separate light points in the figure) with the data of spatial–statistical models developed in Bondur and Sharkov (1982, 1990) and described in Chapter 3. This comparison shows a good quantitative correspondence. In other words, the value of geometrical parameters lies "inside" the confidence boundaries over the 0.9 level—that is, they can be supposed to relate to the same general set that underlaid the statistical models (Chapter 3). Note that the statistical models of Bondur and Sharkov (1982, 1990) were generated on the basis of experimental results obtained in completely different geographical zones and under quite different hydrometeorological conditions—namely, in the Barents and Black Seas.

Similar conclusions can be drawn from analysis of comparison of the geometrical characteristics of foam (strip) fields with statistical models (Bondur and Sharkov, 1982, 1990) (Figure 4.7). In particular, it can easily be concluded that foam fields (in the sense of their geometrical size and ellipsoidal shape) relate to the same general set as the structure in the Black Sea does; in addition, they possess an ellipsoidal shape with a stable eccentricity (Figure 4.7a).

As noted above, when Bondur and Sharkov (1982, 1990) formed their statistical models (see Chapter 3), all life stages of crest and strip types of foams were included in the models, since observations were carried out according to type II measurements (Section 4.1). Figures 4.6 and 4.7 present the values of geometrical parameters

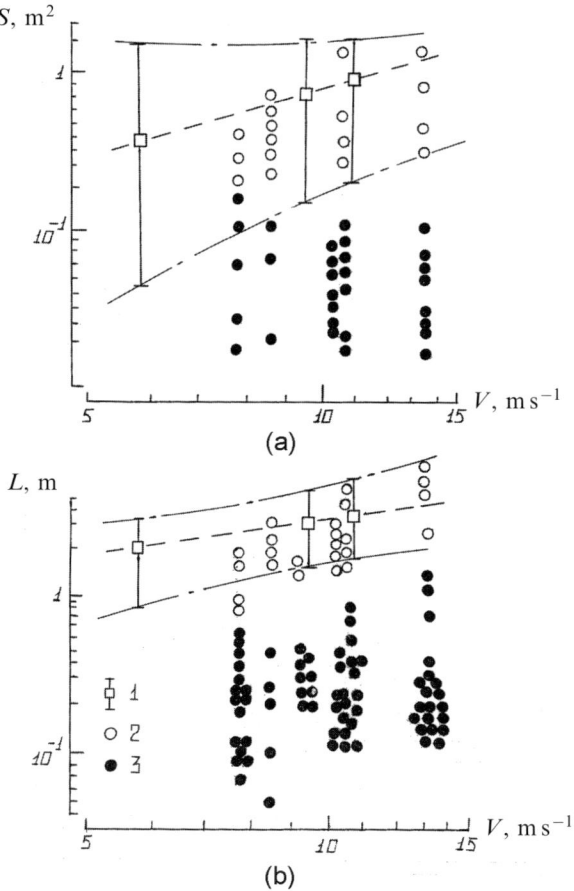

**Figure 4.6.** Wind dependences of the area (a) and linear size (b) of separate samplings of whitecapping foam and mesobreakings: 1—approximated spatial-statistical experimental models for the area (see Bondur and Sharkov, 1982 and Section 3.5) and the linear size (see Bondur and Sharkov, 1990 and Section 3.4) of whitecapping foam with 0.9 confidence limits; 2—experimental data for whitecapping fields; 3—the same for mesobreakings.

obtained at the most developed stage only. Thus, we can consider that—when constructing statistical models—the "age" of foam structures does not play a principal part and is entirely "absorbed" due to natural spatial variability.

## 4.4  SPATIOTEMPORAL CHARACTERISTICS OF MESOBREAKINGS

A high spatial resolution and special photographic conditions made it possible to find, in addition to details about how large waves break (Timofeev and Sharkov, 1992; Sharkov, 1994, 1995), another type of breaking, which was previously

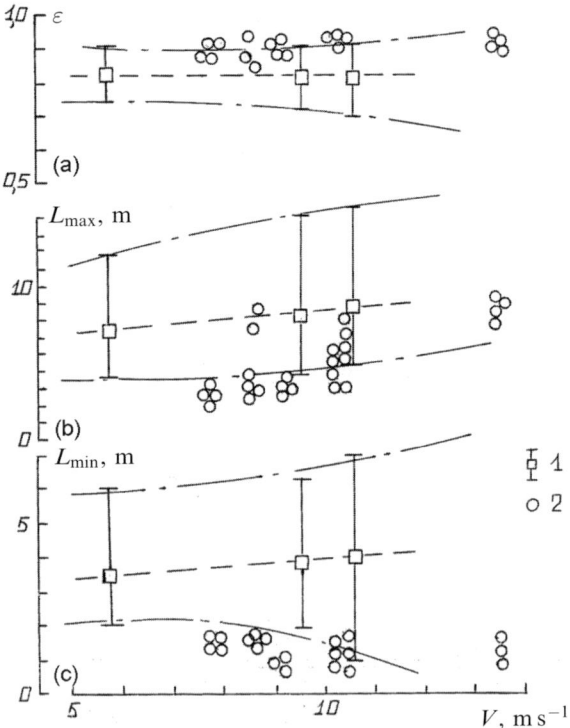

**Figure 4.7.** Wind dependence of the eccentricity (a) and the maximum (b) and minimum (c) linear size of separate samplings of residual foam fields and mesobreakings: 1—approximated spatial-statistical experimental models for the linear size (see Bondur and Sharkov, 1990 and Section 3.4) of residual foam with 0.9 confidence limits; 2—experimental data for residual foam fields.

unknown, and given the name "mesobreaking" (MB). This type of breaking differs from the "usual" breaking of large gravity waves in the following features. First, the disperse foam mass lifetime is rather short (lower than 1 s); second, MB is relatively small in its characteristic size (smaller than 0.5 m); third, MB does not leave behind a spot-like foam on the sea surface or an aerated layer inside the water medium; fourth, MB possesses a weak optical contrast and cannot be identified from aerial photographic data or from perspective pictures taken from onboard the RHV. We can regard MB as swells on the wave crest (or close to it) that immediately dissipate without generating either foam mass on the water surface or a turbulent rotor-type flow (along with an aerated layer) inside the water medium.

Let us analyze the results of measuring the parameters of this specific type of breaking, which, in our opinion, should constitute a separate class—namely, the class of mesobreakings (MB). It is expedient to do this not only because of their specific physical properties (listed above), but also because MB-type breakings can make a noticeable contribution to total foam coverage (see below), and, accordingly, they

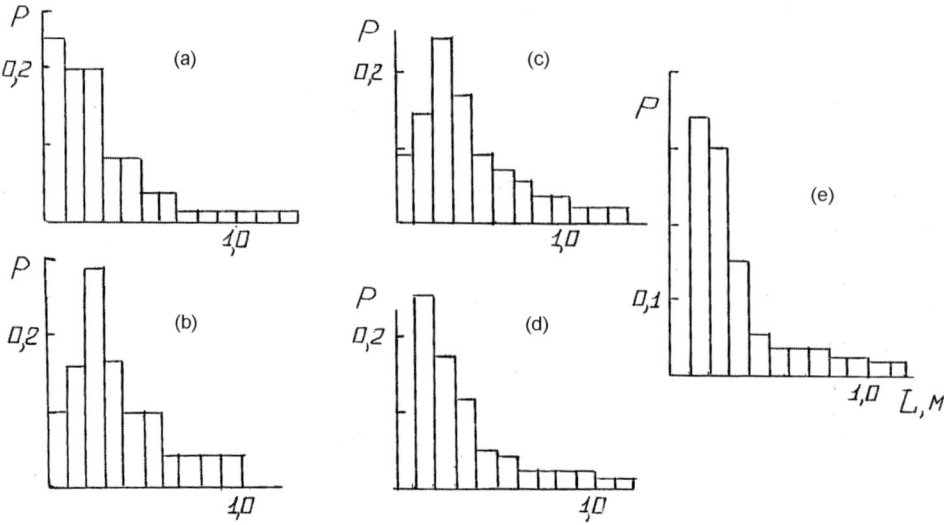

**Figure 4.8.** Distribution histograms of maximal sizes of separate mesobreaking-type random events in a 74-m$^2$ spatial window under the following hydrometeorological conditions: (a) Ia, sampling volume is $n = 31$ (mesobreakings); (b) Ib, sampling $n = 35$; (c) Ic, sampling $n = 85$; (d) Ic, sampling $n = 75$; (e) Ie, sampling $n = 65$.

should be taken into account when developing radiophysical models of radiation and scattering (see Chapter 6). For visual (standard) estimations of the state of disturbed sea surface (e.g., using the Beaufort Scale), though, large breakings that leave behind optically contrasting foam fields and turbulent "calming" fields of slick type or the turbulent wake of breakings are certainly more indicative.

As to the spatial characteristics of MB distribution, field observation has shown that the spatial field studied, inside which the breaking took place, contained a considerable number of mesobreakings of both neighboring waves (in relation to a breaking wave) and, in some cases, of a studied wave itself, after a major large breaking had taken place (Figure 4.2). The MB appearance was also recorded on a studied wave before its major large breaking. In other words, the mesobreaking field represents a kind of noise field in which large breaking takes place, but it may not necessarily be associated with this field directly.

Analysis of Figure 4.8, where histograms of the distributions of the maximum size of MB-type breakings are presented, clearly indicates that approximating distributions belong to the gamma distribution with rather stable parameters and weakly vary as wind strength grows. For all wind velocities the mean values of MB lengths lie within the limits of 0.3–0.4 m (see Table 4.2), and, so, no deterministic dependence on wind velocity exists.

Comparison of the linear and area geometry of large breakings and MBs (in the bi-logarithmic scale in Figure 4.6) evidently shows that the fields of existence of geometrical values of MB are essentially (up to 90%) greater than the confidence

**Table 4.2.** Average characteristics of breaking zone geometry.

| Type of HMC | Parameters of large wave breaking | | | Parameters of mesobreakings | | |
|---|---|---|---|---|---|---|
| | $S^W$ (m$^2$) | $S^S$ (m$^2$) | $\tau$ (s) | $L$ (m) | $S$ (m$^2$) | $n_0$ |
| Ia | 0.4 | 0.6 | 5.0 | 0.44 | 0.06 | 1.2 |
| Ib | 0.5 | 0.5 | 2.5 | 0.37 | 0.05 | 1.8 |
| Ic | 0.7 | 1.6 | 3.1 | 0.40 | 0.05 | 5.1 |
| Id | 0.7 | 1.8 | 1.9 | 0.30 | 0.06 | 6.8 |
| Ie | 1.0 | 5.5 | 3.2 | 0.35 | 0.04 | 6.3 |

*Note*: $S^W$ and $S^S$ are the average areas of whitecappings and foam structures; $\tau$ is the average lifetime of foam structures; $L, S$ are the average length (along the wave front) and area of mesobreaking; and $n_0$ is the average intensity of breaking in the 74-m$^2$ sea surface area. The root-mean-square errors are 0.1–0.2 m for $S^W$ and $S^S$, 0.1 m for $L$, and 0.02 m$^2$ for $S$.

boundaries of statistical models of large breakings (Bondur and Sharkov, 1982, 1990), thus forming their own, separate statistical set.

By virtue of the short lifetime of MB-type breaking, we consider it expedient to introduce a parameter that characterizes the intensity of birth (and decay) of MB-type whitecaps per unit of time (s) and within a certain spatial window (in our case of 100 m$^2$). The birth intensity parameter determined in such a manner showed (see Table 4.2) that there exists a doubtless tendency of MB birth intensity to grow from 1.2 (for $V = 7.7\,\mathrm{m\,s^{-1}}$) up to 6.3 (for $V = 13.5\,\mathrm{m\,s^{-1}}$). We can easily also estimate the full (instantaneous) contribution of MB-type breakings to the relative foam coverage of MB type. It constitutes a considerable value: from 0.1 to 0.6% (at a spatial window of $100 \times 100\,\mathrm{m^2}$).

As far as the physical nature of formation of MB-type breakings is concerned, in some theoretical works (e.g., Banner and Phillips, 1974; Phillips, 1988) there are indications (though rather fragmentary) of the possible existence of small-scale breakings of capillary gravity waves. This type of breaking is called "microbreaking" in the science literature (Banner and Phillips, 1974; Phillips, 1988). It is actively investigated under laboratory conditions by means of sensitive IR panoramic systems (infrared imagers) (Zappa *et al.*, 2004). However, comparison of these theoretical predictions and laboratory data with the field experimental results presented in this chapter represents a separate problem, because the features of the IR-radiation field related to violations of the thermal regime of a surface skin-layer are recorded under laboratory conditions, and, so, the problem of physical identification of individual microbreaking events arises (Zappa *et al.*, 2004).

Nevertheless, some physical considerations can be stated on the basis of analyzing the results presented in this chapter. Since the MB birth intensity only weakly depends (and average physical dimensions do not depend at all) on wind velocity and energy-carrying disturbance parameters, the nature of this formation is associated, most likely, with small-scale, but intensive turbulent wind rushes (wind field inhomogeneity), and are most likely initiated by parasitic capillary waves on the crests of gravity waves. In this case the absence of intensive turbulent intermixing in a water volume "does not include" physical and chemical factors (surface-active substances) in the process, and, accordingly, the foam structures accompanying a large breaking are not formed.

These considerations are qualitative in character; thus, to obtain specific results purposeful investigations are necessary.

As to the remote aspect of the problem, we should note that considerable time variations in the foam coverage of individual structures are naturally included in the detailed statistics of the intensity of radio emission of a disturbed sea surface, which undoubtedly should be taken into account when forming appropriate statistical radiophysical models. Though data on the radiative characteristics of MB-type breakings are no longer available, it is quite obvious (e.g., from the experience of studying foam system radio emission—see Chapter 6) that MB emissivity will be rather low. However, we should keep in mind that MB-type breakings can make a noticeable contribution to total foam coverage (see above) and, thus, most likely make their own noise-wise contribution to the intensity of radio emission of a disturbed sea surface.

## 4.5  SPECTRAL CHARACTERISTICS OF AN AERATED LAYER

Experimental data from studying the disperse structure of an aerated layer in the breaking process, obtained photographically from a semi-immersed sealed box, are presented in Figure 4.9. This figure shows a picture of the size distribution (in diameters) of air bubbles formed by turbulent intermixing at a depth of about 5 cm directly under the breaking zone. Analysis of Figure 4.9 indicates that the density of distribution of the diameters of captured bubbles is close to the normal

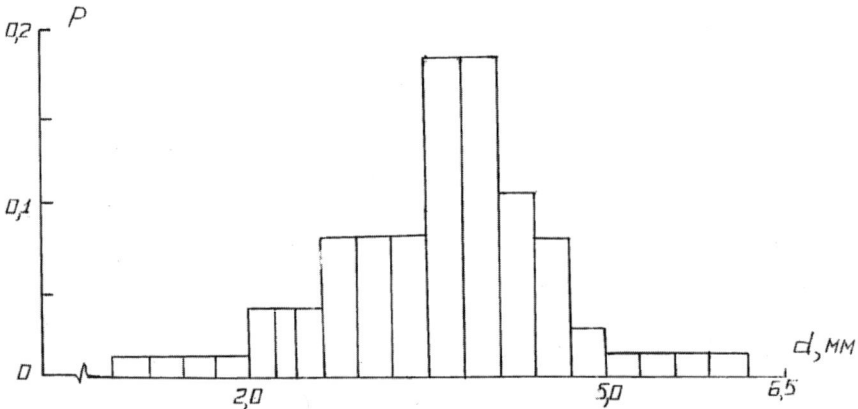

**Figure 4.9.** Distribution histograms of air bubble diameters in the breaking zone (below the water surface) with sampling volume $n = 38$ .

distribution with a mean value of 3.68 mm and a root-mean-square deviation of 0.4 mm.

The total region of spatial dimensions in which experimental data were obtained constituted values from 1 to 6 mm, which does not contradict literature data (Raizer and Sharkov, 1980) on the disperse structure of a foam mass "lying" on the water surface. In this case a rather wide range in the lifting velocities of bubbles was observed—from 0.15 to 0.8 m s$^{-1}$. Quantitatively, the obtained results correspond well with the data of other works (Bortkovskiy and Timonovsky, 1982; Bezzabotnov, 1985).

## 4.6  CONCLUSIONS

1   Three characteristic stages can be distinguished in the temporal dynamics of foam formation breaking: (a) generation, development, and dissipation of a foam wave crest (the source of aeration of the sea's upper layer); (b) generation of a rather stable foam spot behind the wave crest; (c) destruction and diffusion of a foam formation. The latter stage takes up a noticeable (more than a half) part of the life of foam coverage.

2   The area of a foam spot, irrespective of its size, decreases in time according to the exponential law with the half-decay constant not depending (on average) on wind–wave conditions.

3   The geometrical size statistics for crest and strip foams, generated from experimental data—obtained in the Pacific Ocean, Barents Sea, and Black Sea basins—belong to the same general sets.

4   Two groups can be distinguished among various types of breakings: "macro" and "meso" breakings. Geometry statistics of the former can be combined into a single general set, whose mean characteristics rather adequately describe the state of wind–wave conditions. MB-type breakings (existing at all wind velocities) possess some specific properties: short lifetime, small size, absence of an accompanying foam field. Geometry statistics are close to the gamma distribution, the average size not being dependent on wind–wave conditions.

5   Statistics on the size of air bubbles existing in an aerated layer under the breaking zone possess a near-normal distribution with a mean size of 3.68 mm.

# 5

# The drop-spray phase over a rough sea surface

This chapter presents the results of experimental remote and *in situ* investigations of the state of the drop-spray phase that forms in the ocean–atmosphere system in moderate sea and breaking gravity sea wave conditions. Information about the physical mechanisms behind drop-spray phase generation is presented, and comparative analysis of literature data on the physical characteristics of a disperse medium is carried out. These characteristics include: the water content of a medium, the spectrum of dimensions, the altitude profiles of water content, the temporal and spatial dynamics of drop-spray clouds. The model presentations on water content of drop-spray clouds generated on this basis are presented according to the degree of sea state roughness and wind velocity.

## 5.1 PHYSICAL MECHANISMS OF DROP-SPRAY PHASE GENERATION

A short observation of the sea surface in windy conditions shows that the appearance of a small-drop-spray phase in an airflow is associated with the breaking of steep crests of wind gravity waves. However, this generally trivial conclusion is far removed from understanding what actually is the physical mechanism involved in production and separation of drops from the water surface. This process, widely used in technological facilities and called suction, has been repeatedly studied in laboratory installations in tanks, where an airflow is generated over a surface film or thin layer of liquid (Levich, 1962; Wallis, 1969; Kutateladze and Styrikovich, 1976). Though the overall objective of these investigations was to discover the empirical dependencies for solution of purely applied problems, the physical processes determining suction were studied simultaneously as well. A much more complicated and less studied process seems to be generation of sprays on the sea surface in moderate and strong winds. Similar investigations have been performed right up to the present time, basically under laboratory conditions and in wind–wave tanks (Wu, 1973; Lai and Shemdin,

1974; Tedesco and Blanchard, 1979; Cipriano and Blanchard, 1981; Koga, 1981; Monahan *et al.*, 1982) and to a much smaller extent under field conditions (Monahan, 1968; Ruben, 1977; Egorov, 1977; Belov, 1978).

It should be noted that even the phenomenon of liquid suction from a thin film (which is simpler since there is no breaking of large wind waves on a surface film) remains incompletely explained despite numerous experimental investigations. Qualitatively, it is described as follows: at a high gas velocity the resistance forces acting at the top of wave crests turn out to be large enough to tear off the liquid droplets from them; the appearance of whitecaps usually precedes the beginning of suction. This qualitative description can be considered as a possible explanation of suction, and the corresponding mechanism can be called the direct "tearing-off" of droplets from the surface.

However, there exist some other mechanisms of spray phase generation, which differ in their significance. One is associated with the liquid surface instability that arises when there is a high-velocity airflow near the interface boundary. Optimal conditions for this mechanism can be observed both when there are sharp crests of wind waves, where the vertical derivative of wind velocity apparently reaches the greatest values, and on windward slopes of breaking waves, where the reverse counter airflow arises over the whitecap mass rapidly moving along a slope.

However, as many specialists (Blanchard and Woodcock, 1957; Toba, 1962; Monahan, 1968; Koga, 1981; Bortkovskiy, 1983) believe, the overwhelming mass of droplets is born not by direct "tearing-off" of water particles from the sharp crests of gravity waves by wind, but several seconds later by bursting air bubbles in the emulsion structure and the rupture of cellular-type foam structures (see Chapter 6). The process of bursting a bubble and forming spray droplets has been studied in detail in numerous laboratory experiments (Blanchard and Woodcock, 1957; Wu, 1973; Lai and Shemdin, 1974; Tedesco and Blanchard, 1979; Cipriano and Blanchard, 1981; Koga, 1981). This process was also thoroughly described and analyzed principally in connection with the problem of studying sea aerosols as a major factor of electrification of the atmosphere (Blanchard and Woodcock, 1957; Blanchard, 1963; Bortkovskiy, 1983). This process can be qualitatively described as follows. When the top of a buoyed bubble happens to be over the interface surface, the liquid begins to flow down from a dome, and the bubble's shell near the top rapidly thins. At last, when a hole is formed, the unbalanced forces of surface tension cause its expansion. The film, moving at high velocity, collides with the liquid's bulk and forms a circular elevation around the hollow. When the circular wave closes itself, a liquid column (i.e., a jet) is formed at the cavity center, and then one or several droplets are separated from the jet and fly up vertically (i.e., "reactive" or "jet" droplets). The size and energy of such "jet" droplets are determined by the size of a collapsed bubble. If the bubble surface is polluted with surface-active substances, the viscosity of the film grows rapidly. In this case the film continues thinning only until its thickness reaches a critical value; after this the liquid ceases to flow down from the upper part of the dome and continues to flow down from thicker film sections. As a result, the thickness of almost all film composing the over-water part of a bubble becomes small and the film bursts at many places simultaneously. The droplets formed in such a process are much smaller than the drops separated from the central jet and fly in all directions rather than just upwards.

This scheme, which is presented in many works, is rather simplified. Nevertheless, investigation of the various phases of destruction of bubbles of emulsion structure and films of cellular foam structure by means of rapid photography, carried out in many works (Hayami and Toba, 1957; Day, 1964; MacIntyre, 1972), indicates the validity of such an approach. According to the conclusions of these works, it is possible to distinguish (conventionally enough) two phases in bubble bursting (collapse) in which water droplets get into the air medium:

(a) bursting (or tightening) of a hemispherical shell medium and formation of many small droplets (film drops);
(b) after a bubble's upper shell bursts, a narrow jet erupts from the bubble bottom that results in formation of one to five large drops ("jet" drops), which are ejected vertically to altitudes depending on the size of the initial bubble.

At the first stage, the initial bursting and subsequent expansion of a hole occur at a high rate of about $8\,\mathrm{m\,s^{-1}}$, so that the full rupture of a shell lasts a few microseconds, and at the instant of its termination small droplets ("film" drops) tear off from the circular thickening that surrounds a cavity. And then the droplets fly almost horizontally at a velocity close to $8\,\mathrm{m\,s^{-1}}$. Since the size of bubbles in the sea environment varies from fractions of a millimeter up to a centimeter, the corresponding area of the film of a bubble's surface can reach values from $10^{-4}$ up to $1\,\mathrm{cm^2}$, which eventually determines the potential energy of ejection of a bubble's disperse phase. So, when a bubble 6 mm in diameter (surface area about $0.3\,\mathrm{cm^2}$) bursts, it can result in the formation of up to 1000 droplets. A few hundred droplets arise from a bubble 2 mm in diameter (surface area about $0.005\,\mathrm{cm^2}$), and no film droplets are recorded from bubbles smaller than 0.3 mm in diameter. The diameter of film drops varies in the range from 5 to 30 microns, and the height they can reach due to viscous friction in still air is low (less than 1 cm).

The central jet at the second stage can also produce a chain of vertically moving drops ("reactive" drops), rather than a single drop. Thus, even in the absence of pollution the bursting of bubbles can be a source of not only large, vertically moving drops, but also of the majority of small drops directed horizontally or obliquely. These conclusions are based on the results obtained by filming in a laboratory (Day, 1964). Small droplets appearing when large bubbles burst (with $r = 1.5$–$2$ mm) in still air were found to be rising by 2–3 cm, and the droplets from bubbles with a radius $r < 0.25$ mm rose by just a few millimeters.

The study of drops formed from a jet, arising when a circular wave closes, showed that there exists a statistical dependence between the size of the drops formed and the size of bubbles and between the size and initial velocity of ejecting droplets. The radius of jet drops is about $1/10$–$1/15$ the bubble radius (Hayami and Toba, 1957). Despite the high velocity of ejection (up to $10\,\mathrm{m\,s^{-1}}$), the maximum height of bubbles rising in still air is rather low: up to 15–20 cm for bubbles of about 2 mm in diameter and decreases as bubble diameters grow further.

Ascending turbulent airflows include large drops up to altitudes of 10–15 m, and aerosol particles of 10–30 microns in diameter are encountered at altitudes of 1000 m and higher.

Experimental data show that the character of bubble bursting—that is, the size and dynamics of drops arising in the process—essentially changes as temperature, salinity, and concentration of surface-active substances (SASs) change. As already mentioned, the presence of quite insignificant quantities of SASs in bubble shells highly influences the character of bursting (MacIntyre, 1972). Hence, we can assert the instability of the drop formation mechanism described with respect to small disturbances and presume that in any given area of the sea surface bubble bursting occurs simultaneously according to various types; the observed picture can be described as a whole only by means of averaging. We can also expect the size distribution of sprays, generated on the sea surface as a result of simultaneous action of several mechanisms (i.e., the spectrum of sprays), must differ from the well-studied distribution of the size of drops formed at breaking of liquid jets, which is described by the gamma-type distribution. This type of distribution, while having a rather general character, is successfully applied for describing the size distribution of various particles and may be applicable for describing the size distribution of bubbles in the upper layer of the sea. However, in this case the gamma-type distribution will also describe only a part of the spectrum of drops—namely, jet drops formed when "pure" (i.e., not containing SASs) bubbles burst. The size of small droplets that form at film rupture weakly depend on bubble size and, consequently, their appearance will shift the distribution maximum to the side of small drops. In exactly the same manner the action of direct tearing-off of sprays from wave crests and the formation of sprays near the wave front should show up in the size distribution observed.

As already noted, the basic mechanism of bubble generation in the near-surface layer of the sea surface is related to gravity wave breaking (Thorpe and Humphries, 1980; Thorpe, 1982; Koga, 1982). However, other mechanisms of bubble generation also exist: precipitation of rainfall in the form of rain, snow, dust, and aerosols. A large drop of rain can form up to 100 or sometimes even 400 small and rather stable bubbles on the sea surface (Blanchard and Woodcock, 1957). Another important mechanism consists in forming the microbubbles by sorption of gases on the surface of various suspensions or in biological processes, such as photosynthesis and respiration of small sea organisms (Horne, 1969). From the hydrodynamical point of view, one important mechanism of bubble appearance in the near-surface layer is the bursting of capillary waves on reaching an extreme configuration of the wave surface (Phillips, 1977).

## 5.2   DISPERSE CHARACTERISTICS OF THE DROP-SPRAY PHASE

To study the fluxes of heat, water, and momentum during a storm (Bortkovskiy, 1983), as well as to construct electromagnetic models of spray phase fields (Sharkov, 2003), it is necessary to know those spray field parameters that depend on altitude and vary along the wind–wave profile—namely, vertical flux and concentration, size distribution (spectra) of sprays, three-dimensional components of the velocity vector of spray clouds. Such a complex of disperse parameters has still not been studied completely enough under field conditions. The reasons for this are crystal clear: the performance of detailed field experiments in a zone directly adjoining the sea surface, under rough sea conditions, meets natural technological and methodological difficulties (see Plate 2 in the Prelims). Ideas about the properties of spatiotemporal fields of spray clouds were based until recently mainly on the results of measurements in coastal zones, on ships, and in laboratory experiments. However, measurements carried out in a zone of surf and on ships cannot be considered strictly representative, not only because of the deforming effect of a ship, beacon, platform, etc. but also because of the fact that there exists a complicated altitude stratification of spray clouds, which can be sub-divided conventionally into a zone of drop injection and a zone of drop suction by a stratified airflow. As laboratory experiments and theoretical calculations have shown, the height reached (injection) of most sprays under quiet wind conditions is low (see Section 5.1), and measurements for the injection zone should be performed as close to the water surface as possible.

### 5.2.1   Laboratory measurements of the characteristics of a drop-spray phase

As to the results of measurements (Lai and Shemdin, 1974; Toba, 1962; MacIntyre, 1972; Hayami and Toba, 1957; Okuda and Hayami, 1959; Blanchard, 1963; Day, 1964; Wu, 1973, 1979; Wang and Street, 1978) carried out under laboratory conditions (without an airflow) and under wind–wave tank conditions in the presence of an airflow, they obviously differ from field measurements, mainly in the character of a rough sea state and, second, in the character of interaction between artificial and natural windflow. The latter circumstance is directly related to the design features of a wind–wave tank, such as the value of fetch, the depth of the water layer, the height and type of formation of the airflow, the shape of the wind–wave tank's roof, the presence of a mechanical wave-producer, the system of recording, and many other features. Obviously, detailed analysis of measurements under the conditions of various wind–wave tanks—and, even more so, under field experiment conditions—is a complicated problem in itself. Attempts at such a kind were undertaken by Wu (1979) and Bortkovskiy (1983). Here we shall briefly analyze only the results of laboratory experiments, basically from the viewpoint of the disperse characteristics of spray clouds (i.e., size spectra and concentration of particles).

Apparently, measurements of the flux of sprays over water were first carried out by Okuda and Hayami (1959) in a wind–wave tank 14.7 m long, at a water depth of 0.43 m, and at an air layer height over the water of 0.57 cm. Sheets of filtering paper $5 \times 5$ cm in size, impregnated with paint, were exposed for 5–10 s at heights from 5 to 30 cm over the water surface. The plane of the paper was normal to the longitudinal axis of the tank. Note that under the experimental conditions sprays began to tear off from the surface at a wind velocity in the air corridor of about $10 \, \mathrm{m \, s^{-1}}$, and then they amplified with growing velocity, while horizontal flux very rapidly decreased with altitude. Data on the size spectrum of drop clouds are no longer available.

In Toba (1962) the dependence of the intensity of spray system generation at the surface (i.e., of their vertical flow) on wind velocity was investigated. Experiments were carried out in a wind–wave tank, having a length of 21.6 m, a width of 0.7 m, and a height of 1.0 m; the tank was half-filled with water. For a flow velocity interval of $15 \, \mathrm{m \, s^{-1}} \leq U_{10} \leq 23 \, \mathrm{m \, s^{-1}}$ (the measured velocity was reduced to the level of 10 m using a logarithmic extrapolation) the following dependence was obtained:

$$q = q_0 \exp[0.4(U_{10} - 15)] \tag{5.1}$$

where $q$ is the vertical flux of drops in $\mathrm{cm^{-2} \, s^{-1}}$; and $q_0 = 1 \, \mathrm{cm^{-2} \, s^{-1}}$. The appearance of drops was noticed at $U_{10} = 13 \, \mathrm{m \, s^{-1}}$ with a wind fetch of 13 m. Toba considers the basic mechanism of spray phase generation to be the bursting of bubbles formed when the steep slopes of gravity-capillary waves close over their hollows (see Section 5.1). Such a process, in Toba's opinion, intensively proceeds at the crests of wind waves as they break. The distribution of drops over radii ($r$), obtained in the range of $0.0025 \, \mathrm{cm} \leq r \leq 0.015 \, \mathrm{cm}$, can be approximated by the dependence $n(r) \sim r^{-2}$.

The results of determination of the size and quantity of sprays over fresh water in a wind–wave tank 14 m long and 1.5 m wide, at a water depth of 1.2 m obtained using a laser are presented in Wu (1973). The diameter of the laser beam, directed horizontally across the tank, was 0.08 cm. Wind velocity ($U_{0.3}$) was measured over the upper boundary of the logarithmic sub-layer (i.e., at $z = 30 \, \mathrm{cm}$), with an overall air layer thickness over the water of 35 cm. The dependence of the horizontal flux of drops ($q \, \mathrm{cm^{-2} \, s^{-1}}$) on wind velocity in the range of $9 \, \mathrm{m \, s^{-1}} \leq U \leq 13.5 \, \mathrm{m \, s^{-1}}$ and altitude in the range of $0.1 \, \mathrm{m} \leq z < 0.25 \, \mathrm{m}$ has an exponential effect on wind velocity

$$q = 4 \cdot 10^{-4} \exp(-4.9z) \exp[2.5(U_{0.3} - 7.5)] \tag{5.2}$$

It is important to note that the "explosive" trigger character of spray cloud generation was experimentally recorded at all heights at once in the logarithmic sub-layer up to heights of 24 cm at the critical wind velocity of $8.5$–$9.5 \, \mathrm{m \, s^{-1}}$. The paper noted that the physical mechanism of spray structure generation was closely associated with the appearance of gravity waves breaking. The size distribution has a maximum at $r$ varying from $7.5 \cdot 10^{-3}$ to $10^{-2} \, \mathrm{cm}$; the modal size grows slightly as wind strengthens. As to large sizes, the spectrum rapidly decreases as $n(r) \sim r^{-8}$. The size distribution does not depend on height within the logarithmic sub-layer limits. The dependence of the average size on wind velocity is characterized by slow growth from $r = 8.3 \cdot 10^{-3} \, \mathrm{cm}$ at $U_{0.3} = 11.7 \, \mathrm{m \, s^{-1}}$ up to $r = 10^{-2} \, \mathrm{cm}$ at $U_{0.3} = 13.4 \, \mathrm{m \, s^{-1}}$. With some additional assumptions, the estimated intensity of drop generation at the water surface for similar wind velocity values was equal to a value about 10 times lower than that given by Toba (1962).

Lai and Shemdin (1974) present the results of measurements of the disperse properties of spray systems in a wave–wind aerohydrodynamic tank having a length of 45.7 m and a width of 1.83 m, the depth of water was 0.9 m, and the height of the air layer over the water was 1.0 m. Using a mechanical wave-productor it was possible to produce both a regular and an irregular disturbance; it was measured by a string wavegraph. The size of drops in the range of $0.0025\,\text{cm} \leq r \leq 0.08\,\text{cm}$ was determined from the change in voltage output of the film thermoanemometer. The sensor in this instrument was small (length 1 mm, diameter 51 microns), which excluded the possibility of simultaneously hitting two drops and ensured minimum disturbance of an airflow. The dependence of a horizontal flux of drops on height was obtained in the form of a logarithmic profile. As to the disperse structure, it was found that at a height of 13 cm (over sea water) the distribution has the form of $n(r) \sim r^{-2}$, and the distribution density grows as the wind strengthens: at $r = 10^{-2}\,\text{cm}$ and $U_{10} = 15.0$, 16.7, and $18.0\,\text{m}\,\text{s}^{-1}$ the spectrum value equals 0.04, 0.15, $0.60\,\text{cm}^{-4}$, respectively. Moreover, an interesting hydrodynamical effect was observed in the work: by suppressing spray cloud generation by exciting gravity waves using the mechanical wave-producer (called "swell" waves by the authors)—at an amplitude of 5 cm due to partially suppressing wind wave breaking, and at a "swell" wave height of 15 cm— the generation of spray drastically increases by an order of magnitude in distribution density values.

Some distinctions were also revealed in the spray field for fresh and salt water. Measurements carried out over fresh water under the same conditions as over salt water showed that the size distribution for large drops ($r \sim 1.5 \cdot 10^{-2}\,\text{cm}$) has the form $n(r) \sim r^{-3}$, and maximum values of $n(r)$ are noticed in the region of $r \sim 0.003$–0.008 cm (over sea water the modal radius is obviously lower than minimally measured values, $r_{\min} > 0.0025\,\text{cm}$). These distinctions in size distributions of drops over fresh and sea water are most likely associated with distinctions in the size distribution of bubbles.

With the purpose of estimating the spray effect on the energy exchange between water and air in laboratory experiments (Wang and Street, 1978), the water in a tank was heated such that the temperature drop between water and air reached 12.5°C and the drop in specific humidity was 20%. The length of the tank was 35 m, the width 0.9 m, overall height was 1.93 m; the tank was half-filled with water. Wind was produced by a fan located at the windward end of a tank. Measurements were carried out at four values of wind velocity (from 12.5 to $14.5\,\text{m}\,\text{s}^{-1}$), measured at a height of about 30 cm over the water (free-flow mode), and at a fetch (from the beginning of the working part of the tank) from 3.0 to 12.3 m. The quantity of drops and their size were determined by a sensor consisting in a piece of nichrome wire 63 microns in diameter and 6 cm in length, to which a voltage (1800 V) was supplied. A drop striking the wire causes a voltage fall in the instrument's output that was proportional to the drop size. The sensor made it possible to determine the size (diameters) of drops in the range from 25 to 1250 microns and the rate at which they fell (up to $35\,000\,\text{s}^{-1}$). The authors noted that there probably exist drops of much greater size that are not recorded by this system. The paper also noted that the "threshold" generation of a droplet phase takes place when gravity waves having crest heights of 5 cm begin to break at wind velocity in a free flow exceeding $12.5\,\text{m}\,\text{s}^{-1}$. Experimental data demonstrate the high growth in horizontal flux of drops and the height they reach as "fetch" $X$ increases. This testifies to the influence of wind waves on the intensity of spray generation and dynamics. The size distribution, on the contrary, keeps its form throughout the working section of a tank (i.e., $3\,\text{m} \leq X \leq 12.3\,\text{m}$).

### 5.2.2    Field measurements of drop-spray phase characteristics

Most informative from the viewpoint of revealing disperse characteristics of a spray phase in the entrainment zone is a paper by Preobrazhensky (1972), who obtained the vertical profiles of the water content and the size distribution in a spray phase in the atmospheric layer up to 7 m (from a still surface). This work was carried out in the Northern Atlantic basin from onboard a drifting research vessel using samples on oil plates at heights of 1.5, 4, and 7 m. The whole set of data was sub-divided into two groups corresponding to the cases of slack wind (7–12 m s$^{-1}$) and strong wind (15–25 m s$^{-1}$). In slack wind the greatest concentrations are observed for drops with diameters of 5–30 microns, and the content of particles of such a size in a volume unit (the water content) varies weakly with height. As the size of particles increases, a decrease in their concentration with distance from the surface becomes noticeable. So, in light wind the largest drops, observed at the level of crests of breaking waves, had a diameter of 130–150 microns. In strong wind the maximum size of drops reaches 0.5 to 2 mm. According to the experimental data, drops of such size, however, are not carried away to higher-lying layers of the atmosphere. Particles with diameter lower than 200 microns reach a level of 4 m, and those with a diameter of 90 microns a level of 7 m.

The same work showed that the vertical profile of the water content of a spray phase $W(z)$ can be adequately described by the exponential law:

$$W(z) = W_0 \exp[-\beta(z - H)] \tag{5.3}$$

where $z$ is the height of the level of measurements; and $H$ is the height close to the mean wave amplitude at the surface. In slack wind the parameters in (5.3) can be estimated as $W_0 \sim 10^{-4}$ g m$^{-3}$ and $\beta \approx 0.35$ m$^{-1}$, and at strong wind the values of these parameters essentially change to $W_0 \sim 10^{-2}$ g m$^{-3}$ and $\beta \approx 1$ m$^{-1}$.

Let us consider in more detail the study of a drop-spray phase under storm conditions. Since complete and reliable data on parameters of drop-spray clouds at storm were no longer available, Egorov (1977) and Belov (1978) suggested investigation at the initial stage of drop-spray cloud formation under strong surf. These disperse systems can be taken as physical models of drop-spray formations flying over the sea surface at strong wind, but their disperse parameters are too extreme for drop-spray clouds at high seas. There certainly exist many peculiarities and distinctions in the disperse properties of drop-spray clouds formed under surf conditions and at high seas. Note that—whereas during a storm a virtually continuous droplet sheet with homogeneous enough characteristics "flies" in the near-water layer—in the surf zone we have a drop-spray cloud that is localized in space and rapidly evolves in space and time. From the viewpoint of obtaining quantitative characteristics, such a drop-spray cloud in the surf zone can be reliably recorded both in space and in time and, accordingly, its disperse parameters can be measured. The drop-spray cloud parameters that need to be investigated include the following: the lifetime of a drop-spray cloud, the length of its flight, the total water content of a cloud, and the spatial and altitude distribution of water drops in a cloud of a given size. As it is difficult—in theoretical investigation of the processes of droplet production and drop-spray cloud propagation under storm conditions—to take into account all the factors influencing these processes, the problem of experimental investigation of drop-spray cloud parameters under storm conditions (or under conditions simulating them) becomes especially important. Let us analyze the technique of experimental investigation of drop-spray formations. To study total water content we used samplers in the form of bent glass test tubes, the openings to which were placed perpendicular to the flow, and the quantity of liquid of interest accumulated at the bottom of the test tube. The size distribution of drops was studied by precipitating them on an absorbing and self-painted substrate. The distribution of drops could be determined by letting the drops be caught by an immersion medium that was a mixture of soot and viscous oil. However, the distribution of drops obtained by the second technique was clearly overestimated in the small-drop region, which can be explained by crushing large drops at stroke about a substrate. In such a case of studying the size distribution of drops correct graduation of a substrate is of most importance; this was performed by means of a special technique (Belov, 1978).

Determination of the parameters of drop-spray clouds that form under strong surf was carried out in the hydrometeorological station area at the Schmidt Cape (off the Chukotsk Sea coast) in the autumn period (Egorov, 1977). During strong surf, which was accompanied by strong droplet production, the length of flight of a spray cloud, its lifetime, and water content were measured, and its size was estimated approximately. The size distribution of drops in a cloud was studied at various distances from the zone of direct droplet production and at various heights. For this purpose, samplers at heights of 1.5, 2, 3, and 4 m were fixed on a mast. In its turn the mast was placed at distances of 5 to 50 m from the droplet production zone. In such a way about 240 spectra of drops in a spray cloud were obtained.

As expected, the disperse characteristics of a drop-spray cloud evolve in a complicated manner depending on its lifetime. Thus, at the instant of spray cloud formation the prevalence of larger drops in its upper part is noticeable, and its water content can be higher than at lower heights. The maximum size distribution of drops at the 4-m height occurs with a drop diameter equal to 0.44 mm. In contrast, at the end of existence of a drop-spray cloud, drops of smaller diameter prevail in its upper part. At a height of 4 m at $l = 50$ m the maximum spectrum of drops occurs with a diameter equal to 0.23 mm, and when approaching the surface the maximum spectrum shifts to the higher values: at $h = 3$ m $d_m = 0.29$ mm, at $h = 2$ m $d_m = 0.33$ mm, at $h = 1.5$ m $d_m = 0.37$ mm, where $h$ is the height, and $d_m$ is the drop diameter at which the size distribution of drops reaches a maximum.

The results obtained by Egorov (1977) showed that the size distribution of drops in a drop cloud at various stages of its evolution is stable enough and can be described by the well-known gamma distribution:

$$n(d) = b_1 d^2 \exp(-b_2 d) \tag{5.4}$$

where the distribution parameters $b_1$ and $b_2$ depend on various heights and on various effective distances to the droplet production zone.

It was also found from the experiment that small drops prevail at the front of a flying cloud, and larger drops at the rear. In this case, as larger drops fall from spray clouds, the water content of a cloud decreases drastically (almost by an order of magnitude) at the initial part of flight (at a distance up to 10 m) and then changes more slowly—this in fact agrees with the exponential law.

From many measurements of spray cloud lifetimes, Egorov (1977) and Belov (1978) managed to construct confidence intervals for the lifetime of an individual spray cloud. The lifetime of spray clouds was considered to be a function of breaking wave height during surf. As breaking wave height grows from 1 m to 6 m, the mean lifetime of spray clouds linearly grows from 1 to 4 s with confidence intervals of 2 to 5 s.

Also interesting is the study of the surf zone under weak sea wave conditions. We need to find the maximum possible contribution of the drop-spray mechanism to energy transfer under weak sea wave conditions. A specially designed electronic system of recording the velocity and size of a drop flux, successively intersecting two illuminated planes, allowed Ruben (1977) to determine the integral function of the distribution of drops in the size range from 0.05 to 0.3 mm. The experiment was carried out in the coastal zone of the Atlantic Ocean at breaking surf with a wave height of 0.7 m and at a wind velocity of $3.5 \, \mathrm{m \, s^{-1}}$ at three heights above sea level: 25, 38, and 56 cm.

Of doubtless interest is comparison of the disperse properties of drop-spray systems on the basis of experimental data obtained in the entrainment zone and in the surf zone by various authors (Preobrazhensky, 1972; Egorov, 1977; Belov, 1978; Ruben, 1977). These authors used various dispersion characteristics in their experiments. In view of this fact, the author of the present book recalculated these data using well-known techniques (Sharkov, 2003) and reduced the obtained data to a single type in the form of water content and differential density of particle size (the spectra). Figure 5.1 presents the profiles of water content of drop-spray clouds at slack and strong wind at heights of 2 to 7 m from the mean sea surface level at high seas, as well as the profiles of water content of spray formations at strong and weak surf. Note that the profiles at high seas take on an exponential character—see formula (5.2). The disperse properties of spray clouds in the surf zone can be complex as a result of the complicated kinetics and aerodynamics of small and large drops. Their water content can exceed by two to three orders of magnitude the water content of clouds at high seas at the same levels. As already noted, at the instant of formation a cloud can have a homogeneous disperse character and no prominent profile in water content ($l = 0$). At the beginning of motion, large particles can prevail in the upper part of a cloud, and, hence, a cloud's water content can be higher than at lower horizons ($l = 10 \, \mathrm{m}$). As the cloud moves, large drops precipitate and thereby the profile acquires an exponential character ($l = 50 \, \mathrm{m}$). In weak surf the drop-spray cloud possesses an essentially lower (by two orders of magnitude) water content, but, nevertheless, it is comparable with the water content of spray formations in the entrainment zone at storm wind (Figure 5.1). In general, we note that the range of water content of spray systems can be very wide—from $10 \, \mathrm{g \, m^{-3}}$ as large waves break up to $10^{-4}$–$10^{-5} \, \mathrm{g \, m^{-3}}$ at slack wind in the entrainment zone.

Analysis of Figure 5.2, which shows the different densities in the size of spray structures at surf, allows us to conclude that such systems possess a narrow spectrum

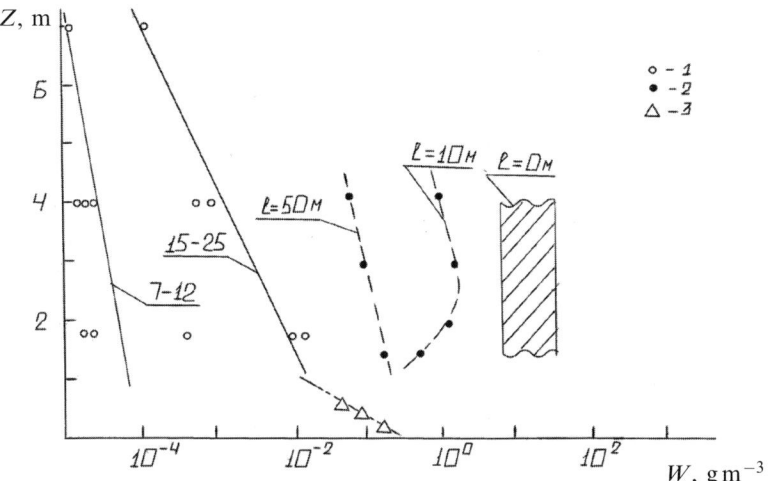

**Figure 5.1.** Experimental level profiles of droplet cloud water contents during light and strong winds from 2 to 7 m above sea level on the high sea (North Atlantic basin ) and water content profiles of spray clouds in strong and slight surf: 1—experimental data (Preobrazhensky, 1972 ) (numbers near the appoximating straight lines correspond to wind velocity at 10-m level ranges; 2—water content profiles in strong surf determined by the author using data from Egorov (1977) ($l$ is the distance in meters from the surf zone); 3—water content profiles in slight surf determined by the author using data from Ruben (1977).

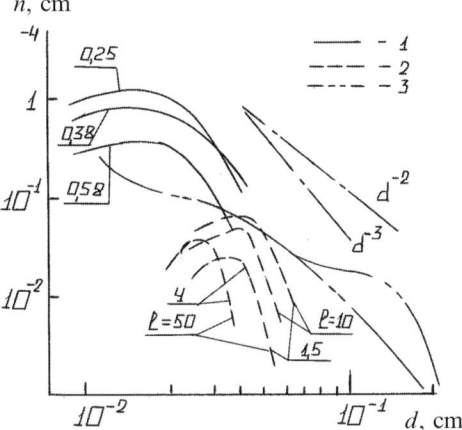

**Figure 5.2.** Experimental droplet distributions of spray clouds at wave breaking in surf and in rain precipitation from warm clouds: 1—spray spectra determined by the author using data from Ruben (1977) obtained in slight surf; 2—spray spectra determined by the author using data from Egorov (1977) obtained in strong surf (numbers near the approximating curves correspond to measuring levels; ($l$ is the distance in meters from the surf zone); 3—drop spectrum for rain precipitation at a 30-mm/hour intensity (data by Takahashi, 1978).

both in weak and strong surf. It is interesting to compare these data with the spectrum of particles for rainfall precipitating from "warm" clouds (Takahashi, 1978). We can clearly see that, compared with spray drops, the spectrum of rain drops is extremely broadband with strong variations in the content of large drops.

Correct measurements of spray field characteristics in the injection zone over the sea surface under field conditions were first performed by Monahan (1968). The experimental setup consisted of a high-speed photo-camera and electronic flash lamp installed on a special buoy. High buoyancy kept instrument height over the water surface during disturbance at a constant level. By means of an anchor the buoy was oriented relative to the wind in such a way that sprays flew between the lamp and the camera directed at the light beam. As a result of system focusing, distinct images were acquired only of drops in a strictly definite volume (about $77 \, \mathrm{cm}^3$). The image of a spherical drop, that flew through an illuminated section, looked like a bright disk with a dark point at the center. The number and size of drops were determined within the limits of $4 \cdot 10^{-3} \, \mathrm{cm} \le r \le 7 \cdot 10^{-2} \, \mathrm{cm}$. The lamp produced pulses lasting $12 \, \mathrm{ms}$ with intervals of $1.5 \, \mathrm{s}$. Each series consisted of 800 pictures. Optical axis height over the surface was $13 \, \mathrm{cm}$. Photography of sprays was accompanied by measuring the wind velocity with an anemometer, installed at a height of $47 \, \mathrm{cm}$ on the same buoy, and gathering salt particles on the wire frames installed at levels of 0.5 to 1.1 m.

Observations were carried out in the coastal and high sea areas of the Atlantic Ocean at rough sea fetches from 1 to 900 km. Functions of the size distribution of sprays are presented in Figure 5.3. This figure presents experimental data on the density of distribution of sprays obtained under wind–wave tank conditions (Toba, 1962; Lai and Shemdin, 1974). All results relate to the 13-cm level. Wind velocities are

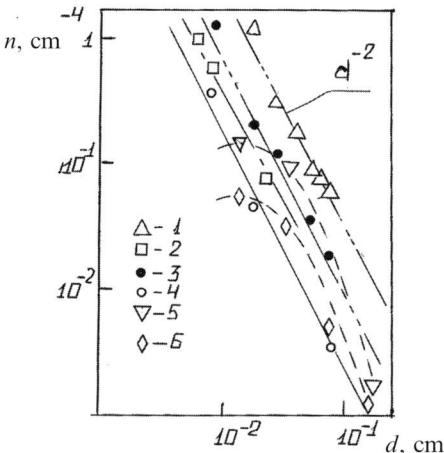

**Figure 5.3.** Experimental droplet distributions of spray clouds (laboratory measurements in wind–wave tanks): 1—at wind velocity $U_{10} = 21.4 \, \mathrm{m/s}$ using data from Toba (1962); 2—at $U_{10} = 23.9 \, \mathrm{m/s}$, 3—at $U_{10} = 21.4 \, \mathrm{m/s}$, 4—at $U_{10} = 17.6 \, \mathrm{m/s}$ using data from Lai and Shemdin (1974). Natural measurements using data from Monahan (1968): 5—at $U_{10} = 15.8 \, \mathrm{m/s}$; 6—at $U_{10} = 12.6 \, \mathrm{m/s}$ .

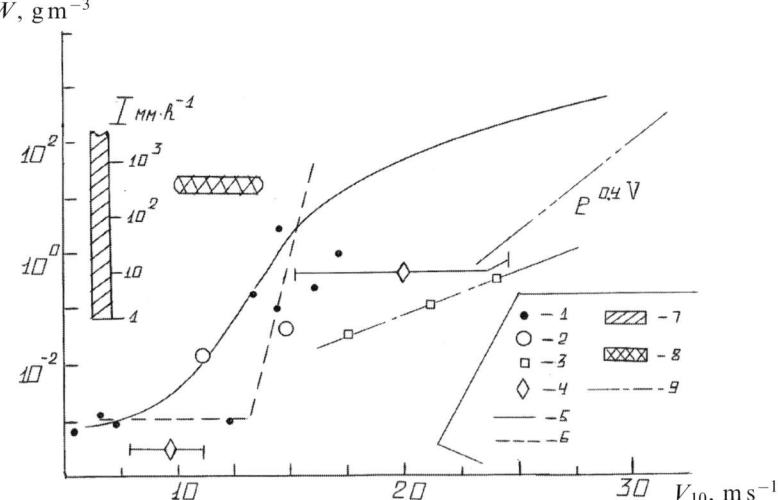

**Figure 5.4.** The dependence of water content for spray clouds at the 13-cm level above the sea surface as a function of wind velocity reduced at the 10-m level: estimation using experimental data (Monahan, 1968), performed by Bortkovskiy (1983) (1) and the author of this book (2). Estimation of water content using data obtained in a wind–wave tank (Lai and Shemdin, 1974) (3) and using natural measurements (Preobrazhensky, 1972), reduced at the 13-cm level (4). The numerical approximations of droplet-spray clouds as a function of wind velocity in accordance with Borisenkov and Kuznetsov (1978) (5), Wu (1979) (6), and Toba (1962) (9). (7) The water content zone of rain precipitation as a function of its intensity (Shiotsuki, 1976). (8) The water content zone of spray clouds determined using data from Egorov (1977) under conditions of wave breakings in strong surf.

reduced to the level of 10 m using logarithmic extrapolation. Note that, despite the principally different character of spray phase generation in wind–wave tank and in field conditions, drop distribution densities were found to be close in qualitative character. Quantitatively, at equal wind velocity the number of drops under field conditions was noticeably (an order of magnitude) greater than under laboratory conditions. Also, under natural conditions there are drops of large size (with diameter of about 1 mm), whereas under wind–wave tank conditions many small drops with diameters less than 80 microns are met. These distinctions are explained not only by the effect of large waves and strong wind—factors principally unmodeled under laboratory conditions—but by obvious distinctions in drop recording systems.

The dependence of water content (the mass of drops in a volume unit) at the 13-cm level on wind velocity, reduced to the level of 10 m using logarithmic extrapolation, is presented in Figure 5.4. Data on water content at the level of 0.5–1.1 m, calculated from salt sedimentation on wire frames, indicate that most drops do not reach this level. This once again emphasizes the fact that injection zones and flying-away zones principally differ in their disperse properties. However, at strong wind—according to data by Bortkovskiy (1983)—the injection zone can drastically increase up to 1 m and more.

Figure 5.4 also presents calculations of the water content of spray clouds, obtained under field conditions (Preobrazhensky, 1972) and reduced to the 13-cm level by means of the exponential law (5.3), as well as calculations obtained under wind–wave tank conditions (Lai and Shemdin, 1974). This drawing further presents data about water content of spray formation arising directly over a breaking wave (Egorov, 1977) and representing some kind of extreme (in saturation) spray formation (dependence 8 in Figure 5.4).

An important element in the interpretation of experimental data is formation of an acceptable approximation dependence of spray formation water content on wind velocity. However, principally different positions exist here. Thus, Wu (1979) considers the appearance of spray systems and their saturation to be an explosion-like process at wind velocities of $12$–$15\,\mathrm{m\,s^{-1}}$ (dependence 6 in Figure 5.4). But, Borisenkov and Kuznetsov (1976, 1978) believe that the process takes on the character of an exponential law (dependence 5 in Figure 5.4) of type

$$\log W = aU^2/(U^3 + C) \tag{5.5}$$

where $a$ and $C$ are normalizing constants: $a = 94.6$ and $C = 172$.

It is interesting to note that data obtained under wind–wave tank conditions— dependence 9 in Figure 5.4 (Toba, 1962) and dependence 3 in Figure 5.4 (Lai and Shemdin, 1974)—also are clearly exponential in character.

Comparing the water content of spray systems with the water content of liquid rainfall, presents a comparison plot of water content and intensity of liquid rainfall (Shiotsuki, 1976—dependence 7 in Figure 5.4). It follows from analysis of these curves in Figure 5.4 that at wind velocities of $15$–$20\,\mathrm{m\,s^{-1}}$ the water content of a spray formation may correspond to rainfall of moderate and high intensity of $50$–$100\,\mathrm{mm\,hr^{-1}}$, and the water content of a spray formation arising directly over a breaking wave may correspond to catastrophic rainfall in tropical cyclones with an intensity up to $1000\,\mathrm{mm\,hr^{-1}}$.

The next step in studying the injection zone comprised investigations of spray structures by means of a specialized setup for remotely gathering drops at consider- able distance from a research vessel (RV) in the tropical zone of the Pacific Ocean and in the Northern Atlantic during expeditions in 1975–1980 (Bortkovskiy, 1977, 1983).

The technique adopted in the setup represents a method used in cloud physics investigations: this involves gathering drops on glass plates greased with an oil mixture. In the setup applied to remotely gather the drops a decrease in disturbance zone is achieved by the horizontal disposition of the receiving plate. The number of drops that settle on such a plate is proportional to their vertical flux. The receiving part of the setup represents a thin disk—a cartridge with 16 slots located on its periphery to which rectangular glass plates $28 \times 14\,\mathrm{mm}$ in size are fixed. The cartridge is placed in a dismountable casing, in the bottom surface of which a window is cut that is equal in size to the glass plate. The whole setup is installed on a specially designed buoy. Buoyancy allows the buoy to retain its vertical position in both the shortest and steepest wind waves; owing to this the receiving device is not likely to be flooded and the distance from it to the water remains virtually constant. During installation from a vessel drifting at considerable speed, the buoy is connected to the vessel by a cable, which moves in the windward direction to distances in excess of 100 m; the cable draft and the stabilizer turn the buoy in such a manner that the casing's window is on the windward side of a buoy. In such a way, minimum distortion of a flow in the spray-gathering zone is provided. After exposing all plates and lifting the setup onboard the RV the cartridge was extracted from its casing, and plates were photographed through the microscope.

Work undertaken in the winter of 1979–1980 in the Northern Atlantic, in the international ocean station "C" area ($52°N$, $35°W$) where the vessel was situated for a month, was characterized by low water and air temperatures (about $5$–$6°C$) and high average wind velocity. The latter circumstance allowed 25 measurements to be made with such a setup, covering the range of wind velocity $U_z$ (at the level $z = 26$ m) from 10 to $19.5 \, \mathrm{m \, s^{-1}}$. In 20 cases the receiving device was installed at a height of 15 m over the water, in 3 cases at 45 cm, and one measurement was made at heights of 6 and 10 cm. Along with measurements of wind velocity $U_{26}$ onboard the RV, the wind velocity $U_{0.3}$ was also measured during the setup operation by an anemometer installed on the buoy at a height of 0.3 m over the water. After performing measurements and photographing the sections of plates through a microscope, the plates were photographed by a camera with a headpiece. Using a number of photographic methods ensured getting data on spray size within a wide range of diameters: $1.25 \cdot 10^{-3} \, \mathrm{cm} \le d \le 0.2 \, \mathrm{cm}$. The measurements made it possible to construct the dependence of a vertical flux of drops from the surface at a height of 15 cm on wind velocity, thus confirming the nonlinear (virtually, square-law) growth of a flux value as wind strengthens. Results of investigation of disperse structures showed that the size distribution of drops in a spray cloud is stable and can be described by the well-known gamma-distribution function (5.4). Such a type of distribution is sometimes called the Nukiyama–Tanasawa distribution (Bortkovskiy, 1977, 1983), widely accepted to describe the field of sprays over the sea. The modal size found in field measurements ($d_m = 2.4 \cdot 10^{-3}$ cm) turned out to be an order of magnitude less than supposed. Bortkovskiy believes that the tearing-off of a sharp breaking crest by a wind generates large ($d \ge 10^{-2}$ cm) but not numerous drops, and the bursting of bubbles provides numerous small ($d \le 1.2 \cdot 10^{-2}$ cm) drops. In the region of $2.4 \cdot 10^{-3} \, \mathrm{cm} \le d \le 1.5 \cdot 10^{-2}$ cm the distribution function can be approximated by the expression $n(d) \sim d^{-k}$, where $k \approx 2$–$3$. In laboratory investigations (Lai and Shemdin, 1974), as well as in field measurements (Monahan, 1968), a result similar to this was obtained for drops at $6 \cdot 10^{-3} \, \mathrm{cm} \le d \le 7 \cdot 10^{-2}$ cm. The shift in size distributions, obtained at measurements under winter conditions, in the small-drop region is probably explained (according to Bortkovskiy, 1983) by the effect of low water temperature on bubble bursting.

## 5.3   CONCLUSIONS

1   Comparative analysis of literature data on the physical characteristics of a disperse spray medium that forms in wind wave breaking mode—such as the medium's water content, the size spectrum, the altitude profiles of water content,

the temporal and spatial dynamics of drop-spray clouds—generally demonstrates the satisfactory qualitative picture of this physical phenomenon. The tentative model presentations of water content of drop-spray clouds are generated on this basis depending on roughness in sea state and on wind velocity.

2   It should be pointed out, however, that all experimental work emphasizes that the technology currently in use does not correctly record the disperse characteristics of drop-spray formations in full volume and, especially, in very rough sea. In addition, virtually nothing is known about the spatial and temporal variability of spray fields both on the scale of World Ocean basins, and on the scale of the World Ocean as a whole. At present, efforts are being undertaken to produce the *in situ* and remote experimental apparatus to record drop particles under field conditions where high numbers of specimens are sampled and where ranges in variation of their size (from 20 up to 1000 microns) and motion velocities are wide. It is important to record possible changes in the phase structure of the flight of drop clouds, as well as their temperature regime. Searches are ongoing in various directions: from updating the mechanical collectors to applying holographic apparatus equipped with electronic counting devices.

# 6

# Electrodynamics of a rough, disperse, closely packed media

This chapter presents the results of theoretical and experimental investigations directed at finding the most adequate method of description and formation of electrodynamic models of multicomponent, closely-packed disperse structures, which form in the ocean–atmosphere system during the gravity wave breaking process. These investigations were carried out within the framework of the USSR Academy of Sciences' series of works on remote sensing the sea surface structure (Raizer *et al.*, 1976; Militskii *et al.*, 1976, 1977, 1978; Bordonskii *et al.*, 1978; Raizer and Sharkov, 1980, 1981; Vorsin *et al.*, 1984). Analysis of the results of active and passive microwave experiments, carried out between 1976 and 1980, made it clear that the mechanism of intensive radio emission of disperse systems is incorporated in the special absorbing properties of an emulsion monolayer and polyhedral cells. We shall try to consider in this chapter the nature of this phenomenon, which does not lend itself to explanation within the classical, static electrodynamic approximation framework. In our opinion, in this case it is necessary to use an approach that takes into account the effects of electromagnetic wave interaction with disperse structure inhomogeneities— that is, to solve the diffraction problem with simultaneous detailed examination of the dispersed structure of foam systems and their complex hierarchical composition.

## 6.1 FOAM AS A COLLOIDAL SYSTEM: PHYSICAL AND STRUCTURAL PROPERTIES

Rough disperse media (foams) belong to a class of colloidal systems containing gas (atmospheric air) and liquid as phase components (Bikerman, 1973; Tikhomirov, 1975; Weaire and Hutzler, 2000). The aggregate state of foam is determined by its steadiness (stability in time) and disperse structure. Since colloids are heterogeneous (inhomogeneous) systems with a huge internal interface surface, foam is essentially unstable. Its natural tendency is to separate the disperse phase from the disperse medium, since in this case the interface surface decreases. Therefore, the stability of foam is always limited, and the "stability" notion is relative in itself. Foam lasting for a few seconds can be considered as unstable, and foam lasting for minutes or hours as stable.

The internal structure of foam (i.e., the relative geometry of its phase components) is determined by the ratio between the volumes of phases in a given disperse system. As a whole, rough disperse media are sub-divided into two classes: the so-called gas emulsions (the aeration layer) and cellular-type foam, or, in other words, foam in the true sense of the word (Bikerman, 1973; Tikhomirov, 1975; Weaire and Hutzler, 2000). However, as we shall see below, there exist many intermediate disperse systems, comprising an entire hierarchical collection of foam colloidal structures, which play an important part in constructing radiophysical models of electromagnetic wave interaction with such structures.

A disperse system's property of retaining particles that are size and shape invariable in time ("aggregation" stability) depends mainly on properties inherent in the liquid films of structures. For foam with a known lifetime to be formed, the liquid should contain dissolved surface-active substances (SASs). Typical examples of SASs are albumins, spirits, and fatty acids.

The following parameters are introduced to characterize the colloidal state of foam (Tikhomirov, 1975):

(1) The foam-forming ability of a solution—the quantity of foam expressed either by its total volume $V$, or by the height of its column $h$.
(2) The foam ratio $\beta$—the ratio of the total foam volume $V$ to the volume of solution spent for its formation $V_2$:

$$\beta = \frac{V}{V_2} = \frac{V_1 + V_2}{V_2} \qquad (6.1)$$

where $V_1$ is the volume of gas in the foam. Then the volume concentration of gas in foam is $\varphi = V_1/(V_1 + V_2)$. Obviously, the foam ratio is related to the volume of gas concentration by the relationship

$$\beta = \frac{1}{1 - \varphi} \qquad (6.2)$$

(3) Foam mass dispersity is specified by the type of spectrum (or by the function of distribution in size) of bubbles in a liquid or by the type of foam structure elements. Dispersity can also be specified by the moments (the mean value, the dispersion) of the distribution function. In some cases we use such parameters as the mean volume of a bubble or the specific area of its surface.
(4) Foam stability—this is the lifetime of an element of foam structure (a film, a bubble) or of some particular volume (column) of foam.

(5) "Aggregation" structure steadiness $\tau$—that is, the time of spontaneous destruction of a foam column to half its initial height $h_0$. Quantity $h_0$ is sometimes called the whitecapping degree (Bikerman, 1973). Several types of temporal dispersion of a foam column are distinguished—in particular, linear, logarithmic, and stepwise.

(6) The "foam factor", by which is meant the quantity equal to the product $F = h_0\tau$. This parameter is often applied to characterize whitecapping under natural sea conditions.

As noted earlier, the internal hierarchical structure of foam systems is mainly determined by the ratio between the volumes of phases in the disperse system. Concentrated "gas emulsions" consist of spherical bubbles of various size (polydispersiveness), which are distributed in a liquid medium more or less chaotically. In this case the greatest close packing is theoretically achieved at a gas phase concentration of $\varphi = 74\%$. Gas bubbles are permanently in motion under the gravity field effect. During motion, bubbles can collide with each other, be kept together by attraction forces (Van der Waals forces), or stick together. This process is called coagulation or flocculation. Floccules can reach a considerable size—up to several diameters of the original bubble. Large aggregates, obtained by bubbles sticking together, float up at considerable velocity and precipitate from the emulsion system (the sedimentation process). It should be noted that the sedimentation rate is determined by the nature and character of gas bubble size distribution. Besides coagulation, the full merging of particles—coalescence—can occur in an emulsion. It is for this reason that a gas emulsion is unstable as a whole: it quickly breaks up into two volume phases (liquid and gas) with a minimal interface surface.

The bubbles of a concentrated emulsion can disappear or enlarge as a result of another mechanism as well. We are talking about gas diffusion through the liquid layers of smaller bubbles into larger ones under an effect caused by the difference in Laplace pressures inside the bubbles (Tikhomirov, 1975). The diffusion rate is proportional to the difference in Laplace pressures, as well as to the permeability of liquid layers that allows bubbles of various size to separate.

At gas concentrations higher than 74% the emulsion is highly concentrated. The shells of its bubbles are contiguous with each other and, at sufficient stability, form a cohesive (continuous) structure that does not coalesce. Cellular foam arises when the gas phase volume exceeds between 10 and 20 times the liquid volume (Tikhomirov, 1975). In this case the lower the ratio between the volumes of gas and liquid phases, the greater the foam film thickness.

At concentrations of 80–90%, bubbles, squeezing each other, are deformed, turning into a closely packed system of spheroidal bubbles and, partially, into polyhedra. Highly concentrated emulsion acquires elasticity in its form—that is, it begins to exhibit noticeably extreme shear stress. Such a system is just foam in its true sense.

The foam films separating bubbles join together to form thickenings that are called Plateau triangles. Three foam films (three films of bubbles) are always associated at the same place, making angles of $\simeq 120°$ with each other. The foam films represent parallel liquid layers, whereas the surface of a liquid in a Plateau triangle is concave. As a result of this, capillary pressure arises causing suction of liquid from the films toward these thickened joints and consequent thinning of foam films. Simultaneously, under the effect of its own weight, liquid runs off from the upper parts of foam to lower parts.

When gas concentration in the foam approaches 100%, the disperse medium assumes the form of many very thin liquid films, and the bubbles turn into many-sided (polyhedral) cells. This is the so-called "tracery" foam (or "dry" foam). The ratio between the volumes of gas to liquid phases equals a few tens, sometimes even hundreds (Tikhomirov, 1975).

The state of foam with many-sided cells is close to equilibrium; as a result, such foams possess greater stability than foams with spherical cells.

The number of sides in a cellular foam "bubble" can vary from 8 to 18, and their shape can be quite different: square, pentagonal, hexagonal, and heptagonal (Tikhomirov, 1975).

It was qualitatively shown (Tikhomirov, 1975) that, owing to the gas diffusion process, bubbles have successively assumed the shape of a parallelepiped, rectangular prism, and tetrahedron, irrespective of their original structure. The process of breakup of cellular-type foam usually begins from above, where the thinnest films are torn one by one, and the "tracery" skeleton gradually falls down.

Summarizing, we distinguish the following structural gradation of a rough disperse medium containing gas and liquid as phase components:

(1) The mono- or polydisperse system of ideally spherical particles chaotically distributed in the liquid medium—a weakly concentrated (diluted) gas emulsion. The ratio of this system is $\beta < 4$.
(2) The continuous structure of (mono- or polydisperse) spherical bubbles, closely packed to each other—a concentrated gas emulsion. The ratio is $\beta \sim 4$.
(3) The cellular (honeycomb) system of particles that takes on an irregular many-sided shape—the transition type from emulsion to foam. The ratio is $\beta = 10–20$.
(4) The foam as such—the structure consisting of thin liquid films, which are fastened in a common rigid skeleton and form polyhedral cells. The ratio of such a structure is $\beta = 20–100$. Foam having a ratio $\beta > 100$ is also called "dry" foam or "tracery" foam.

It is important to note that each type of disperse structure produces its particular geometry of the region bordering the atmosphere. So, in the case of gas emulsion in water this surface represents a continuous film of liquid, whose local radii of curvature are determined by the packing and size of particles of an external layer. The "tracery"-type structure causes considerable roughness in foam surface. The geometrical size of individual cells and of their combinations continuously change because of ruptures in external films of the structure. As a result, the configuration of the external boundary of a foam layer can assume diversified shapes. It should be noted that "tracery" foam always arises on top of the structure of closely packed bubbles, which are at the point of contact with the water surface.

In nature, there apparently exists a certain hierarchy in foam layer structure: the upper part is always formed by polyhedral cells, below which are bubbles being deformed in shapes of spheroids and irregular polyhedrons which then settle down as spherical particles. In such a case the notion of the "foam–water" interface boundary becomes conventional to some extent, since it does not represent a sharp transition from one medium to another. It seems that such a boundary resembles a kind of "brush" formed of water capillaries inverted upwards and penetrating the emulsion layer of bubbles. As a visual example of the hierarchical structure of an inhomogeneous layer of foam systems between the water surface and atmosphere, we shall consider the results of microphotography of the cross-section of a layer of foam systems prepared according to techniques specially developed by specialists from the Space Research Institute (SRI) of the Russian Academy of Sciences for performing laboratory experiments (Figure 6.1a) (Bordonskii *et al.*, 1978). Analysis of these microphotographs shows that the laboratory sample contains all the hierarchical types considered above. Thus, at the boundary with the water surface the emulsion monolayer is situated, which contains a "brush" of water capillaries penetrating the transition region of several layers of closely packed spheroidal bubbles. And, finally, the upper part of the foam layer is formed by polyhedral cells, whose sides represent thin liquid films fastened to each other.

The capability of any geometrical foam structure to retain particles invariable in size ("aggregate" stability) is mainly determined by the specific properties of its liquid films.

It turns out that completely pure liquids virtually do not form a foam—it instantly collapses when the liquid ceases to be agitated or stirred. For foam with a given lifetime to be formed, the liquid should contain dissolved surface-active substances (SASs). Typical examples of SASs are albumins, spirits, and fatty acids. SAS molecules are easily adsorbed by the interphase surface at the interface, thus forming adsorption layers. When a film extends, the adsorption SAS layer at the interphase surface becomes more rarefied, the surface tension as a result of extension grows and promotes the reverse contraction of a film. This is the so-called kinetic factor of foam stability (Tikhomirov, 1975). There exist many hypotheses of physical and chemical nature to explain the relative stability of foam cover (Weaire and Hutzler, 2000).

## 6.2  PHYSICAL AND CHEMICAL PROPERTIES OF SEA FOAM

The whitecapping process on the sea surface is one of most complicated physical and chemical transformations in the ocean, which are related not only to the hydro-dynamic instability of gravity waves, but with the colloidal properties of real sea water and with the biological activity of a sea medium as well. Atmospheric air, which in some way or another gets into the water (e.g., by the jet-turbulent impact mechanism as wave crests break—see Deane and Stokes, 2002), generates many bubbles within a wide range of sizes. The surface-active substances contained in a sea medium are adsorbed on the surface of these bubbles. In such a way, the bubbles are

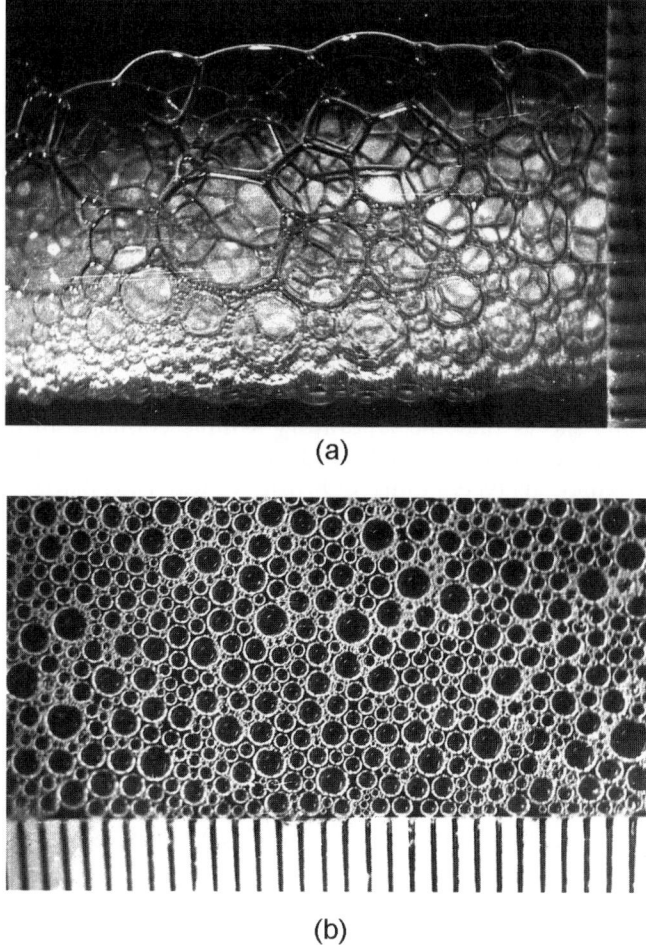

(a)

(b)

**Figure 6.1.** Scaled model optical images of a dispersive foam medium, performed using microphotography: (a) vertical section view of inhomogeneous tracery foam; (b) emulsion mono-layer, top view. The scale bar, vertical in (a) and horizontal in (b), is equal to 0.1 cm.

enveloped by an adsorptive SAS layer, which impedes gas from exiting the bubble cavity (i.e., prevents its reverse dissolution). Rising up to the surface, the bubbles entrain this film behind them. The structure in the near-surface layer, consisting of bubbles "sewn together" by an adsorptive film, just represents the birth of foam. The foam mass accumulates as jets overturn and impact a sea surface having a high SAS concentration.

Other important sources of gas bubble formation are oxygen and hydrogen sulphide dissolved in the water depth (Zaitsev, 1970). Oxygen bubbles, rising due to temperature fluctuations, continuously grow in size, migrate to the surface, and carry an adsorptive film.

Only SASs whose adsorptive layers possess high mechanical strength can form stable foam. These SASs include insoluble and soluble organic substances. Insoluble organic substances are hydrophilic colloids and semi-colloids arising as a result of lifetime and posthumous secretions of animals and plants populating a sea medium. It is believed that plankton is the basic source of soluble organic substances in the open sea (Zaitsev, 1970).

It was found (Abe, 1957, 1962, 1963) that foam powder (dry foam sediment) contains a huge quantity of microscopic fragments of marine algae and phytoplankton albumin. The presence of a very small quantity of foam powder in filtered sea water essentially changes whitecapping characteristics. Thus, the addition of 0.05 g of foam powder to 100 cm$^3$ of filtered sea water increases the lifetime and foam factor by a factor of 100, and the addition of 0.08 g of the same powder increases these characteristics as much as 1500–2000 times (Abe, 1962, 1963). According to Abe's data, the concentration of organic substances in the sea water of a near-surface layer equals 0.099 equiv. mg l$^{-1}$, whereas the concentration of these substances in foam reaches a value hundreds or even thousands of times greater.

When the hydrochemistry of an ocean surface is homogeneous, foam fields do not form a continuous uniform cover even during strong sea disturbance (Figure 6.2), but possess some peculiar spatial–statistical characteristics (see Chapters 2, 3). However, in the presence of a complex hydrochemistry, foam fields can take on strange shapes: even when the sea is rough, stable foam can accumulate in flow convergence zones as linear structures; when river water intrudes into sea water, foam can assume complicated spiral and mushroom-like shapes; it can frame large slick areas as petal-shaped structures and assume other forms. Large foam masses form in coastal zones under the effect of seasonal winds. Such foam is very stable; sometimes it accumulates as huge clots (about one meter in size) and is ejected onshore (Figure 6.3). In particular cases, such as the northeast coast of Japan, the mass of foam thrown onto the coast can damage electric transmission lines and even impede railway and automobile traffic. Under calm conditions the foam mass breaks up generating slick fields on a smooth water surface, which are clearly seen (both in the optical range, and on radar images) because the gravitation-capillary effect of sea waves is suppressed.

Table 6.1 presents the physical constants of stable sea foam and foam liquid,[1] obtained by Abe (1957, 1962, 1963) and systematized in Raizer et al. (1976). We shall consider them in more detail.

(1) The surface tension of foam liquid equals the values $\gamma = 27$–34 dynes cm$^{-1}$ at temperature $t = 12°$C. This is approximately two to three times lower than the surface tension of normal sea water. Such a low value in surface tension of foam liquid is probably caused by the presence of the many organic substances in it.

It should be noted that whitecapping in the regions of congestion of surface organic film (which represents, in essence, the "foam liquid") could be

---

[1] By "foam liquid" we mean the liquid that remains after the complete diffusion of sea foam (Abe, 1963).

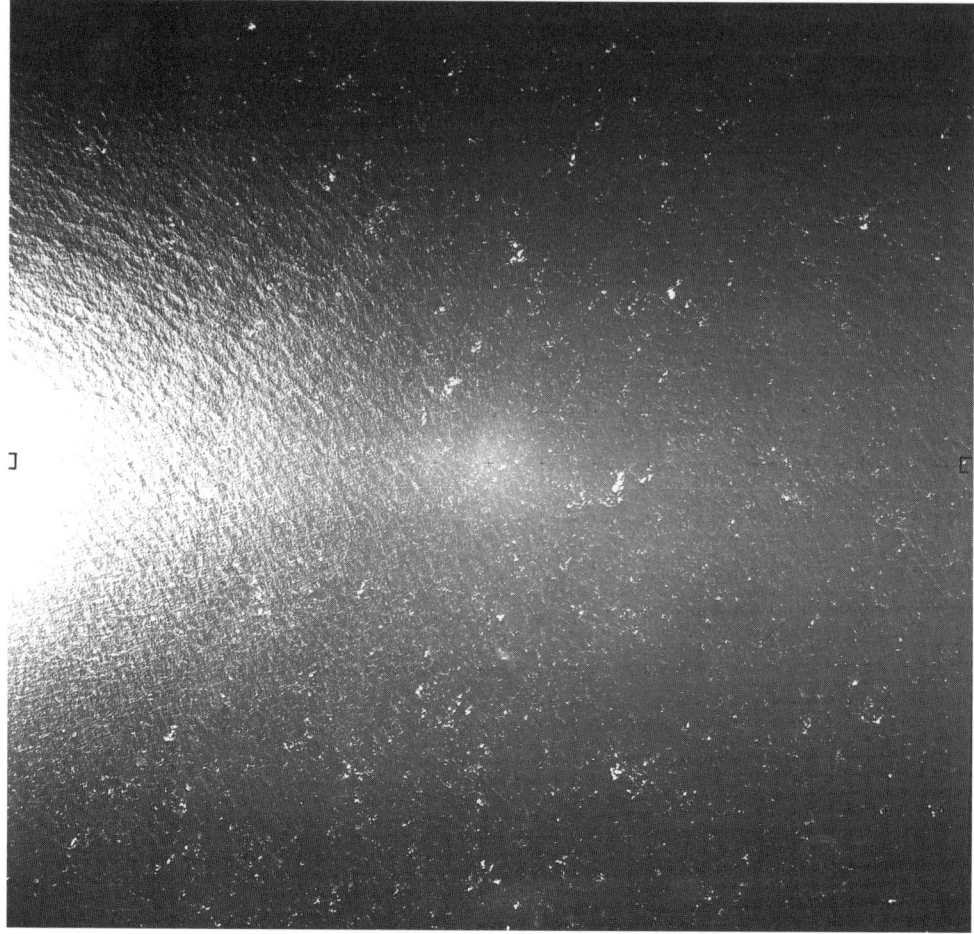

**Figure 6.2.** Aerial photograph of a disturbed sea surface under a shimmering sun. The flight altitude of the Russian aeroplane laboratory IL-14 was 1000 m. Image dimensions are 1800 × 1800 m. Sea conditions reached 5 on the Beaufort Scale. The sea basin is the Sea of Japan (Pacific). Observation date and time was March 25, 1975, 14:10 (local time).

    accomplished much more efficiently than whitecapping on a sea surface free of such a film.

(2) The (molecular) viscosity of foam liquid equals the value $\eta = 0.63 \, \text{g cm}^{-1} \, \text{s}^{-1}$ at temperature $t = 11.5°C$, which is about 60 times greater than the viscosity of normal sea water. Viscosity counteracts thinning of a bubble's film, determines its size, and essentially influences foam stability. We note that low surface tension is still not sufficient for forming stable foam; a specific molecular viscosity of foam-forming solution is also required in this case (Tikhomirov, 1975).

**Figure 6.3.** A photograph of the coastal zone of the northeast seaboard of Japan at the advent of stable foam mass after passage of a severe local storm. Scale of the vertical black bar is 100 cm (Abe, 1963).

**Table 6.1.** Some physical parameters of sea water and foam structures.

| Parameter | Substance | Quantity | Notes |
|---|---|---|---|
| Surface tension $\gamma$ (dyne/cm) | Foam fluid | 27–34 | $t = 12°C$ |
| | Sea water | 67–74 | Natural, $t = 12°C$; $S = 35‰$ |
| Dynamic viscosity $\eta$ (g/cm s) | Foam fluid | 0.63 | $t = 11.5°C$ |
| | Sea water | 0.0137 | Natural, $t = 10°C$; $S = 35‰$ |
| | Sea water | 0.0189 | Natural, $t = 0°C$; $S = 35‰$ |
| Heat conductivity $q$ (cal/cm s grad) | Fresh water | 0.0131 | $t = 10°C$ |
| | Fresh water | 0.0179 | $t = 0°C$ |
| | Foam fluid | 0.0076 | $t = 25°C$ |
| | Stable foam mass | 0.0081 | $t = 20°C$ |
| | Sea water | 0.00135 | $t = 17.50°C$; $S = 20‰$ |
| | Fresh water | 0.00172 | $t = 20°C$ |
| | Graphite | 0.0377 | $t = 20°C$ |
| Electrical conductivity $\sigma \cdot 10^2$ (ohm cm)$^{-1}$ | Stable foam mass | 0.18 | $t = 12°C$, $f = 1\,kHz$ |
| | Stable foam mass | 0.13 | An hour later $t = 12°C$, $f = 1\,kHz$ |
| | Foam fluid | 2.9 | $t = 12°C$, $f = 1\,kHz$ |
| | Sea water | 3.8 | Natural, $S = 35‰$ |
| Density $\rho$ (g/cm$^3$) | Stable foam mass | 0.0113 | $t = 12°C$ |
| | Foam fluid | 0.977 | $t = 12.6°C$ |
| | Sea water | 1.0255 | $t = 12°C$; $S = 35‰$ |
| | Fresh snow | 0.05–0.08 | |

(3) The heat conductivity of foam liquid equals the value $q = 0.0076$ cal $(\text{cm s deg})^{-1}$ at temperature $t = 25.2°C$. The heat conductivity of foam liquid is higher than that of sea water (see Table 6.1). This can affect the radiation regime of ocean–atmosphere interaction processes.

(4) The specific electric conductivity of a stable foam mass turns out to be equal to $\sigma = 0.18 \cdot 10^{-2}$ $(\text{Ohm cm})^{-1}$ at frequency $f = 1000$ Hz and temperature $t = 12°C$. Over an hour its value decreases to the value of $\sigma = 0.13 \cdot 10^{-2}$ $(\text{Ohm cm})^{-1}$ and then remains almost invariable. Remarkable is the fact that the specific electric conductivity of stable foam mass is about 20 times lower than that of foam liquid, which equals the value of $\sigma = 2.9 \cdot 10^{-2}$ $(\text{Ohm cm})^{-1}$.

(5) Density can also serve as a particular characteristic of sea foam. According to data from Abe (1963), stable foam density equaled the value of $\rho = 0.0113$ g cm$^{-3}$ at temperature $t = 12°C$, which is about a hundred times lower than the density of normal sea water, but is of the same order as the density of freshly fallen snow (Table 6.1).

Dependencies of the physical parameters of foam mass on sea water temperature and salinity are summarized in Raizer *et al.* (1976).

## 6.3 DISPERSE STRUCTURE OF SEA FOAM IN THE BLACK SEA BASIN

As noted above, investigations into the characteristics of the disperse foam zone, arising on the sea surface due to wave breaking, are important for constructing electrodynamic models of disperse natural systems and for interpreting the data of remote sensing of the ocean surface (see Chapter 1). However, little is known so far about the physical properties and disperse microstructure of sea foam under natural conditions. The available data relate either to artificial samples, formed from a salt solution equivalent to sea water (Miyake and Abe, 1948; Monahan and Zietlow, 1969), or to a stable foam mass gathered on a coast after a strong storm—a rare phenomenon (Abe, 1963).

This section presents the results of field investigations of a disperse structure of sea foam of natural origin. Observations were carried out in the shore zone of the Black Sea in two highly spatially separated regions (the Blue Gulf bay on the Crimean Peninsula and the Adler Cape on the Caucasian coast) at sea disturbance of about 4 on the Beaufort Scale (Raizer and Sharkov, 1980).

Sections of the sea surface covered by foam were photographed from the pier. Figure 6.4 is a photograph of the surface in the surf zone, which clearly demonstrates two types of structures: so-called breakers and strips (or foam spots). The latter form as a rule on the leeward slope of the wave, or when air bubbles float up to the surface.

To estimate parameters for the disperse structure of whitecappings, micro-photography was used. However, because of the high dynamism of a breaking wave crest we did not manage to obtain a detailed microphotograph of the crest structure, and the data presented in the section relate only to stable (lifetimes of a few minutes) samples of strip foam. Samples of such foam were withdrawn from the surface by means of a special tray; but, by so doing the disperse structure of foam was disturbed. Then the sample in the tray was photographed in the reflected light by a Zenit camera using transition rings and a "Tair-11" lens (with a focal length of 135 mm). Picture scale was achieved by a flag situated at the same distance from the lens as the foam sample. The camera was hand-operated. It should be emphasized that during the 1–2 minutes that it took for the whole cycle of work to be completed, the studied formations at the sample selection place on the sea surface had time enough to be broken. This essentially limited the observation statistics. Nevertheless, the geometrical characteristics of the disperse structure of strip-type foam, being the most stable, was determined.

Figure 6.4b shows an enlarged photograph of a foam strip on the sea surface (the characteristic size of the strip in the real scale was about $2 \cdot 0.5 \, \text{m}^2$), and Figure 6.4c presents the microstructure of the same section. As seen, this strip represents concentrated gas emulsion in the water, whose bubbles are closely packed and form a single layer on the water surface—the "emulsion monolayer". The shape of bubbles varies and depends on their size: the smallest bubbles are spherical, the largest are deformed as irregular polyhedra.

Processing of photographs—like that shown in Figure 6.4c—was performed as follows. All bubbles on a frame were broken down into fractions $n_i$ characterized by external diameters $d$ in the range $d_i \leq d \leq d_i + \Delta d$ ($\Delta d$ is the quantization step used to form the fractions). Then, using a special scale the number of bubbles of various fractions was calculated, and a histogram $n = (n_i/n_0) \cdot 100\%$ was constructed, where $n_0 = \sum n_i$ is the quantity of bubbles of all fractions. Afterwards, the normalized distribution function $f(d) = n_i/(\Delta d \cdot n_0)$ was determined (in accordance with well-known rules—Sharkov, 2003). The mean values of diameter $d$ and of thickness $\delta$ of the liquid shell of bubbles and their volume density $\bar{N}$ were also calculated by the corresponding formulas:

$$\bar{d} = \frac{1}{n_0} \sum_i d_i n_i \qquad \bar{\delta} = \frac{1}{n_0} \sum_i \delta_i n_i \qquad \bar{N} = \frac{n_0}{\bar{d} S_F}$$

where $\delta_i$ is the thickness of the shell of the $i$ bubble (as estimated from a light halo of a bubble), and $S_F$ is the area of the processed section of the photograph. The formula for $\bar{N}$ supposes that the distribution of bubbles in a hypothetical volume that is equal to the product $\bar{d} S_F$ is the same as that in the frontal layer (i.e., in the layer closest to the camera lens). The results of processing microphotographs of two foam samples, selected in various seasons at the Black Sea shore, are presented in Figure 6.5 and in Table 6.2 (these data also include water temperature and salinity at the same place during observations). In Figure 6.6 the experimental data are approximated by linear dependence of the logarithm of a normalized function on the size distribution of bubbles. Based on this dependence, the form of distribution function $f(d)$ itself, averaged over the two realizations, was restored. For both samples the exponential approximation of a normalized function of the particle size distribution was found— namely, $f(d) = a \exp(-ad)$, $a = 10.8 \, \text{cm}^{-1}$.

**Figure 6.4.** Sea surface photographs with foam structures (Blue Bay, Crimea Peninsula, the Black Sea): (a) breakers in the coastal zone (4 on the Beaufort Scale).

These results reveal an amazing fact: the emulsion-type foam of both samples, taken in completely different Black Sea areas (and in various seasons), is identical in structure (an exponential drop in size) and differs in disperse composition only slightly. Good correlation in the distribution of bubbles is observed: in both cases the spectral interval of particle size was $0 < d \leq 0.6$ cm, and the most probable value for bubble diameter lies in the region of $0 < d \leq 0.05$ cm. The distribution maximum for these foam samples lies apparently within the same region.

It is interesting to compare these results with similar data by Abe (1962). Figure 6.5 presents a histogram of the size distribution of bubbles for foam generated artificially from sea water taken in the area of the Myson Beach bay, California (realization no. 3). According to the photograph in Abe (1962), the foam layer also has an emulsion structure. Comparison with the data by Miyake and Abe (1948) and Abe (1963) suggests that the size distribution of bubbles for emulsion-type white-cappings obeys a unique law close to the gamma-distribution.

From the viewpoint of investigation of natural radiation and electromagnetic wave scattering by foam systems, interesting results can be found in laboratory radiophysical experiments (Militskii *et al.*, 1976, 1977, 1978; Bordonskii *et al.*, 1978; Raizer and Sharkov, 1980, 1981), in which an artificially generated emulsion monolayer, similar to sea foam, was studied. Analysis of such histograms indicates

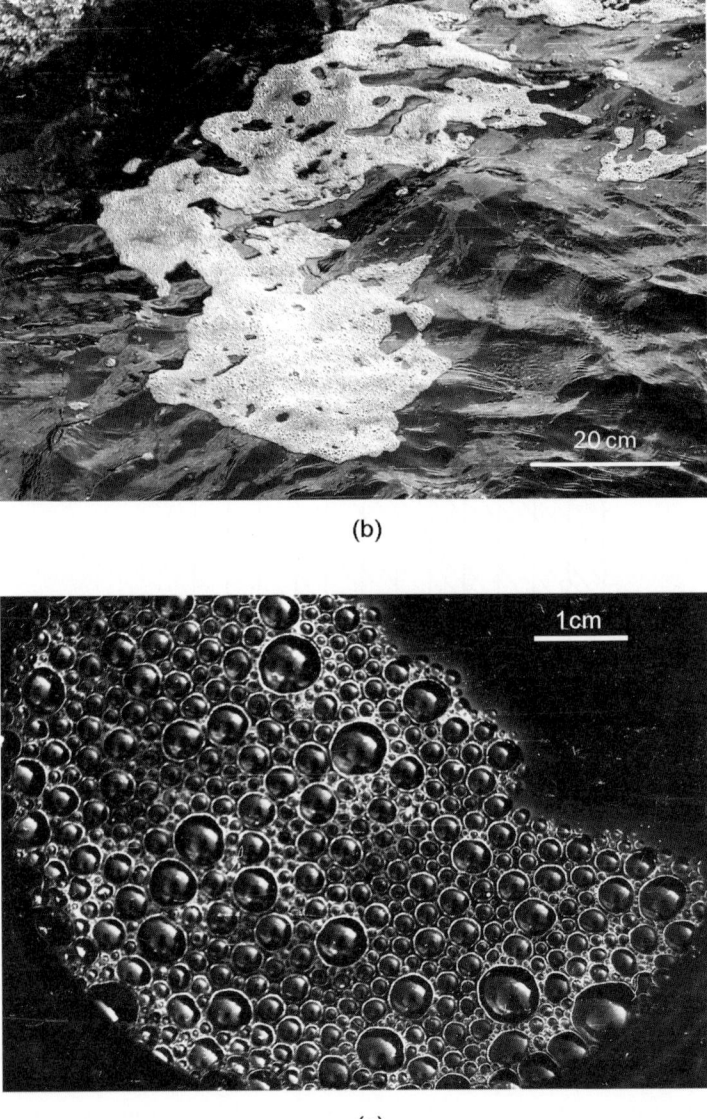

(b)

(c)

**Figure 6.4** (*cont.*). (b) Close-up view of the residual foam field under study; (c) dispersive structure of this residual foam field using microphotography with a magnification of 2.5.

that, in general, the disperse structure of sea foam is close to its laboratory analog prepared using industrial detergents (shampoos). Thus, the results of laboratory radiophysical experiments can reflect, in some sense, the regularities of radiowave propagation and scattering in disperse media of sea origin.

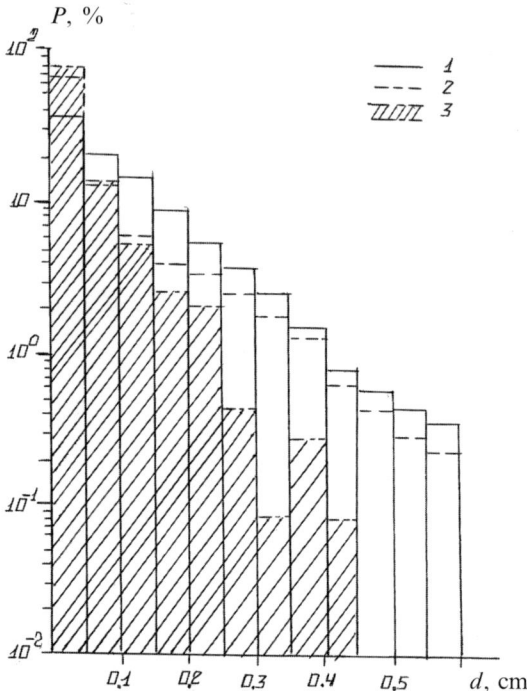

**Figure 6.5.** Experimental probability density histogams of bubble dimensions (diameter $d$, cm) using microphotography for sea foam samples: (1) Blue Bay (Crimea Peninsula, the Black Sea) (Figure 6.4c); (2) Adler Cape (The Caucasus, the Black Sea); (3) Maison Beach Bay, California, Pacific (Abe, 1962).

## 6.4   EARLIER MEASUREMENTS AND "NAIVE" ELECTROMAGNETIC MODELS

Experts in complex media electrodynamics did not pay close attention to studying the electrodynamic properties of roughly disperse systems of sea foam type for a number of years. This was related to the belief, widespread at that time, that hyper-thin film structures were hardly able to deform, let alone absorb, the electromagnetic field of microwave range, except for possible weak scattering caused by the finite radii of films of polyhedral structure. In such a case the emulsion monolayer, because of the small size of its bubbles (in relation to the electromagnetic wavelength), cannot make any significant contribution to the electrodynamics of such systems at all. Detailed investigations, carried out by experts from the Space Research Institute (Raizer *et al.*, 1976; Militskii *et al.*, 1976, 1977, 1978; Bordonskii *et al.*, 1978; Raizer and Sharkov, 1980, 1981; Vorsin *et al.*, 1984; Sharkov, 2003), showed that the situation with the electrodynamic properties of disperse media is quite opposite.

For this reason, revelation of the abnormal increase in the radiative properties of a water surface covered with foam structures and obtained under very high sea disturbance by airborne remote radiothermal systems (Williams, 1969) certainly

**Table 6.2.** Characteristics of the dispersive microstructure of sea foam.

| Number set | Data and site of experiment | Sea surface temperature $t$ (°C) | Water salinity $S$ (‰) | Sampling volume $n_0$ | Mean value of bubble diameter $d$ (cm) | Mean value of bubble envelope $\delta$ (cm) | Mean value of volume density $N$ (cm$^{-3}$) |
|---|---|---|---|---|---|---|---|
| 1 | September 1, 1977 The Blue Bay; the Black Sea; the Crimea Peninsula | 20 | 16 | 796 | 0.135 | 0.036 | $6.9 \cdot 10^3$ |
| 2 | May 16, 1978 The Adler Cape; the Black Sea; the Caucasus | 16 | 18 | 600 | 0.101 | 0.027 | $4.2 \cdot 10^3$ |

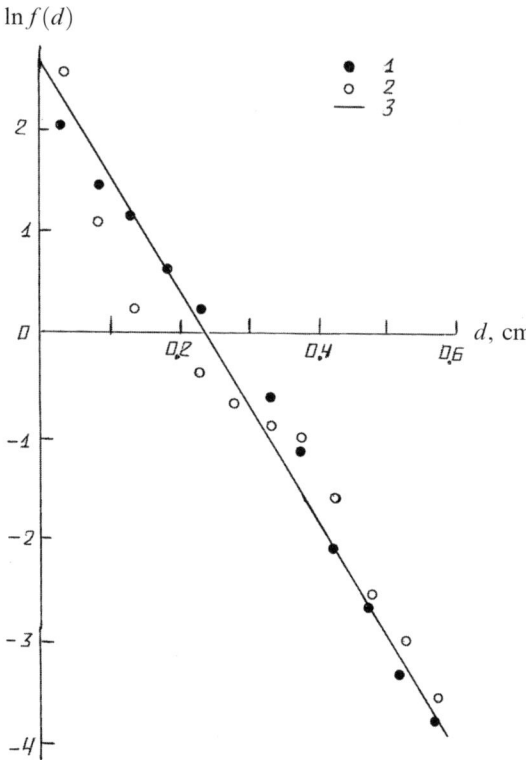

**Figure 6.6.** Approximation of the normalized probability density function of the bubble dimensions for samples 1 and 2 (Table 6.2 and Figure 6.5); (3) exponential function $f(d) \approx 10.8 \exp(-10.8d)$; $f$, cm$^{-1}$; $d$, cm.

attracted a lot of attention. This occurred because it was believed at that time (see, for example, Basharinov et al., 1969; Matveyev, 1971; Shulgina, 1972; Martsinkevich and Melentyev, 1975) that large-scale disturbance did not influence the emissivity of a disturbed sea surface, and the appearance of highly emissive foam fields fundamentally changed the situation with the advent of microwave remote monitoring of the sea surface state. In addition, Williams (1969) advanced an idea about the possibility of measuring the magnitude of the near-surface wind velocity vector based on measuring the increment (which, in his opinion, was virtually linear with wind velocity) of a radiothermal signal using space instruments. This idea stimulated a range of primarily experimental research-type works on observing radiothermal radiation of a disturbed sea surface both from space (Russian satellite "Cosmos-243") and air carriers (Matveyev, 1971; Nordberg, et al., 1971; Ross and Cordon, 1974; Kondratyev et al., 1975; Webster et al., 1976), and from special sea platforms (Holliger, 1971) (see Chapter 8). Though, in general, the results of these observations were extremely contradictory because of serious distinctions in the spatial and temporal resolution of observational systems and because of the obviously unsatisfactory energy sensitivity of space- and air-based instruments, it nevertheless became clear that the problem of determining the radiothermal characteristics of foam structures, as an independent structural element of a disturbed sea surface, acquired great significance.

The effects of such a drastic increase in sea emissivity due to whitecapping certainly demanded a prompt physical explanation. In this connection some theoretical estimations were made at an early stage (1970–1976) in which foam cover was modeled by a layered structure with sharp separation boundaries. In essence, this structure represented either a homogeneous dielectric layer with parameters adequate for the heterogeneous mixing of water and air (the "porous dielectric"—Droppleman, 1970; Matveyev 1971), or a multi-layered system of thin water films alternating with air layers (Rosenkranz and Staelin, 1972). In both cases, the homogeneously absorbing half-space with dielectric permeability, corresponding to sea water, and a smooth boundary was taken as the substrate. So, the porous dielectric model provides a sharp increase in emissivity—up to unity (the "blackbody" model) for centimeter radiowaves at a thin layer of the mixture and a low ($C \approx 10^{-2}$) water concentration (Matveyev, 1971). This effect clearly represents a typical interference maximum in the natural radiation of a two-layered medium with sharp separation boundaries (in the quasi-coherent approximation) and is rather characteristic on reception of natural radiation of multi-layered structures by microwave radiometers with sharply restricted passbands (Sharkov, 2003). Similar results are also obtained in the case of a multi-layered system of water and air films at a fully arbitrary selection of model parameters (Rosenkranz and Staelin, 1972). Despite the obvious roughness and imperfection of these models (see Section 6.6 for more details), they facilitated the first qualitative estimations and even have been used for quantitative interpretation of experimental data (Matveyev, 1971; Rosenkranz and Staelin, 1972). Thus, Webster *et al.* (1976), interpreting aircraft data of the USSR/USA Bering Sea experiment that took place in 1973, proposed another simple model—namely, a thin (1.5 mm) transition dielectric layer with smoothly varying parameters. The authors believed such a model could describe the concordant effect of the whole drop-spray zone on the sea's radio emission, since no sharp boundaries of media separation probably exist under intensive whitecapping conditions. Hence, the characteristic feature of the transition model is the fact that it does not produce interference effects in spectral dependencies of system emissivity that are inherent in layered models. However, the application of such a simple model did not yield any serious results.

So, first-stage (1969–1976) models were based on rather intuitive ideas about the structure of foam formations, which, as we shall see below, are far from physical reality. As a result, the choice of model parameters often—to be blunt—took on a "fitting character" that could not be substantiated physically in any way. See the reviews of first-stage works in Stogryn (1972) and Raizer *et al.* (1976). Such "fitting character" approaches were shown to be ineffective because they did not take into account the specificity of a disperse structure as a set of material particles interacting with electromagnetic radiation in which an allowance was made for diffraction effects. It was these effects that turned out to be principally important in the interaction between microwave range electromagnetic waves and roughly disperse systems. The issues surrounding construction of "diffraction" models that take into account these features will be considered in subsequent sections. However, construction of a new type of model required detailed experimental data not only on the electrodynamic characteristics of a foam cover and its model analogs but on its microstructure as well (see Section 6.3).

## 6.5 EXPERIMENTAL INVESTIGATIONS OF CHARACTERISTICS OF ROUGHLY DISPERSE SYSTEMS BY RADIOPHYSICAL METHODS

This section presents the results of radiospectroscopic active–passive microwave experiments, carried out under the scientific guidance of the author of this book, on the radiophysical characteristics of laboratory analogs of disperse systems by carefully monitoring the parameters of the samples studied (Raizer *et al.*, 1976; Militskii *et al.*, 1976, 1977, 1978; Bordonskii *et al.*, 1978; Raizer and Sharkov, 1980, 1981; Vorsin *et al.*, 1984; Sharkov, 2003).

### 6.5.1 Laboratory analogs of foam systems and their disperse characteristics

In choosing the laboratory analogs of sea foam we proceeded from information obtained as a result of field observations. The main task was to reproduce a sample that was certain in its structure, and then to check its parameters carefully. To do this we also applied a microphotography method.

The model media were generated on the water surface in two ways: by mechanical medium intermixing using blade-rotation instruments (the profiles of the blades were specially selected) at a controlled rotation rate, and by the method of air blowing (with a controlled blow-down rate) through a porous plate immersed in the water medium. An industrial shampoo, containing chemical detergents, was used as a surface-active substance. The first method allowed us to obtain a thick foam layer of mixed structure—polyhedral and emulsion—and the second method a mono- or polydisperse (unitary) layer of bubbles, depending on the volume rate of air feeding through the porous plate. The chemical foam was highly stable: for example, the lifetime of emulsion monolayer bubbles reached some hours, which is sufficient to carry out a cycle of detailed measurements. It was also important that—as the foam aged—the size and shape of bubbles and, accordingly, their disperse structure virtually did not change.

Typical photographs of model media—showing a thick (tracery, cellular) foam layer of inhomogeneous structure and an emulsion monolayer—are presented in Figure 6.1. The first sample, as already noted, included a great hierarchy of structures. Directly on the water surface boundary was situated the emulsion monolayer (Figure 6.1b), with its closely packed spherical bubbles. The diameters of bubbles were different and equal to $d = 0.2$–$1.5$ mm with the most probable value of $d_M = 0.3$ mm. The monolayer thickness was $h \approx d$. The transition region consisting of several layers of deformed bubbles (the polydisperse region was pierced by water capillaries—the "brush") and the region of cellular structure were situated above the

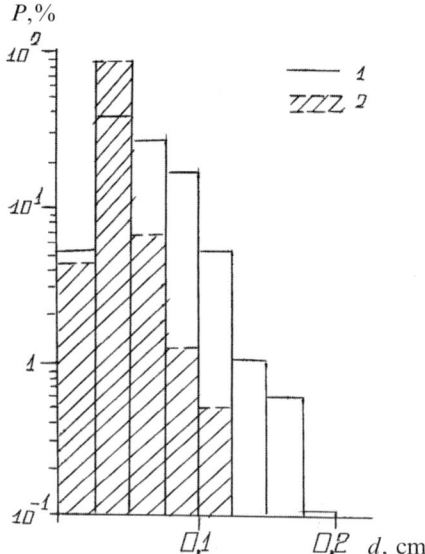

**Figure 6.7.** Experimental probability density histograms of bubble dimensions for an emulsion monolayer test specimen: 1—polydispersive structure of the monolayer (Figure 6.1a); 2—mono-dispersive structure (Figure 6.1b). $d$ is bubble diameter.

monolayer. The latter consisted of many multi-sided (polyhedral) solid cells of various shape. The sides of cells comprised liquid films with a linear size $d = 0.5$–$3.0$ mm and thickness of $0.1$ mm (Figure 6.1a). The characteristic size of cells of the upper floor of tracery-type foam reached $d = 0.5$–$0.6$ cm; the thickness of liquid films bordering the atmosphere can be estimated as $\delta \leq 0.005$ cm. The mean height of an inhomogeneous emulsion and tracery foam layer (Figure 6.1a) was $h_0 \simeq 1.0$–$1.5$ cm. Experimental histograms of the size distribution of bubbles for artificial samples of a polydisperse structure and for a monodisperse layer are presented in Figure 6.7.

The emulsion monolayer was a direct analog of sea foam of emulsion type (see Section 6.4 and Figure 6.5). A comparatively small distinction consisted in the disperse composition of these formations: the spectrum of sizes of bubbles of an artificial sample was much narrower than that in the natural case; however, the most probable value of particle diameter remained the same. The mean concentration of bubbles also corresponded to real samples and was $N \sim 3.8 \cdot 10^3$ cm$^{-3}$ for a poly-disperse structure (compare this with the data of Table 6.2). The bubbles had shells of thickness $\delta \simeq (1$–$5) \cdot 10^{-3}$ cm. The mean height of the monolayer (Figure 6.1b) was $h \simeq 0.1$ cm.

To estimate the volume distribution of phase components over the thickness of a disperse layer of polyhedral structure, we made use of a microphotograph in its cross-section (Figure 6.1a). For the emulsion monolayer, however, we used a hypothetical model in the form of a two-dimensional compact-hexagonal grid of hollow spherical

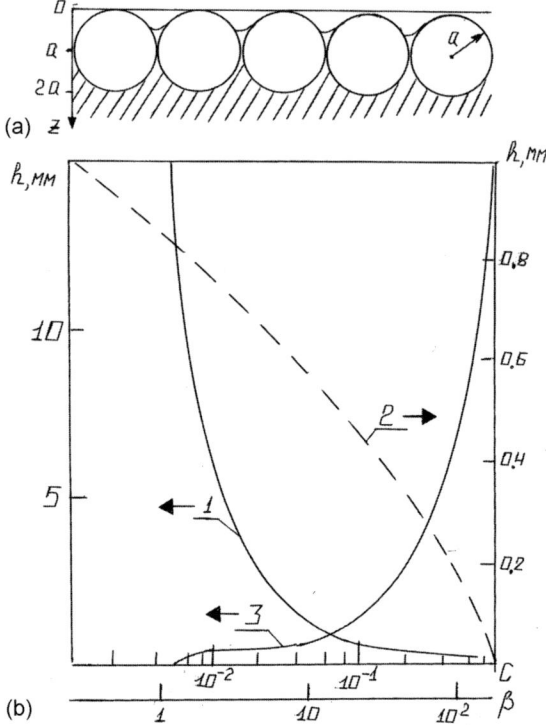

**Figure 6.8.** Experimental profiles of water content in the polydispersive structure of an inhomogeneous layer that includes emulsion and tracery foam (see Figure 6.1): (a) geometrical diagram for the emulsion monolayer; (b) water content volume $C$ in the dispersive layer. 1—The profile $C(h)$ for an inhomogeneous layer of height $h_0 = 1.5$ cm (Figure 6.1a); 2—the profile for monolayer $C(h)$ with $h_0 = 0.1$ cm; 3—the profile of the ratio parameter $\beta(h)$ for an inhomogeneous layer with $h = 1.5$ cm.

particles of identical radius $r = a$ ($a = 0.5$ mm) and infinitely thin shells (Figure 6.8a). To do this we considered that only the intervals between particles in the range $a < z \leq 2a$ were filled with water. In general, this accords with reality (Figure 6.1b). Then, the $C(z)$ profile can be found from geometrical considerations and takes the following form:

$$C(a < z \leq 2a) = 1 - \frac{\pi}{2\sqrt{3}} + \frac{\pi}{2\sqrt{3}}\left(1 - \frac{z}{a}\right)^2 \qquad (6.3)$$

If we now smoothly interpolate formula (6.3) into the region $0 < z \leq a$ (as though describing the diffusion of a disperse system's boundary in such a manner), then we can obtain the continuous distribution of $C(z)$ concentration over the whole "monolayer". Further estimations of the concentration value for an inhomogeneous tracery layer were performed according to the layered technique using a series of microphotographs, one of which is presented in Figure 6.1a.

Semi-empirical profiles of the volume concentration of water and of the ratio parameter in the disperse system of the two types—the monolayer and the inhomogeneous emulsion-tracery layer—are constructed in Figure 6.8. The volume concentration $C(h)$ in a monolayer (curve 2) drops according to the near-exponential law from 1 down to values of about $10^{-3}$ (the conventional boundary with the atmosphere). The volume concentration $C(h)$ in the inhomogeneous emulsion–tracery layer (curve 1) is distributed (as expected) non-uniformly: first, its value drastically (according to the near-exponential law) decreases from 1 down to 0.05 in the height range of $h = 1$–$1.5$ mm, and then it decreases more smoothly down to values of about 0.005 at the conventional boundary of foam and atmosphere media separation at $h = 15$ mm. Comparison of Figures 6.1 and 6.8 reveals the clear correspondence between the ratio value of foam and the character of its structure: in the emulsion monolayer region the ratio equals $\beta \simeq 4$–$8$, for the polydisperse region $\beta \simeq 10$–$50$, and for the cellular structure $\beta \simeq 100$–$200$. Thus, this ratio can clearly be considered as a stable numerical parameter characterizing the type of foam formations.

These media were the basic models on which the physics of the interaction between microwave radiation and the "water-disperse layer" system was investigated. It should be added here that a unique technique of foam formation under laboratory conditions allowed us to obtain, each time it was used, new samples that are not only identical in type, but close in their disperse structure as well. Thereby, a high reproducibility of results of radiophysical measurements was achieved, which added to confidence in their reliability and facilitated the processing of experimental material.

### 6.5.2   Spectral and polarization properties of the radiothermal radiation of disperse systems

The spectral and polarization characteristics of microwave radiation of disperse media have been measured by means of high-sensitivity radiometric systems with wavelengths $\lambda = 0.26, 0.86, 2.08, 8$, and $18$ cm, and fluctuation sensitivity thresholds of, respectively, $\Delta T = 1, 0.1, 0.03, 0.05$, and $0.3$ K at a time constant $\tau = 1$ s (Bordonskii et al., 1978). The experimental installation consisted of a special rack with a rotary device, which provided remote measurement in the wave zone of antennas at each wavelength (Figure 6.9a), and of a laboratory tray of size $60 \times 60 \times 15$ cm$^3$ (Figure 6.9b). The entire measurement complex was mounted between May and June, 1976, on the roof of a tall building at the Space Research Institute of the Russian Academy of Sciences (in Moscow) for the purpose of providing reliable and delicate radiothermal measurements (Figure 6.10). In addition to the internal absolute calibration of instruments, "outdoor" calibration of radiothermal complexes and antenna systems was accomplished according to the absolute measurements technique that was described in detail in Sharkov (2003). The technique takes into account the geometry of calibration and working measurements using absolutely absorbing ("black" body) coating and reflecting (metallic) (for "artificial" sky illumination) sheets for the ranges of electromagnetic radiation studied. The measurements were carried out in clear weather at minimum water

(a)

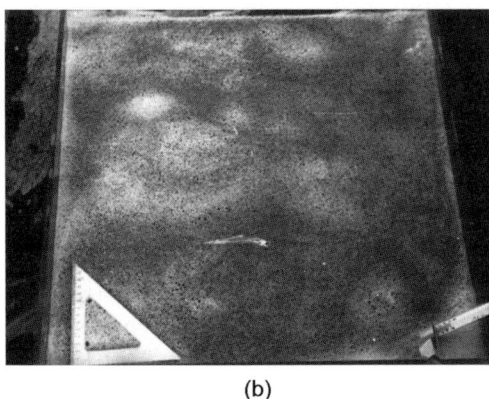

(b)

**Figure 6.9.** General view of experimental instruments for laboratory multi-channel radio-thermal measurements: (a) the device turned so that the antennas of radiothermal instruments are visible (the antenna of decimeter radiometer R-18 that works in the 18-cm wavelength is in the foreground); (b) the dispersive structure of a laboratory tray for carrying out radiothermal measurements.

**Figure 6.10.** General view of the experimental area installed on the roof of a tall building at the Space Research Institute (Moscow). The author (1976) of the book near a laboratory tray during the preparation process of the dispersive structure.

vapor content in order to decrease as much as possible the contribution of sky illumination variations. The reliability of all measurements was checked on a smooth water surface, whose emissive characteristics were calculated according to the modern version of the Debye model (Sharkov, 2003) and sky radiation in a clear atmosphere. The geometry of calibration and working measurements was identical for all ranges, which made it possible to take sufficiently correctly the transition function of the antenna system into account in determining the magnitudes of brightness temperatures. In this case the error in absolute radiothermal measurements reached a record value $\delta T_B = \pm 0.2\,\mathrm{K}$ (or 0.2%) for the range $\lambda = 0.8\,\mathrm{cm}$ (see Table 6.3) in relation to the calculated theoretical values of water emissivity using modern data on the dielectric properties of water (Sharkov, 2003).

The samples of disperse media were stationary and comprised a foam layer of cellular structure and an emulsion monolayer. Using high-sensitivity radiometric instruments, the following experimental data were obtained by Militskii *et al.* (1976, 1978), Bordonskii *et al.* (1978), and Raizer and Sharkov (1981): data about the spectral (1) and polarization (2) characteristics of the radiation of stationary disperse systems of various structure, and (3) about variations in emissive properties caused by non-steadiness of a disperse layer that is polyhedral in structure. In other

**Table 6.3.** Experimental values of emissivities for dispersive media over the waveband 0.26–18 cm.

| | | | | Emissivity | | |
|---|---|---|---|---|---|---|
| Wavelength $\lambda$ (cm) | 0.26 | 0.86 | 2.08 | | 8 | 18 |
| Incidence angle $\theta$ (deg) | 35 | 35 | 35 | | 35 | 20 |
| Surface type | VP | HP | VP | HP | HP | VP |
| Cell structure $h = 1$–1.5 (cm) | $0.985 \pm 0.015$ | $0.905 \pm 0.005$ | $0.977 \pm 0.005$ | $0.800 \pm 0.018$ | $0.560 \pm 0.015$ | $0.460 \pm 0.017$ |
| The emulsion monolayer $h \leq 0.1$ (cm) | $0.950 \pm 0.015$ | $0.815 \pm 0.005$ | $0.970 \pm 0.005$ | $0.634 \pm 0.017$ | $0.335 \pm 0.016$ | $0.396 \pm 0.017$ |
| Smooth water surface: | | | | | | |
| Experiment | $0.655 \pm 0.012$ | $0.390 \pm 0.006$ | $0.515 \pm 0.006$ | $0.330 \pm 0.017$ | $0.335 \pm 0.016$ | $0.396 \pm 0.016$ |
| Theory | 0.660 | 0.389 | 0.511 | 0.333 | 0.306 | 0.380 |
| Surface thermodynamic temperature $T_0$ (Kelvin) | 300 | 296 | 296 | 285 | 285 | 294 |

*Note:* VP and HP are emissivities at V polarization and H polarization.

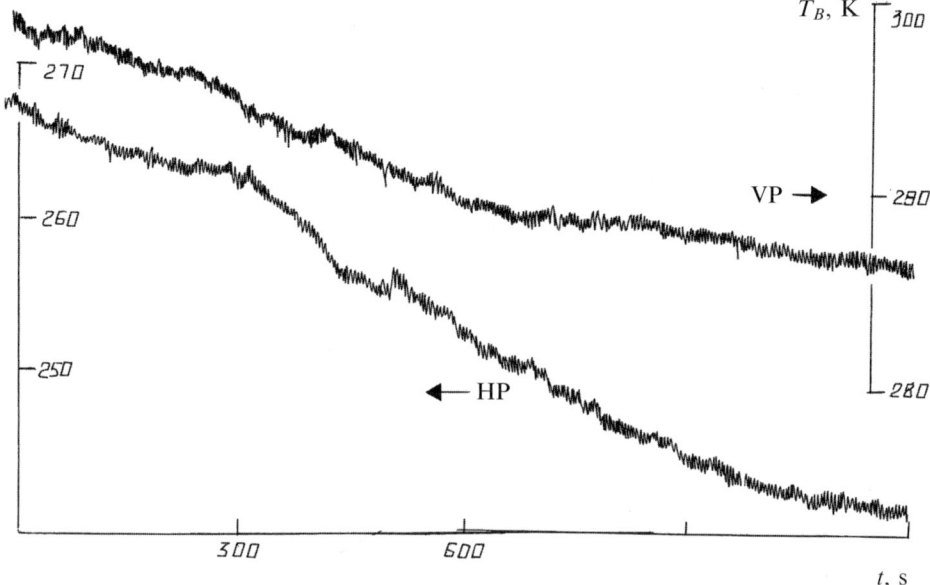

**Figure 6.11.** Experimental synchronous signal registrations during the temporal decay of tracery foam at the 0.86-cm wavelength and two polarizations (incidence angle 35°, structure temperature 292 K). VP and HP are the vertical and horizontal polarizations.

words, the process of slow, temporal, spontaneous diffusion of an original sample of cellular foam into the emulsion monolayer was investigated.

The results of measurements are presented in Figures 6.11–6.14 and in Tables 6.3–6.4. The essence of the experiment on studying the emissive properties of a disperse layer at its temporal diffusion was as follows: an unstable layer of polyhedral structure, whose initial thickness was $h \sim 1$–2 cm, was produced in a laboratory tray. In due time the layer spontaneously destroyed itself, monotonously decreasing in height, until the stable emulsion monolayer remained on the water surface. A radio-thermal signal was continuously recorded by the radiometer's recorder at two polarizations during the disperse layer diffusion process (Figures 6.11–6.13). Simultaneously with radiometric measurements, the disperse structure parameters were monitored by means of microphotography (Figure 6.12). As a result of repeating the experiment several times, the exponential law of time variation in cellular layer height $h(t)$ at destruction (under laboratory conditions) was established for all frequency ranges mentioned above:

$$h(t) = h_0 \exp(-k\tau) \tag{6.4}$$

where $h_0 = 2$ cm and $k = 0.090$–0.095 min$^{-1}$ at $\tau \leq 25$ min and medium temperature of 20–25°C.

In general, the disperse layer destruction process was accompanied by a decrease in brightness temperature; however, the character of signal variation and the

**Table 6.4.** Experimental values of emissivities for dispersive media on the water surface at $\lambda = 0.86$ cm; $T_0 = 300$ K.

| Surface type | Emissivity, polarization | Incidence angle $\theta$ (deg) | | | | | | | | | | |
|---|---|---|---|---|---|---|---|---|---|---|---|---|
| | | 10 | 15 | 20 | 25 | 30 | 35 | 40 | 45 | 50 | 60 |
| Polyhedral structure | VP | 0.973 | 0.973 | 0.973 | 0.974 | 0.974 | 0.974 | 0.976 | 0.978 | 0.980 | 0.983 |
| | HP | 0.973 | 0.973 | 0.972 | 0.970 | 0.968 | 0.967 | 0.965 | 0.965 | 0.964 | 0.965 |
| The emulsion monolayer | VP | 0.872 | 0.865 | 0.871 | 0.879 | 0.884 | 0.892 | 0.900 | 0.902 | 0.809 | 0.931 |
| | HP | 0.875 | 0.855 | 0.852 | 0.850 | 0.844 | 0.826 | 0.796 | 0.777 | 0.762 | 0.729 |
| Smooth water surface | VP | 0.438 | 0.450 | 0.460 | 0.483 | 0.489 | 0.501 | 0.517 | 0.566 | 0.598 | 0.697 |
| | VP (theory) | 0.437 | 0.453 | 0.463 | 0.474 | 0.490 | 0.510 | 0.533 | 0.562 | 0.597 | 0.690 |
| | HP | 0.449 | 0.426 | 0.417 | 0.406 | 0.405 | 0.383 | 0.356 | 0.334 | 0.309 | 0.276 |
| | HP (theory) | 0.447 | 0.431 | 0.422 | 0.411 | 0.397 | 0.380 | 0.361 | 0.338 | 0.313 | 0.253 |

*Note*: VP and HP are emissivities at V polarization and H polarization.

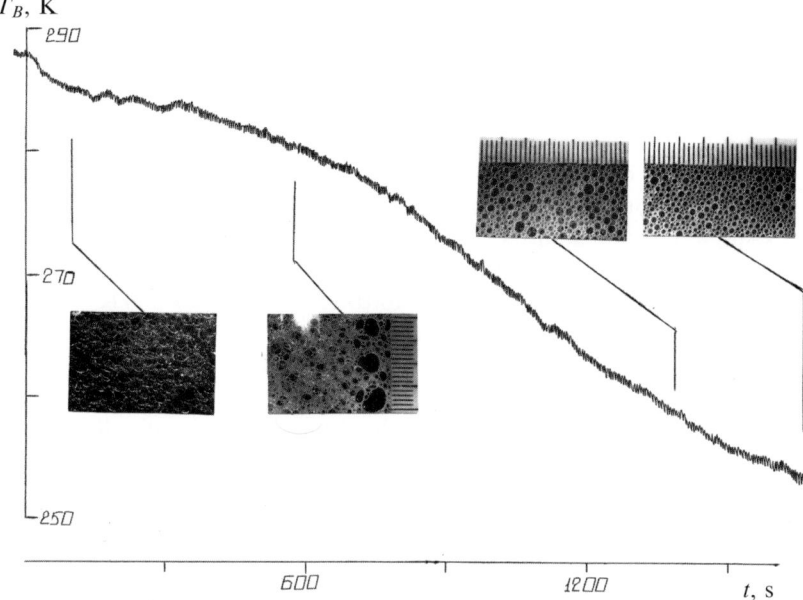

**Figure 6.12.** Experimental signal registrations during the temporal decay of tracery foam at the 0.86-cm wavelength and vertical polarization in synchronism with photoframes of the surface structure (incidence angle 35°, structure temperature 292 K).

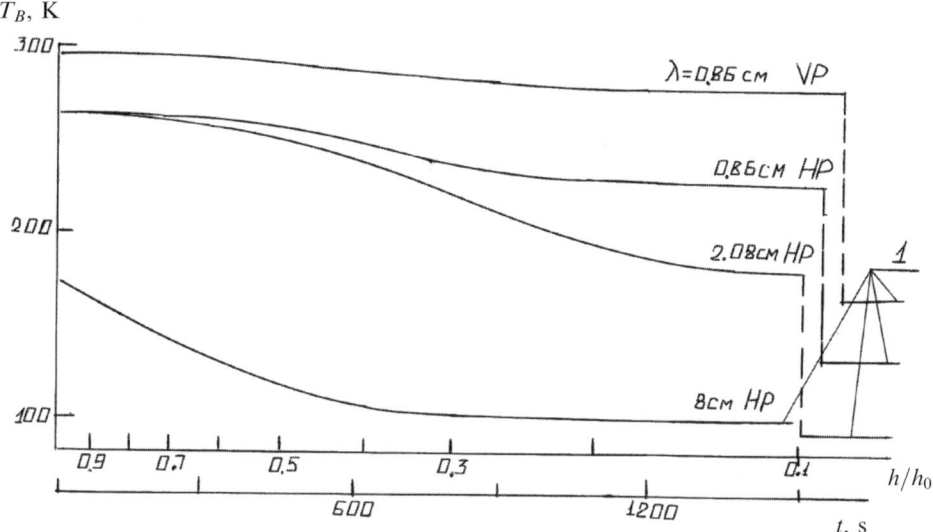

**Figure 6.13.** Experimental synchronous signal registrations during the temporal decay of tracery foam at the 0.86-cm, 2.08-cm, and 8-cm wavelengths and two polarizations (incidence angle 35°, for structure temperature see Table 6.3). VP and HP are the vertical and horizontal polarizations. 1—The emission of a smooth water surface at the corresponding wavelengths.

radiation intensity level essentially depended on sensing wavelength used. So, in the short-millimeter range ($\lambda = 0.26$ cm) the transformation of one (cellular) structure into another (monolayer) resulted in brightness temperature decreasing by just a few Kelvins. In this case the intensity of emission of a disperse system remained high: emissivity of transformed disperse structures remained at the level of 0.98–0.95 (see Table 6.3) and virtually did not depend on received signal polarization or on observation angle. In other words, some kind of "diffusive" emission mode was observed. Even when a single-layer structure of bubbles (the emulsion monolayer) appeared on the water surface, the brightness temperature of the system was equal to $T_B = 295$ K (at the thermodynamic temperature of an emitting structure of 300 K).

The situation essentially changed, however, in the long-wave millimeter range. Figures 6.11 and 6.12 present typical radiometric records of a signal received at disperse layer destruction at a wavelength of $\lambda = 0.86$ cm. Figure 6.11 shows the registration of radio emission of a part of the temporal cycle of cellular structure diffusion, obtained at vertical and horizontal polarizations simultaneously. Figure 6.12 presents a registration of radio emission (at vertical polarization) during a nearly total cycle of temporal decay of a cellular structure into a monolayer. Several synchronous photographs illustrate the transformation of disperse structures on the water surface in due time. As seen from analysis of Figures 6.11–6.12 and Table 6.3, the decrease in brightness temperature while observing cellular foam layer decay into a monolayer reached 30–40 K. When a stable emulsion-type monolayer is formed, the increment in brightness temperature (which had values $T_{B_0} = 172$ and 132 K for vertical and horizontal polarizations, respectively, see Figure 6.11) with respect to water surface radiation was recorded—namely, $\Delta T_B \simeq 110$–115 K.

Figure 6.13 demonstrates the character of synchronous variation in brightness temperatures at wavelengths $\lambda = 0.86$, 2.08, and 8 cm at spontaneous decay of a foam layer of cellular structure ($h = 1.5$ cm) into the emulsion monolayer ($h = 0.1$ cm) for a period of $t = 25$ min. To better visualize the results the "background noise zigzag" in radiothermal signal registrations in Figure 6.13 was eliminated. Brightness temperature is plotted here both as a function of observation time $T_B(t)$, and as a function of the relative height of the disperse layer $T_B(h/h_0)$. It follows from Figure 6.13 that the indicated wavelength range is highly sensitive to structural transformation of the disperse layer. So, at diffusion of a cellular layer the values of brightness temperatures decreased by $\Delta T_B = 85$ and 65 K for wavelengths $\lambda = 2.08$ and 8 cm, respectively. Note that at $\lambda = 0.86$ cm the change in brightness temperature value was essentially lower at $\Delta T_B = 30$ K. While a stable emulsion-type monolayer formed, the increment in brightness temperature (which had the value $T_{B_0} = 95$ K for horizontal polarization) at a wavelength of 2.08 with respect to water surface radiation was recorded as $\Delta T_B \simeq 86$ K. At a wavelength of 8 cm no signal from a monolayer was recorded (Figure 6.13 and Table 6.3).

Table 6.3 and Figure 6.14 present the experimental values of emission factors (emissivities) of the structures and contrasts studied, as well as the calculation data for a smooth water surface. As already noted, the latter allow us to judge the reliability of results of multifrequency absolute radiothermal measurements. Table 6.3 and Figure 6.14 demonstrate the essential dependence of the frequency spectrum

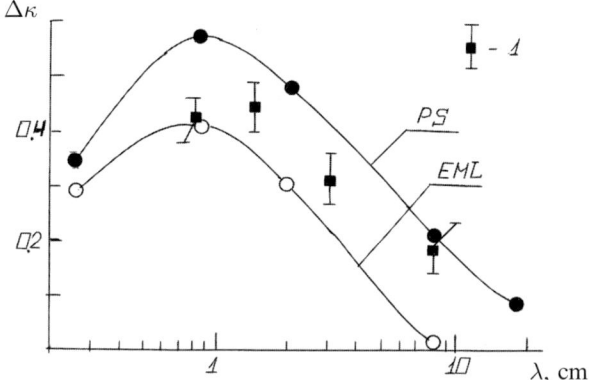

**Figure 6.14.** Experimental spectral increment in emissivity of a water–foam structure at the expense of the dispersive layer (incidence angle 35°; horizontal polarization). PS is the polyhedral structure. EML is the emulsion monolayer. 1—Data of natural experiments onboard a research vessel (Matveyev, 1978).

of increments of "water surface–foam" system emissivity on disperse system type in the range from short millimeters up to decimeters. The spectral dependence of emissivity increments has some kind of weakly resonant character both for a cellular structure and for an emulsion monolayer. Here, the wavelength range of 0.8–2 cm possesses the highest sensitivity to the structural transformation of disperse layers (Figure 6.14). Increments in emissivity values, in the presence of an emulsion monolayer, can reach values of 0.3–0.35 in the range of $\lambda = 0.26$–2 cm and values of 0.5–0.55 for a cellular structure in the range of 0.8–2 cm.

It is important to emphasize that—in this range—the main contribution (more than 60%) to radio emission of a water surface covered with a disperse layer is made by the lowermost emulsion monolayer. This is especially noticeable in the millimeter range: the monolayer contribution to emission of the total system exceeds 80% in this case. In the longwave part of the centimeter range and in the decimeter range (at wavelengths of 8 cm and 18 cm) the emissive properties of the surface change only when rather thick layers of disperse formations of cellular structure are present. So, the brightness contrast value for the decimeter range (at a wavelength of 18 cm) for a cellular foam structure layer of height 4 cm was $\Delta T_B = 14$ K. The upper part of this layer's structure was formed by polyhedral cells a few centimeters in size. Much later (in 2003) similar results were obtained by Camps *et al.* (2005) at a wavelength of 21 cm.

The experimental polarization dependencies of emissivities of disperse structures of two contrasting types $\kappa_{V,H}(\theta)$ are presented in Figure 6.15 for wavelength $\lambda = 0.86$ cm. It is important to note that—qualitatively—the character of $\kappa_{V,H}(\theta)$ dependencies for an emulsion monolayer only slightly differs from similar dependencies for a smooth water surface. However, noticeable smoothing of polarization properties in the emission of a polyhedral structure is observed. In the presence of an emulsion monolayer the polarization degree does not exceed $p < 15\%$, and in the

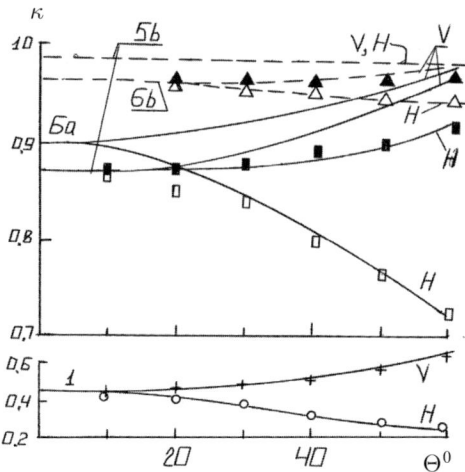

**Figure 6.15.** Experimental and theoretical emissivity dependences of the foam–water system as a function of incidence angle at the 0.8-cm wavelength. Experimental data for vertical (V) and horizontal (H) polarizations: crosses and circles correspond to a smooth water surface; dark and light boxes correspond to the emulsion monolayer; dark and light triangles correspond to the polyhedral structure. Theoretical dependences: solid curves correspond to the emulsion monolayer; dashed curves correspond to the polyhedral structure. 1—A smooth water surface. The numbers near curves correspond to theoretical models (Table 6.8).

presence of a thick layer of foam of cellular structure it is less than 1% (for a smooth water surface $p \simeq 50\%$). In the latter case the emission intensity virtually does not depend on observation angle, yet emissivity is high enough ($\kappa_{V,H}(\theta) \simeq 0.98$).

### 6.5.3  Reflective properties of disperse systems in the microwave range

The purpose of experiments, whose results are described in the present section, was determination of the spatial and energetic distribution of intensity of an electromagnetic field diffused into the upper hemisphere by the disperse layer of various structures, or so-called bidirectional reflectance (Militskii *et al.*, 1977).

Measurements were carried out by the bistatic method in wavelength ranges of $\lambda = 0.43$, 0.83, and 3.06 cm. A skeleton diagram of the measuring device is shown in Figure 6.16. A planar, linearly polarized electromagnetic wave of specified frequency fell on the studied surface at angle $\theta_0$. Diffused radiation of the same frequency was recorded in the incidence plane. The sector of observation angles was $\theta = 0–60°$. In transmission–reception mode were used horizontal (HH) polarization (when vector **E** was perpendicular to the incidence plane), vertical (VV) polarization (when vector **E** was situated in the incidence plane), and cross-polarization (VH and HV). Disperse structures were produced on the water surface in laboratory tray 1 (Figure 6.16). The disperse composition of samples strictly corresponded to those used in radiothermal experiments (see Section 6.5.1). The tray was made of a radiotransparent material

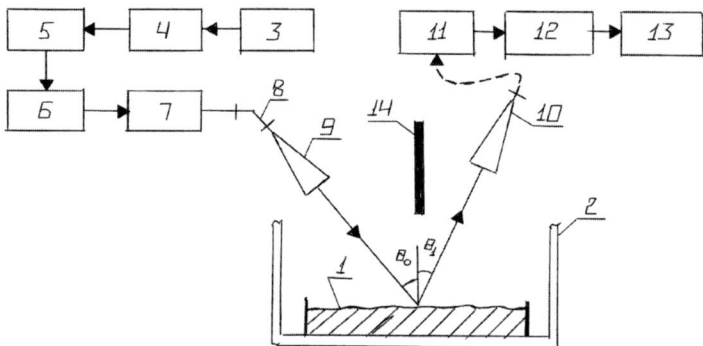

**Figure 6.16.** Skeleton diagram for bistatic scatterometric measurements: 1—a laboratory tray with dispersive structure on the water surface; 2—microwave chamber with absorptive blocks; 3—signal generator; 4—isolator; 5—dissipative attenuator; 6—resonant wavemeter; 7—measurement attenuator; 8—waveguide; 9—transmit antenna; 10—receive antenna; 11—isolator; 12—detector block; 13—direct current indicator; 14—absorptive block.

(plexiglas) and had a transversal size of $600 \times 600 \, \text{mm}^2$ and a depth of $150 \, \text{mm}$. Samples together with the tray were placed in a special microwave echo-free chamber of rectangular cross-section 2 (Figure 6.16). The geometrical volume of the chamber was about $1 \, \text{m}^3$. The sources of microwave power were klystron generators that emit continuously in the ranges of a few millimeter and centimeter wavelengths, as well as a sweep generator with a frequency deviation higher than $4 \, \text{GHz}$ (Figure 6.16). The "echo-free" factor of the chamber was less than $40 \, \text{dB}$, and accuracy in determination of the reflection coefficient was better than 1%. Identical pyramidal horns with an opening cross-section of $26 \times 33 \, \text{mm}^2$ and $80 \times 80 \, \text{mm}^2$ (for millimeter and centimeter ranges, respectively) were used as transmitting and receiving antenna systems. According to direct measurement data, the angular width of the main petal was $10° \times 12°$ for a wavelength of $0.4 \, \text{cm}$, $18 \times 22°$ for a wavelength of $0.8 \, \text{cm}$ and $24°$ for a wavelength of $3 \, \text{cm}$.

It should be noted that the reliability of all measurements was checked by comparing the experimental and calculated values of the reflection coefficient for a smooth water surface (Table 6.5).

The results of measurements—shown as polar scattering diagrams (PSDs)—of the types of structures considered are presented in Figure 6.17 in polar coordinates. The zero power level ($0 \, \text{dB}$) corresponds to the diagram of reflection of a metal sheet constructed during calibration. The experimental values of reflection coefficients in power $|R_{\text{HH}}|^2$ and $|R_{\text{VV}}|^2$ (and calculated values for a smooth water surface) along with their errors are presented in Table 6.5.

We now distinguish the basic regularities found during the experiments. First, the presence of a disperse layer always results in a sharp decrease in surface reflectivity. The reflection coefficient lowered with growing radiation frequency, and its value essentially depended on the disperse layer structure. So, in the millimeter range in the case of a cellular foam layer the reflected signal was not recorded at all—its power

**Table 6.5.** Experimental values of the mirror reflection coefficients for dispersive media.

| | 69.9 | 36.2 | | 9.8 | |
|---|---|---|---|---|---|
| Frequency $f$ (GHz) | | | | | |
| (wavelength $\lambda$, cm) | (0.43) | (0.83) | | (3.06) | |
| Incidence angle $\theta$ (deg) | 24 | 34 | | 34 | |
| | **Reflection coefficient, polarizations** | | | | |
| Surface type | $|R_{VV}|^2$ | $|R_{HH}|^2$ | $|R_{VV}|^2$ | $|R_{HH}|^2$ | $|R_{VV}|^2$ |
| Water surface experiment | $0.43 \pm 0.01$ (18°C) | $0.50 \pm 0.01$ (15°C) | $0.50 \pm 0.01$ (22°C) | $0.61 \pm 0.01$ (23°C) | $0.66 \pm 0.01$ (18°C) |
| Theory | 0.425 | 0.473 | 0.491 | 0.623 | 0.670 |
| The emulsion monolayer $h = 0.1$ cm | $0.025 \pm 0.005$ (18°C) | $0.040 \pm 0.001$ (13°C) | $0.032 \pm 0.001$ (22°C) | $0.063 \pm 0.001$ (23°C) | $0.63 \pm 0.01$ (18°C) |
| Polyhedral structure $h \sim 1$ cm | 0.005 (18°C) | 0.005 (13°C) | 0.005 (22°C) | 0.005 (23°C) | $0.160 \pm 0.003$ (18°C) |

*Note*: VV and HH are the polarization types for radiation and reception. The media temperature is shown in parentheses.

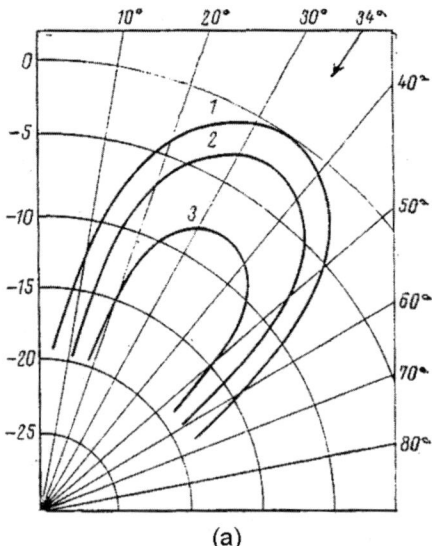

(a)

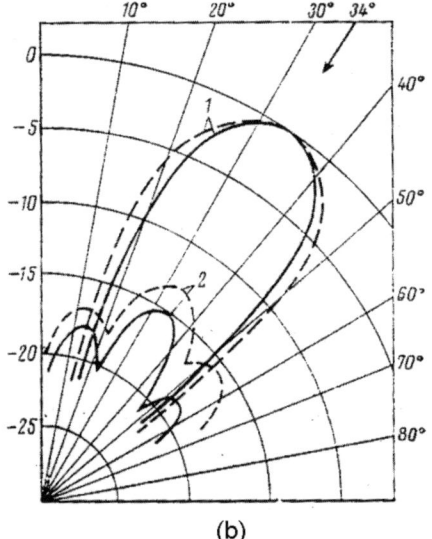

(b)

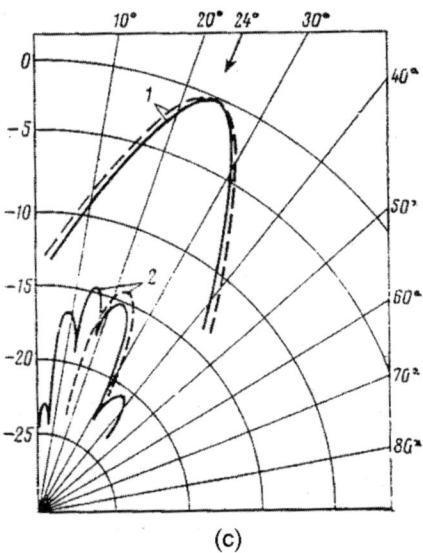

(c)

**Figure 6.17.** Experimental scattering diagrams in the forward reflection regime for two foam structures: (a) working frequency 9.8 GHz, incidence angle 34°; (b) working frequency 36.2 GHz, incidence angle 34°; (c) working frequency 69.9 GHz, incidence angle 24°. 1—A diagram of the reflection from a metal sheet; 2—diagrams of reflection from the emulsion monolayer structure; 3—diagrams of reflection from the tracery foam structure. Solid curves correspond to vertical–vertical polarization for the radiation–reception regime (VV); the dashed curves correspond to horizontal–horizontal polarization for the radiation–reception regime (HH).

level was less than 30 dB. Thus, it follows that the reflection coefficient in power is $|R|^2 < 0.005$.

Second, the spatial distribution of intensity of a field reflected at the given frequency was essentially determined by the surface layer structure. Whereas in the centimeter range (at a wavelength of 3 cm) the disperse structures of both types virtually did not change the smoothness (mirror character) of the surface, in the millimeter range (at wavelengths of 0.83 and 0.43 cm) scattering diagrams took on a completely different (multilobe) character (Figure 6.17).

In the presence of a polydisperse layer of bubbles on the water surface, the scattering of millimeter-range radiowaves outside the specular direction was observed. This is testified to by some broadening (1.5–2 times at the 5-dB level) of the scattering diagrams obtained as compared with those related to a metal sheet. The main feature consisted, however, in the fact that scattering indicatrices in the milli-meter range had maxima and minima typical of the interference picture. This was in no way associated with the appearance of receiver sidelobes: their level was not recorded in the experiments at all, which is testified to by the smoothness of the calibration reflection diagrams. An increase in incidence radiation frequency resulted in an interference picture of higher order arising. The relative amplitudes of maxima could reach 3–4 dB. At the same time, when vertically polarized radiation (VV) was used the irregularity of scattering indicatrices was observed more distinctly than when using horizontally polarized radiation (HH). In the latter case the maxima and minima were on occasion fully blurred, and then the scattered signal was only localized within mirror direction limits.

The appearance of interference features in the scattering indicatrices indicates that the reflecting surface was perceived by an electromagnetic wave as a spatial periodic structure. Indeed, the considered surface resembles a kind of two-dimensional grid consisting of floating bubbles. Using experimental data, it is possible to determine the characteristic period of the grid on which radiowave diffraction took place. The condition of the diffraction maximum of the $m$th order is as follows:

$$\sin \theta_0 - \sin \theta_m = m(\lambda/\Lambda) \qquad (6.5)$$

where $M = 0, \pm1, +2, \ldots$; $\theta_0$ and $\theta_m$ are the incidence and observation angles; $\lambda$ is the length of a radiowave; and $\Lambda$ is the period of a grid. Estimations made using the data of Figure 6.17 (vertical polarization) for wavelengths $\lambda = 0.43$ and 0.83 cm led to the values of $\Lambda = (3–4)\lambda$. Hence, the radius of correlation of a structure's effective inhomogeneities has the order of several emission wavelengths and exceeds at least ten-fold the mean size of a bubble. A special experiment with the "monodisperse" layer showed that an interference picture is completely absent: the reflection coefficient was higher than that for a polydisperse layer and was equal to $|R_{HH}|^2$ and $|R_{VV}|^2$ (compare with Table 6.5).

So, we can draw the following conclusions on the experimental results of investigating the electromagnetic properties of the disperse structures of foam systems:

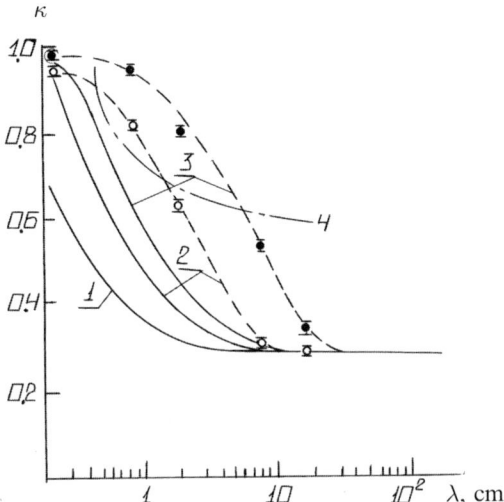

**Figure 6.18.** Spectral dependences of foam structure emissivity for a model of the inhomogeneous dielectric layer on parameters being in agreement with the heterogeneous mixture of air and water. System temperature $T_0 = 293$ K; water salinity $S = 0‰$; incidence angle $\theta = 35°$; horizontal polarization. 1—A smooth water surface; 2—the emulsion monolayer with $h_0 = 1$ mm; 3—polyhedral structure with $h_0 = 10$ mm; 4—spectral emissivity approximation using data from Stogryn (1972). Solid curves correspond to calculated data. Dashed curves and points with confidence intervals correspond to experimental data.

(1) The spectrum of radio emission of a disperse system essentially depends on its geometrical microstructure and qualitatively differs from the emission spectrum of a smooth water surface (Figure 6.18).

(2) The influence of an emulsion monolayer on a system's thermal radiation in the millimeter and centimeter ranges dominates. The existence of a rich structural hierarchy of a cellular foam layer does not have any essential effect on radiation characteristics.

(3) The wavelength range of $\lambda = 0.8$–2 cm possesses the greatest sensitivity to the structural transformation of a disperse system. In the short millimeter range, disperse systems are close to an "absolute blackbody": their emissivity is of the order of unity. In the long-wave part of the centimeter range and in decimeter ranges, variations in a system's radio emission can be observed only for sufficiently thick $h/\lambda > 1$ disperse layers of mixed or polyhedral structure.

(4) The emission of disperse systems in the millimeter range is close to that of diffusion. Polarization degree decreases to zero as layer height increases to a centimeter.

(5) In the millimeter range at vertical and horizontal polarizations nearly identical sensitivity of brightness temperature to a system's disperse structure variations is observed. In sensing layer destruction processes, angles close to the nadir are optimum.

(6) The spectral dependence of increases in emissivity as a result of the presence of a disperse layer is quasi-resonant in character, possesses a prominent maximum in the wavelength range of $\lambda = 0.8$–$1$ cm, and qualitatively differs from the spectral dependence of an absolute value of the emissivity of the total surface–disperse layer system (compare Figures 6.14 and 6.18).

(7) The intensity and spatial distribution of radiation scattered from a disperse system essentially depend on the height and structure of a layer. In the millimeter range the emulsion system (a monolayer) can be interpreted as a dielectric amplitude-phase grid which causes diffraction maxima to arise that are comparable in intensities between each other and with a mirror component.

(8) Cellular structure is a kind of "black" body for microwave radiation whose absorptivity increases with growing radiation frequency. This is associated with repeated re-reflection of electromagnetic waves by the liquid walls of closed cavities. Cavity configurations themselves do not play any essential part. No reflected signal was recorded in cross-polarization mode at the 35-dB level in the frequency range indicated.

## 6.6 THE THEORY OF NATURAL RADIATION OF DISPERSE CLOSELY PACKED SYSTEMS

From the electrodynamic point of view, it is quite natural to represent the disperse system as a planar, layered, inhomogeneous structure with some law of dielectric permeability variation with depth. Since this law is unknown *a priori*, it is necessary to have a calculation technique that allows determination of a system's emissivity for an arbitrary profile of the complex dielectric constant $\dot{\varepsilon}(z)$ of a medium. The analytical solution of such a problem, as shown by Brekhovskikh (1957), can be obtained only for a limited number of particular (and specific) cases. This is the reason we shall make use of the numerical method in this section. For this purpose the inhomogeneous—regarding its dielectric properties—half-space is broken down as a system of plane-parallel layers, and each elementary layer is presented as dielectrically homogeneous. The quantization step is chosen from considerations of smoothness of function $\dot{\varepsilon}(z)$. The theory of thermal radiation of an inhomogeneous, non-isothermal half-space, in its representation as a multilayered structure with sharp boundaries, was studied in detail by Sharkov (2003).

### 6.6.1 An inhomogeneous dielectric layer adequate for heterogeneous mixing of water and air

This model represents a further development of the model by Matveyev (1971), which was based on two-layer presentation of a disperse (foam) two-component (binary) structure in the form of a homogeneous dielectric with smooth boundaries. However, taking into account the statistical character of distribution of the mixture's phase components in each elementary layer, we shall use Odelevskii's formula (see

Odelevskii, 1951) for heterogeneous mixtures

$$\sum_{i=1}^{2} \frac{\dot{\varepsilon}_i - \dot{\varepsilon}_C(z_k)}{\dot{\varepsilon}_i + 2\dot{\varepsilon}_C(z_k)} C_i(z_k) = 0 \tag{6.6}$$

where $\dot{\varepsilon}_i$ are the dielectric permeabilities and $C_i$ the volume concentrations of phases; $\dot{\varepsilon}_C(z_k) \equiv \dot{\varepsilon}_k$ is the dielectric permeability of a heterogeneous mixture of the $k$th layer. It is important to note that relation (6.6) is strictly symmetric with respect to choice of mixture component—carrying phase or inclusion phase.

Solving (6.6) with respect to $\dot{\varepsilon}_C$ and attributing indices $i = 1, 2$ to air and water, respectively, we shall have for $\dot{\varepsilon}_1 = 1$ and $C = 1 - C_2$:

$$\dot{\varepsilon}_C = \tfrac{1}{4}[\dot{Z} + (\dot{Z}^2 + 8\dot{\varepsilon}_2)^{1/2}]$$

$$\dot{Z} = 3C_2(\dot{\varepsilon}_2 - 1) - (\dot{\varepsilon}_2 - 2) \tag{6.7}$$

Obviously, $\dot{\varepsilon}_C = 1$ at $C_2 = 0$ and $\dot{\varepsilon}_C = \dot{\varepsilon}_2$ at $C_2 = 1$.

By virtue of the strong dependence of the dielectric properties of water on frequency $\dot{\varepsilon}_2(f)$ in the wavelength range considered (as is known, the relaxation maximum of water as a polar dielectric only falls in the centimeter range—Sharkov, 2003), the spectral dependencies of total multilayered system emissivity were calculated by mathematical modeling applying special techniques (Bordonskii et al., 1978; Sharkov, 2003).

The volume distribution of phase components in the thickness of a disperse layer of polyhedral and emulsion structures was described in detail in Section 6.5.1 (Figures 6.1 and 6.8). The monolayer was represented as a two-dimensional compact-hexagonal grid formed by spherical particles of identical size (Figure 6.8a). The profile of water content in the intervals between spheres can be presented by formula (6.3). Then, smoothly interpolating formula (6.3) into the region $0 < z \leq a$ (just like describing the blurring of a disperse system's boundary), we can obtain the continuous distribution of concentration $C_2(z)$ throughout the "monolayer". Semi-empirical profiles of the volume concentration of water in the tracery structure are constructed in Figure 6.8. As can be seen, function $C_2(z)$ is close to exponential one in the case of a thick foam layer of polyhedral structure. The value of concentration increases here within the limits of $0.5\% < C_2 < 7$–$8\%$ for $1.5 < h < 0.1$ cm (polyhedral structure) and up to unity (emulsion structure).

The statistical model of a disperse structure—Figure 6.8 and formula (6.3)—was used to calculate the spectral dependencies of emissivity for systems of two types according to the multilayered technique. In so doing, the quantization step of an inhomogeneous half-space varied from $1/50$ to $1/150$ of the total thickness of a disperse layer. Theoretical curves and the results of radiothermal experiments are presented in Figure 6.18. As can be seen, the experimental points lie well above the calculated values throughout the spectral range—from 0.2 to 18 cm. The discrepancy between experimental values and those calculated by the model reaches 50–60% in the range of $\lambda = 0.8$–2.0 cm. Therefore, the model considered (or, in essence, the transition dielectric layer), which takes into account the inhomogeneity of the volume distribution of phases in the disperse layer, is neither qualitatively nor quantitatively adequate for the experimental data. It is obvious that models of the form of a homogeneous dielectric layer (see, for example, Matveyev, 1971), with parameters corresponding to real values of the volume concentration of phases in a mixture, are hardly likely to be more reliable because of the presence of sharp boundaries in a two-layer structure.

Figure 6.18 also gives the approximate frequency dependence of foam system emissivity presented in Stogryn (1972) and widely used in Western scientific literature up to the present time. It can easily be concluded from analysis of Figure 6.18 that the approximation considered does not describe, either qualitatively or quantitatively, the experimental spectral dependencies of foam system radiation and cannot be used as a reliable scientific source. This is most likely related to the fact that in 1972 approximation was generated generalizing individual and uncoordinated experiments that were carried out under completely uncontrollable conditions.

### 6.6.2 Transition and layer-inhomogeneous models

Such types of dielectric models can be based on combinations of discrete and continuous distribution of the function of a structure's complex dielectric constant profile $\dot{\varepsilon}(z)$. In our opinion, the most reasonable and physically justified versions are presented in Figure 6.19 (above the spectral diagrams with numbers 2–7). The continuous profile $\dot{\varepsilon}(z)$ was specified according to the hyperbolic tangent law:

$$\dot{\varepsilon}_{n,n+k}(z) = \tfrac{1}{2}(\dot{\varepsilon}_n + \dot{\varepsilon}_{n+k}) + \tfrac{1}{2}(\dot{\varepsilon}_n - \dot{\varepsilon}_{n+k})th\frac{mz}{2} \qquad (6.8)$$

where $k > 1$ (the integer) is the "size" of a transition region, and $m > 0$ is the coefficient of "smoothness" of a hyperbolic function. It follows from the diagrams of the spectral characteristics of emissivities (Figure 6.19a, b) that the character of spectral dependencies $\kappa(\lambda)$ is essentially determined by the form of function $\dot{\varepsilon}(z)$. The presence of one or several thin water films with smooth boundaries in the models results in the appearance of well-known resonance effects in thermal radiation in the quasi-harmonic approximation (Sharkov, 2003). We emphasize that monotony in spectral dependencies $\kappa(\lambda)$ can be achieved by just using transition layers with smoothly varying parameters within the limits of $1 < \dot{\varepsilon}(z) < \dot{\varepsilon}_2$. For example, the transition layer (6.8) correlates well and over a wide frequency range with the wave resistances of water and air—that is, it decreases reflections from the separation boundary of these media, which essentially increases system emissivity.

Figure 6.19 presents the results of calculation of the spectral dependence of emissivity for a series of versions of foam structure models. The profile of complex dielectric permeability in the transition region was specified according to the hyperbolic tangent law (6.8). The presence in the model of an emulsion monolayer of one or several alternating water films (their thickness equals $\Delta h = 0.05$ mm for model 2 and $\Delta h = 0.01$ mm for models 3 and 4) in the region bordering the atmosphere drastically changes the spectral dependence of emissivity. In this case interference effects can appear, thus contradictiing the experiment. For a thick layer of cellular structure,

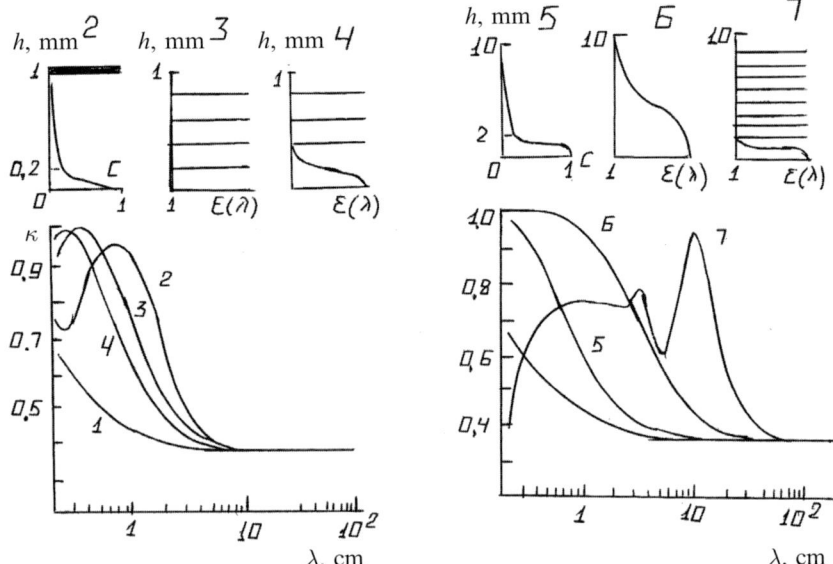

**Figure 6.19.** Theoretical spectral dependences of foam structure emissivity for six models of dielectric structures 2–7. System temperature $T_0 = 293\,\mathrm{K}$; water salinity $S = 0‰$; incidence angle $\theta = 0°$. 1—A smooth water surface. Schematic images of the dielectric structure are displayed over corresponding diagrams (see text for details).

transition dielectric models 5 and 6 essentially increase surface emissivities even in the decimeter wavelength range. Model 7—alternating plane-parallel water films ($\Delta h = 0.1$ mm, cellular structure), situated above the transition layer (emulsion structure)—gives several sharp interference maxima. As in the case of emulsion models, this qualitatively contradicts the experiment. It should be noted that best agreement with experiment for a thick foam layer is achieved by transition model 6.

The results of radiothermal measurements of disperse systems of various types could not be explained within the framework of models that have a discrete-layered structure and transition layers. Such a "traditional" approach, as should be expected, turned out not to be very effective, most likely because of the fact that it did not take into account the disperse structure in the form of material particles with complicated character of scattering and absorption of electromagnetic waves. Issues in the construction of electrodynamic models, which take into account this feature, will be considered below (Sections 6.6.3 and 6.6.4).

### 6.6.3   Electromagnetic properties of a bubble in the microwave range

Moving over to the scattering problem, we should first of all define the types of effective inhomogeneities. The structure of a system in the presence of an emulsion monolayer suggests that the simplest inhomogeneities of radiowave length scale (in the millimeter and centimeter ranges) can be air bubbles floating on the water surface. However, it turns out to be rather difficult to get a theoretical solution even to the quasi-static problem of emission and scattering of a closely-packed medium at the boundary of two half-spaces in a rigorous statement of theory. But, it is undoubtedly meaningful to consider a simplified version of the problem, in which a set of bubbles is suspended in free space; and, as further analysis will show, in such a case it is possible to obtain the results that adequately fit physical reality.

An individual bubble can be represented mathematically as a two-layer concentric sphere, whose core is filled with air and the shell with water. This is the so-called hollow spherical particle. Analytical solution of the problem of electro-magnetic wave scattering on two concentric spheres based on Mie theory was obtained in a general form in Aden and Kerker (1951). A series of numerical results, obtained using the Mie theory for two-layer, hollow spherical particles and for thin-walled coating in the optical range and fragmentary in the microwave range is presented by Rozenberg (1958, 1972) and Kerker (1969). These authors, however, have considered only an insignificant number of cases of varying diffraction param-eters of studied spheres and dielectric properties of particle substances. These cases are necessary for constructing the microwave models. To overcome the calculation difficulties for thin-walled coating structures in the optical range, the so-called phenomenological optics of foam systems is currently being developed (Kokha-novsky, 2004). As far as the microwave range is concerned, Dombrovskiy (1979, 1981) made numerical calculations of absorption and scattering characteristics both for individual hollow spheres, and for a polydisperse structure of emulsion foam. Calculations were performed within a wide range of variation of diffraction param-eters and with due account of the highly prominent frequency dependence of the dielectric properties of water.

Numerical calculations were carried out for five emission wavelength values— 0.26, 0.86, 2.08, 8.0, and 18 cm—used for the aforementioned (see Section 6.5.2) experiments on studying the natural microwave radiation of foam systems. The external radius of bubbles varied from 0.01 to 0.5 cm, and the thickness of shells $\delta$ was assumed to be independent of particle radius and varied in the range of 0.001– 0.01 cm. To calculate polydisperse foam, normalized distribution functions were used that corresponded to the disperse structure of the samples of emulsion-type foam studied (see Section 6.5.1).

The basic characteristics of the Mie scattering theory are the efficiency factors of extinction, scattering, and absorption for separate particles, normalized on a geo-metrical cross-section, whereas for the polydisperse medium the basic characteristics are the spectrum-specific factors of extinction, scattering, and absorption per unit length in the medium studied (Stratton, 1941; Deirmendjian, 1969; Rozenberg, 1972; Hulst, 1981; Ishimaru, 1978; Sharkov, 2003).

Let us now consider the absorption and scattering of electromagnetic waves by individual hollow water spheres with respect to the same characteristics of full aqueous spheres (water drops). It follows from analysis of Figure 6.20—where the efficiency factors of extinction and absorption are presented as depending on the external radius of a hollow sphere—that bubbles with a thin shell scatter

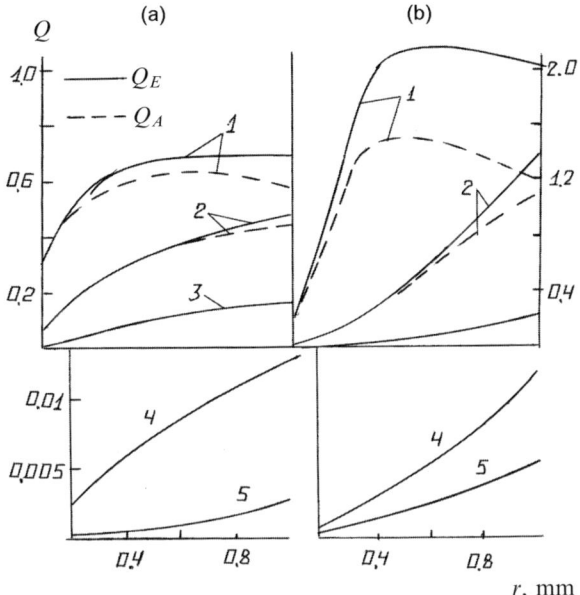

**Figure 6.20.** Theoretical dependences of the efficiency factors of extinction ($Q_E$) and absorption ($Q_A$) for individual hollow water spheres: (a) with the thickness of an envelope $\delta_1 = 0.001$ cm; (b) with $\delta_2 = 0.005$ cm. Working wavelengths: 1—$\lambda = 0.26$ cm; 2—$\lambda = 0.86$ cm; 3—$\lambda = 2.08$ cm; 4—$\lambda = 8$ cm; 5—$\lambda = 18$ cm.

electromagnetic energy very weakly. Throughout the considered range of variation of a hollow sphere's external radius and wavelength, the relative fraction of scattered energy in the total extinction does not exceed 16%, and for small hollow spheres it can be equal to 2–4% and even lower than 1%. Here we note that hollow spheres with a thin-walled shell scatter energy that is essentially weaker than that of spheres with thick-walled shells. Scattering at a wavelength of 0.26 cm is essentially anisotropic and differs highly from Rayleigh scattering. As the wavelength increases, the scattering becomes more symmetric and approaches Rayleigh scattering. The relative value of scattering at centimeter and decimeter waves becomes very small, so that the curves of efficiency factors of extinction and absorption in Figure 6.20 virtually coincide for wavelengths greater than 2 cm.

These distinctions can be traced most explicitly by direct comparison (in magnitudes) of the absorption and scattering characteristics of a homogeneous water sphere and a hollow sphere with an aqueous shell with identical external dimensions (diameters) (Figure 6.21). The figure was constructed by the author of this book using the calculation data of Dombrovskiy (1979, 1981). It presents the dependencies of the efficiency factors of extinction ($Q_E$), scattering ($Q_S$), and absorption ($Q_A$) for homogeneous spherical water drops and a hollow sphere with an aqueous shell at a wavelength of 0.86 cm versus the dimensionless diffraction Mie parameter (or the size parameter) $x = (2\pi a/\lambda)$. The qualitative form of these dependencies for a

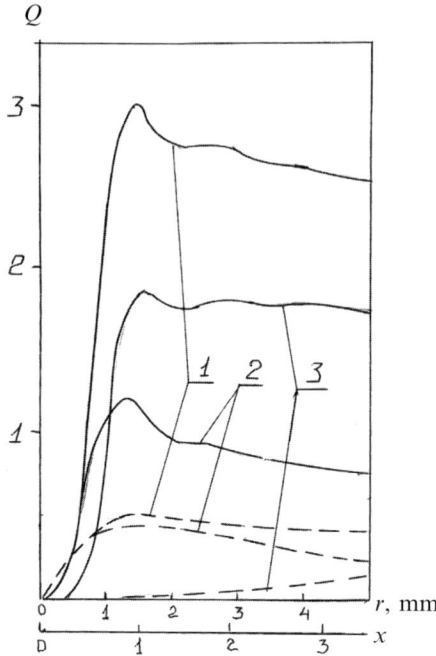

**Figure 6.21.** Theoretical dependences of the efficiency factors of extinction, scattering, and absorption for individual water drops and hollow water spheres with the thickness of an envelope $\delta = 0.001$ cm at the 0.86-cm wavelength as a function of external particle radius ($r$) and size parameter ($x$). System temperature $T_0 = 293$ K; water salinity $S = 0‰$. 1—Efficiency factors of extinction; 2—efficiency factors of absorption; 3—efficiency factors of scattering. Solid curves correspond to water drops. Dashed curves correspond to hollow water spheres.

homogeneous water sphere, consisting of a highly absorbing substance, represents a rather classical picture: all three dependencies have prominent maxima (Mie maxima) at a parameter value of about unity and two (further maxima are "blurred"). Then, as the size parameter increases, elongated falling tails are observed in these dependencies. In this case the extinction factor tends to the value $Q_E \to 2$ (the extinction paradox well known in scattering theory) (Ishimaru, 1978). For diffraction parameter values in the region of $1 < x < 3.5$ a spherical water drop mainly scatters falling electromagnetic energy—its spectral albedo—that is, the ratio of scattering to extinction factors equals the value of 0.65–0.7. The total region from the first maximum to zero was called the Rayleigh region, in which the interaction of electromagnetic waves with small particles ($x < 1$) takes place according to quite certain regularities (so-called Rayleigh scattering) (Hulst, 1981; Ishimaru, 1978; Sharkov, 2003). Principally, we should note here a sharp decrease in electromagnetic energy scattering by a particle: all the energy falling on a particle is absorbed. This is well seen in Figure 6.21: at values $x = 0.5$ the spectral albedo equals 0.1 and then promptly decreases, and at $x < 0.3$ the value of albedo is lower than 0.01. In the optical range the situation is essentially different: small particles ($x < 0.05$), consisting of a weakly absorbing substance, highly absorb electromagnetic energy at Rayleigh scattering, but large particles, consisting of the same substance, at $x > 0.1$ are purely scattering—the albedo of such particles tends to unity (Deirmendjian, 1969; Sharkov, 2003).

We shall now analyze scattering on a water shell (the relative thickness of a shell with respect to its external radius is $\delta/a = 0.003$–$0.01$). It follows from Figure 6.21 that in the whole range of size parameter values studied $0 < x < 3.2$, absorption essentially prevails over scattering: for large particles ($x = 3$) this ratio equals 4–4.5, and for small hollow spheres this ratio drastically increases. Note that in absorption and extinction dependencies there is no prominent maximum at $x \simeq 1$. Scattering (in magnitudes) by a large ($x = 3$) hollow sphere is greater than 22 times suppressed with respect to scattering by a spherical drop of the same size. In this case—since total extinction also decreases with respect to a spherical drop—for large hollow spheres ($x = 1$–$3$) the albedo equals the values of 0.04–0.1. In the Rayleigh region (at $x < 1$) the albedo drastically decreases and tends to zero. Thus, absorption essentially prevails over scattering for hollow particles with thin-walled shells both in the Mie region and in the Rayleigh region.

The aforementioned data related to scattering of aqueous thin-walled shells of the emulsion monolayer, whereas in the foam of cellular structure the appearance of larger hollow particles with thicker walls is possible (see Section 6.5.1). Figure 6.22a–f presents the results of calculation of the efficiency factors for aqueous media depending on an external radius up to 5 mm and for shell thickness values in the range of 0.01–0.1 mm (or for the relative thickness of shells with respect to the external radius $\delta/a = 0.02$–$0.1$). Figure 6.23a–e gives the dependencies of the same values on shell thickness for fixed values of a particle's radius. The results of calculations presented in Figures 6.22 and 6.23 give, in combination, sufficient information on quantities $Q_E$, $Q_S$, $Q_A$ as functions of two variables: the thickness and external radius of a shell (Dombrovskiy, 1981). As follows from analysis of Figures 6.22 and 6.23, the main feature of the electromagnetic interaction of such kinds of particles consists in the fact that in all spectral ranges the scattering efficiency factor monotonously grows both as shell radius and shell thickness increase. For the largest thick-walled aqueous shells in the millimeter region of the spectrum, scattering can exceed absorption. Moreover, we can indicate a critical value in shell thickness at which the character of interaction of a hollow sphere acquires the scattering features of a homogeneous sphere—scattering begins to exceed the value of 0.5. In this case the relative value of shell thickness with respect to the external radius of a particle should exceed $\delta/a = 0.025$–$0.03$. For the millimeter range, unlike the case for scattering, the dependencies of the efficiency factors of absorption and extinction on particle radius are essentially non-monotonous—that is, Mie maxima are revealed (Figure 6.22a–c). The increase in shell thickness in all cases results in increasing the efficiency factor of extinction. At the same time, the efficiency factor of absorption, beginning with a certain shell thickness, noticeably drops due to increased radiation reflection by the particle. The Rayleigh scattering region is small: at $\lambda = 8.6$ mm it propagates up to a radius $a = 1$ mm.

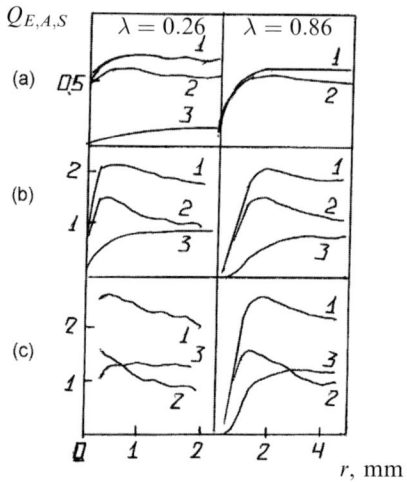
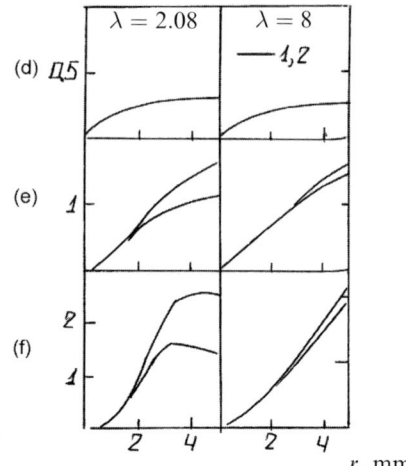

**Figure 6.22.** Theoretical dependences of the efficiency factors of extinction, scattering, and absorption for individual hollow water spheres as a function of external particle radius ($r$). System temperature $T_0 = 293$ K; water salinity $S = 0‰$. 1—Efficiency factors of extinction; 2—efficiency factors of absorption; 3—efficiency factors of scattering. Working wavelengths: (a)–(c) $\lambda = 0.26$ cm and $\lambda = 0.86$ cm; (d)–(f) $\lambda = 2.08$ cm and $\lambda = 8$ cm. The thickness of an envelope of (a)–(d) corresponds to 0.001 cm; (b) and (e) 0.005 cm; (c) and (f) 0.01 cm.

In the centimeter range the aqueous shells under consideration scatter more weakly than they absorb (Figure 6.22c–e). So, for $\lambda = 8$ cm—even at $a = 5$ mm and $\delta = 0.1$ mm—scattering equals only 15% of the extinction value. For particles with a fixed shell thickness the efficiency factors of absorption and extinction monotonously grow as shell radius increases, except in the case of strongest scattering at $\lambda = 20.8$ mm and $\delta = 0.1$ mm, where the absorption and extinction curves have maxima. The dependences of $Q_A$ and $Q_E$ on shell thickness for $A = $ const. take on different characters depending on aqueous shell radius. Whereas for shells with a radius of 1 mm the radiation weakening by a particle monotonously decreases as thickness increases from 0.03 to 0.1 mm, for shells with a radius of 5 mm the extinction monotonously increases throughout the calculation range.

As for the scattering indicatrix for particles with an aqueous shell, investigations (Dombrovskiy, 1981) suggest—despite the fact that the form of scattering indicatrix and the degree of polarization of chaotically scattered, polarized radiation for those particles with a radius of 5 mm at a wavelength of 20.8 mm—(strictly speaking) the inapplicability of Rayleigh formulas to determining the angular characteristics of scattering. However, if we pay attention to the low value of the scattering efficiency factor, then, in spite of the high asymmetry of scattering, we can hope for the possibility of applying Rayleigh formulas to efficiency factors in the case of shell thickness of 0.01 mm and radius up to $a = 5$ mm to an accuracy of 10–20%.

As is known (Stratton, 1941; Deirmendjian, 1969; Hulst, 1981; Ishimaru, 1978), Rayleigh formulas are contained in the first term of the Mie series, so we can obtain

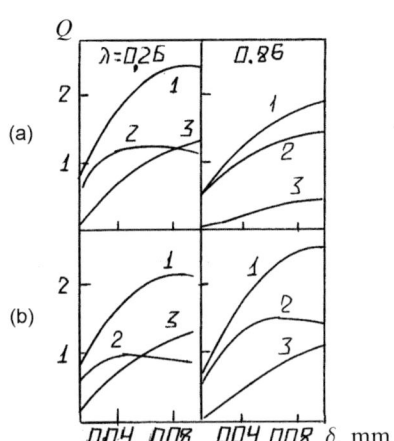
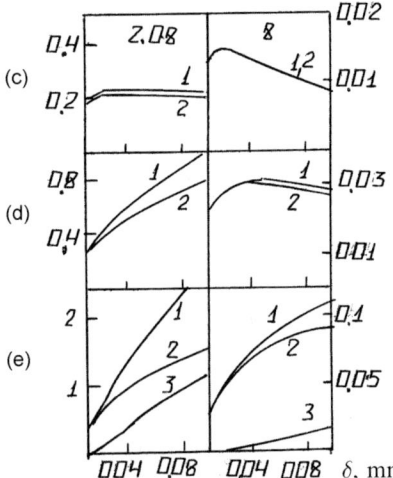

**Figure 6.23.** Theoretical dependences of the efficiency factors of extinction, scattering, and absorption for individual hollow water spheres as a function of the thickness of an envelope. System temperature $T_0 = 293$ K; water salinity $S = 0‰$. 1—Efficiency factors of extinction; 2—efficiency factors of absorption; 3—efficiency factors of scattering. Working wavelengths: (a)–(b) $\lambda = 0.26$ cm and $\lambda = 0.86$ cm; (c)–(e) $\lambda = 2.08$ cm and $\lambda = 8$ cm. External particle radius ($r$): (a) and (c) correspond to 0.1 cm; (b) and (d) 0.2 cm; (e) 0.5 cm.

the efficiency factors of extinction $Q_E$ and scattering $Q_S$:

$$Q_S = \frac{6}{x^2}|\dot{a}_1|^2$$

$$Q_E = \frac{6}{x^2}\operatorname{Re}\dot{a}_1$$

(6.9)

The expression for the Mie series coefficient $\dot{a}_1$ can be found either directly from Mie formulas, or from solution of a corresponding electrostatic problem. If we make use of the latter method, then the expression for $\dot{a}_1$ will be as follows (Landau and Lifshitz, 1957):

$$\dot{a}_1 = \frac{2}{3}jx^3\frac{(\varepsilon - 1)(2\varepsilon + 1)(1 - q^3)}{(\varepsilon + 2)(2\varepsilon + 1)(1 - q^3) + 9\varepsilon q^3}$$

(6.10)

where $q = 1 - (\delta/a)$ is the coefficient for filling a particle's cavity; $j$ is the imaginary unity; $\varepsilon = \dot{\varepsilon}(\lambda)$ is the complex dielectric permeability of a shell's substance (water in the considered case). In the millimeter range the discrepancy between accurate solution of the problem and Rayleigh approximation can reach 10–15% in the case of thin shells ($\lambda = 0.26$ cm) and 20–25% in the case of thick shells ($\lambda = 0.86$ cm). In the centimeter range Rayleigh formulas result in high accuracy.

Let us move on to studying the electrodynamic properties of a polydisperse structure. Detailed calculations of the polydisperse structure of a foam system are labor-intensive, but the final results are substantially similar to the monodisperse case. Dombrovskiy (1979) showed, to a satisfactory degree of accuracy, that calculated values of real foam can be replaced by the data of monodisperse foam that has the effective radius of hollow spheres.

Table 6.6 presents the values of extinction parameters per unit length and integrated-over-distribution extinction $\beta$ (cm$^{-1}$), and the single-spectrum albedo $\omega$ for a polydisperse system of bubbles, whose size distribution function corresponds with the experimental histogram in Figure 6.7 (here $\delta$ is the thickness of a bubble's wall).

Let us now summarize basic results on the characteristics of scattering and absorption of electromagnetic waves in the millimeter, centimeter, and decimeter ranges by an emulsion-polydisperse medium:

(1) In all spectral ranges, absorption in a medium of thin-walled shells essentially prevails over scattering. The relative fraction of scattering in extinction does not exceed 7% for a shell thickness of $\delta \leq 0.001$ cm.
(2) The scattering value increases as the shell thickens and can compose up to 30% in the extinction fraction. At the same time, scattering drastically decreases as emission wavelength grows, and, beginning with a wavelength of 0.8 cm, scattering can be disregarded in the analysis of thin-walled shells.
(3) In virtually all microwave ranges the Rayleigh approximation can be used to determine the absorption by particles with a thin shell ($\delta \leq 0.01$ mm) to an error no greater than 10%. For particles with a thick shell ($\delta \sim 0.05$ mm) the Rayleigh approximation can be used in the centimeter and decimeter ranges only. In the millimeter range the error can reach 25% or more.

Note that for a homogeneous spherical drop of water of the same radius, the albedo of a particle in the millimeter range is approximately ten-fold greater than for a bubble. Drops and bubbles of identical size differ very highly in their properties in the microwave range. This is already testified to by calculations in the Rayleigh approximation.

It is also important to note that in the optical range the situation is directly opposite: a hollow particle absorbs electromagnetic energy much more weakly than a homogeneous particle of the same external radius (Dombrovskiy, 1974); and foam structure as a continuous medium has a reflection coefficient in the optical range close to unity (0.94–1.00) (Kokhanovsky, 2004).

Thus, rigorous solution of the diffraction problem helped us to establish that microwave radiation weakening by bubbles is mainly caused by absorption rather than by scattering. In the extreme case of an infinitely thin shell even a bubble on the electromagnetic wavelength scale will behave as an "absolutely blackbody", absorbing all the incident radiation. At the same time, for bubbles the Rayleigh region turns out to be much broader than for complete drops. This allows us to use Rayleigh formulas not only in the centimeter, but, in some cases, in the millimeter range as well. The special property of hollow spherical particles to absorb a greater part of incident radiation can be taken as the basis for constructing a qualitatively new electrodynamic model of a disperse system.

**Table 6.6.** Spectral characteristics for per unit length extinction ($\beta_0$) and scattering albedo ($\omega$) in a polydispersive bubble media in the microwave range 0.26–18 cm.

| Working wavelength $\lambda$ | Mie theory | | | | Rayleigh theory | | | |
|---|---|---|---|---|---|---|---|---|
| | $\delta = 10^{-3}$ cm | | $\delta = 5 \cdot 10^{-3}$ cm | | $\delta = 10^{-3}$ cm | | $\delta = 5 \cdot 10^{-3}$ cm | |
| (cm) | $\beta_0$ (cm$^{-1}$) | $\omega$ | $\beta_0$ (cm$^{-1}$) | $\omega$ | $\beta_0$ (cm$^{-1}$) | $\omega$ | $\beta_0$ (cm$^{-1}$) | $\omega$ |
| 0.26 | 11.72 | 0.074 | 33.80 | 0.266 | 10.10 | 0.218 | — | — |
| 0.86 | 5.71 | 0.015 | 7.16 | 0.074 | 4.98 | 0.018 | 4.57 | 0.099 |
| 2.08 | 1.57 | 0.003 | 1.08 | 0.018 | 1.41 | 0.003 | 0.86 | 0.020 |
| 8 | 0.12 | $<10^{-3}$ | 0.066 | 0.001 | 0.108 | $<10^{-3}$ | 0.059 | 0.001 |
| 18 | 0.03 | $<10^{-3}$ | 0.015 | $<10^{-3}$ | 0.024 | $<10^{-3}$ | 0.013 | $<10^{-3}$ |

*Note*: Dielectric characteristics of bubble envelopes agree with the relaxation model (Sharkov, 2003). $\delta$ is the thickness of sphere envelopes.

### 6.6.4   Optical model of a disperse medium

The applicability of the radiation transfer theory to electromagnetic radiation in a discrete scattering medium whose particles are closely packed (i.e., realized in foam-like systems) still remains poorly studied, in general (Ishimaru, 1978). In this case so-called cooperative effects (correlations in the radiation field of particles), whose description is outside the framework of the classical transfer theory, can be essential. Nevertheless, it seems expedient to try to use such an approach to solution of a model problem on the thermal radio emission of disperse media in connection with the presence of specific experimental material.

We shall now obtain that solution of the radiation transfer equation that is used for estimating the spectral and polarization characteristics of the radio emission of disperse systems.

Let the plane-parallel layer of hollow spherical particles be situated on the water surface. The scalar radiation transfer equation in the case of axial symmetry has the following form for the intensity of radiation $I(\tau, \mu)$ generated inside the layer (Ozisik, 1973; Sharkov, 2003):

$$\mu \frac{\partial I(\tau, \mu)}{\partial \tau} + I(\tau, \mu) = F(\tau, \mu) \qquad \mu = \cos \theta \tag{6.11}$$

where the spectral function of a source is

$$F(\tau, \mu) = (1 - \omega) I_B[T(\tau)] + \frac{\omega}{2} \int g(\mu, \mu') I(\tau, \mu') \, d\mu' \tag{6.12}$$

where $\tau$ is the optical thickness of a layer; $\omega = \sigma / \beta$ is the spectral albedo of single scattering; $g$ is the indicatrix of scattering of a medium's volume unit; $I_B[T(\tau)]$ is Planck's spectral function; and $T(\tau)$ is a profile of the thermodynamic temperature of a layer.

The boundary conditions for mirror-reflecting an underlying surface and, in the absence of radiation, falling on a layer from outside will be written as

$$I(0, \mu) = \phi(\mu) \qquad I(\tau_0, -\mu) = 0 \qquad \mu > 0 \tag{6.13}$$

where $\phi(\mu)$ is the specified function.

In searching for a solution to equation (6.11), intensity is usually sub-divided into two components (the two-flow approximation): direct $I^+$ and reverse $I^-$. In this case solutions are written in the form

$$I^+(\tau, \mu) = I^+(0, \mu) \exp\left(-\frac{\tau}{\mu}\right) + \frac{1}{\mu} \int_0^\tau F(\tau', \mu) \exp\left(-\frac{\tau - \tau'}{\mu}\right) d\tau'$$

$$I^-(\tau, \mu) = I^-(\tau_0, \mu) \exp\left(-\frac{\tau_0 - \tau}{\mu}\right) - \frac{1}{\mu} \int_0^{\tau_0} F(\tau', \mu) \exp\left(-\frac{\tau - \tau'}{\mu}\right) d\tau' \tag{6.14}$$

under boundary conditions (6.13).

The spectral coefficient of the emission of a system under conditions of isothermality ($T = $ const.) is determined in accordance with Kirchhoff's law as follows:

$$\kappa(\mu) = \frac{I^+(\tau_0, \mu)}{I_B(T)} \tag{6.15}$$

where $I_B(T) = CT$ (the Rayleigh–Jeans law); and $C$ is a constant.

By virtue of the smallness of the spectral albedo of a medium ($\omega < 1$) at all wavelengths, in the radiation transfer theory equation we can neglect the integral term that describes internal re-scattering (6.12), and make use of well-known solutions (Ozisik, 1973; Sharkov, 2003). In such a case the spectral function of source $F$ can be allowed to equal

$$F = (1 - \omega)I_B(T) = \text{const.} \tag{6.16}$$

and then equation (6.11) allows a simple analytical solution satisfying the boundary conditions (6.13).

After simple transformations we obtain from (6.14)–(6.16) the following expression for the spectral emissivity of a system:

$$\kappa(\mu) = (1 - |R(\mu)|^2) \exp\left(-\frac{\tau_0}{\mu}\right) + (1 - \omega)\left[1 - \exp\left(-\frac{\tau_0}{\mu}\right)\right]$$

$$+ |R(\mu)|^2(1 - \omega)\left[1 - \exp\left(-\frac{\tau_0}{\mu}\right)\right]\exp\left(-\frac{\tau_0}{\mu}\right) \tag{6.17}$$

where $R(\mu)$ is the spectral Fresnel (complex) coefficient of reflection over the field of an underlying surface (the water surface in this case).

The physical sense of the obtained relation is rather obvious: the first term of (6.17) describes an underlying surface that is radiation-weakened in the scattering layer, and the second and third terms describe the natural and mirror-reflected radiation of a layer with regard to losses on scattering.

At $\omega = 0$ (where scattering in a layer is completely absent) formula (6.17) is transformed to the well-known form:

$$\kappa(\mu) = (1 - |R(\mu)|^2) \exp\left(-\frac{2\tau_0}{\mu}\right) \tag{6.18}$$

We note that the obtained expression (6.18) was earlier repeatedly used for various estimations of the intensity of emission of cloud-like systems over a smooth surface, including estimations of disperse system emission (Matveyev, 1971). However, in those estimations the question concerned a homogeneous dielectric layer model that had effective parameters corresponding to the matrix system of air and water. Naturally, diffraction effects were disregarded in such a model.

The spectral and polarization dependencies of system emissivities were calculated for two values of the geometrical thickness of a layer: $h = 0.1$ and $1$ cm (Figures 6.15 and 6.24). The optical depth of the layer was assumed to be constant: $\tau = \beta h$. The values of integrated parameters $\beta$ and $\omega$ were taken from Table 6.6 as accurate values of Mie theory that allow for the diffraction phenomena on hollow spheres.

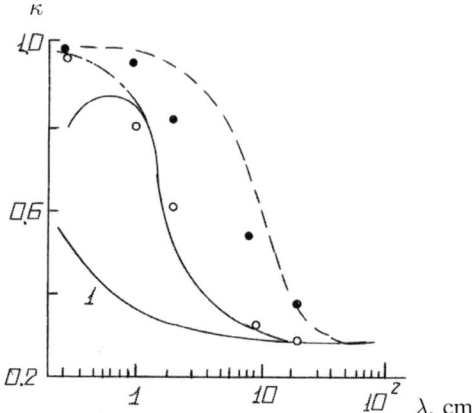

**Figure 6.24.** Spectral dependences of foam structure emissivity for an optical model (model 5b in Table 6.8). Temperature $T_0 = 293$ K; water salinity $S = 0‰$; incidence angle $\theta = 35°$; horizontal polarization; $\delta = 5 \cdot 10^{-3}$ cm. 1—A smooth water surface. Solid curve is estimation of the emulsion structure using equation (6.17), dashed–dotted curve using equation (6.18); dashed curve corresponds to the polyhedral structure. 1—A smooth water surface. Experimental data: circles correspond to the emulsion layer, dark circles to the polyhedral structure.

In the case of a "monolayer" and for $\tau < 1$ the spectral dependencies, calculated and experimental, correspond well with each other (Figure 6.24). However, in the polarization characteristics for $\lambda = 0.86$ cm (Figure 6.15) qualitative distinction is observed: both calculated components $\kappa_H(\theta)$ and $\kappa_V(\theta)$ grow as the observation angle increases (in the figure this is designated as model 5b), which obviously contradicts the experiment (dark and light squares in the figure). In this case the radiation polarization degree turns out to be approximately four times lower than follows from measurements.

In the case of a cellular layer and when the layer's optical depth increases ($\tau > 5$), agreement between calculated and experimental data is violated. In the millimeter range, for example, the polarization dependencies, designated by the dashed line in Figure 6.15 (model 5b), are completely absent, whereas the experiment clearly demonstrates the presence of polarization properties in the cellular structure (dark and light triangles in the figure), though the latter are weakly expressed. This is due to the fact that the theoretical emissivity of a system is only determined by the value of the spectral albedo of the scattering layer's unit of volume:

$$\kappa = 1 - \omega \qquad (6.19)$$

since the exponent-containing terms have an order of $10^{-5}$–$10^{-6}$. The frequency characteristic of a system $\kappa(\lambda)$ turns out to be much steeper, and the calculated values of emissivity exceed experimental ones, in the centimeter range especially (Figure 6.24).

It is also important to note that the inclusion of an integrated term in the transfer equation (6.12), describing multiple scattering in a layer, does not essentially cause any change in spectral characteristics $\kappa(\lambda)$. This is checked by comparison of calculation data with the solution results of a similar problem using the standard method of double spherical harmonics (Dombrovskiy, 1979; Dombrovskiy and Raizer, 1992). However, the latter solution—unlike ours (formula (6.17))—contains hemispherical values of spectral coefficients $\kappa(\lambda)$ and principally disregards polarization of the natural radiation emitted by the medium. This approach is widely developed in heat transfer problems to find a simplified solution to the full equation of transfer theory (Ozisik, 1973). Such an approach, however, is not applicable to remote-sensing problems, since the basic parameter measured in such problems is the spectral intensity of radiation, considered in the infinitesimal solid angle around the observation direction and with regard to polarization of the radiation received (Sharkov, 2003). This is the reason that interpretation of the experimental data of Bordonskii *et al.* (1978), carried out by Dombrovskiy (1979), was purely qualitative in character. Calculations obtained by such a method cannot be used for quantitative analysis of real remote measurements.

Thus, the "optical" model of a disperse system, despite its simplified version (absence of multiple scattering, medium homogeneity in optical depth), turned out to be, in general, more consistent, both qualitatively and quantitatively, than the simplest dielectric models. This is associated with the fact that the "optical" model takes into account the main feature of the disperse structure as a medium consisting of highly absorbing particles—bubbles. However, the optical model possesses certain deficiencies as well. First, it does not describe correctly the polarization dependencies of spectral emissivity of a system in the millimeter range. Second, as quantity $\tau$ increases to a critical value $\tau_C$, the degree of blackness of a system ceases to change. The latter circumstance is not casual in any way, since the vertical distribution of $\beta$ and $\omega$ parameters, caused by changing the geometry of a disperse layer's particles with height, was disregarded in this case. The essential discrepancy with the experiment in the case of $\tau \sim 10$ probably had the same reason.

### 6.6.5  Diffraction models of disperse systems

In this section we turn once again to the macroscopic description of a disperse system, while keeping within the framework of a planar-layered model (see Section 6.6.2). However, the $\dot{\varepsilon}(z)$ profile is chosen here from completely different physical considerations—namely, taking due account of the diffraction effects on hollow particles of the emulsion structure (Raizer and Sharkov, 1981).

As is known, the question about the effective dielectric permeability of a scattering medium belongs to basic issues in the theory of multiple scattering of waves (Born and Wolf, 1999). To estimate the effective microwave characteristics of the emulsion medium $\dot{\varepsilon}_{EF}$ we shall make use of expressions for the quasi-static dipole approximation $\dot{\varepsilon}_{N\alpha}$ and take into account the multipole contribution to scattering $\dot{\varepsilon}_{N\alpha}$ (Hulst, 1981; Born and Wolf, 1999). The expressions are generalized to the case of a polydisperse system of hollow spherical particles (Table 6.7) (Raizer and Sharkov, 1981). The first expression represents the well-known Lorentz–Lorenz formula (formula N1 in Table 6.7), which takes into account the dipole interaction in

**Table 6.7.** Formulae of the dielectric properties for diffraction models.

| Number | The formula for $\varepsilon_{EF}$ | Model parameters | Applicability limits |
|---|---|---|---|
| 1 | $\dot\varepsilon_{Na} = \dfrac{1 + \dfrac{8}{3}\pi\overline{N\alpha}}{1 - \dfrac{8}{3}\pi\overline{N\alpha}}$ | $\overline{N\alpha} = k\displaystyle\int_0^\infty \alpha(a)f(a)\,da\left[\dfrac{4}{3}\pi\int_0^\infty a^3 f(a)\,da\right]^{-1}$ $\dot\alpha = a^3\dfrac{(\dot\varepsilon_0 - 1)(2\dot\varepsilon_0 + 1)(1 - q^3)}{(\dot\varepsilon_0 - 1)(2\dot\varepsilon_0 + 1)(1 - q^3) + 9\dot\varepsilon_0 q^3}$ | $\lambda > 0.8\,\text{cm}$ $0 \le k \le 1$ $a \le 0.1\,\text{cm}$ $\delta < 5\cdot 10^{-3}\,\text{cm}$ | The Rayleigh scattering range |
| 2 | $\dot\varepsilon_{NS} = 1 + j4\pi\left(\dfrac{2\pi}{\lambda}\right)^{-3}\overline{N\dot S_0}$ | $\overline{N\dot S_0} = k\displaystyle\int_0^\infty \dot S_0(a)f(a)\,da\left[\dfrac{4}{3}\pi\int_0^\infty a^3 f(a)\,da\right]^{-1}$ $\dot S_0 = \displaystyle\sum_{n=1}^\infty \dfrac{2n+1}{2}(\dot a_n + \dot b_n)$ | $\lambda < 0.8\,\text{cm}$ $a, \delta$ are optional values $0 \le k \le 1$ | The Mie scattering range |

*Note:* $\dot a_n$ and $\dot b_n$ are the Mie coefficients for hollow spheres.

a closely-packed system (quasi-static approximation). The second expression represents the Hulst formula, which describes the contribution of a particle's multipole moment to effective permeability (electrodynamically discharged medium approximation) (formula N2 in Table 6.7). The parameters that appear in these formulas are as follows:

$$a \text{ and } \delta = \text{external radius and thickness of a hollow sphere's shell;}$$
$$f(a) = \text{normalized function of a size distribution;}$$
$$N, k = \text{volume concentration and coefficient of particle packing;}$$
$$\dot{\varepsilon}_0(\lambda) = \text{complex dielectric permeability of a shell's substance (water);}$$
$$\dot{\alpha} \text{ and } \dot{S} = \text{complex polarizability and amplitude of "forward" scattering of a}$$
$$\text{hollow spherical particle;}$$
$$\lambda = \text{electromagnetic wavelength in vacuum; and}$$
$$q = 1 - (\delta/a) = \text{coefficient of "filling" a particle's cavity.}$$

Let us consider the results of model calculations of $\dot{\varepsilon}_{EF}$ by relations (1) and (2) (Table 6.7). We shall make use of the following experimental (see Section 6.5.1) values of parameters $f(a)$ ($0.01 \leq a \leq 0.1$ cm)—$\delta = 10^{-2}$–$10^{-3}$ cm and $N = 5 \cdot 10^3$ cm$^3$—as well as of the theoretical values of the complex dielectric permeability of water $\dot{\varepsilon}(\lambda)$ (Sharkov, 2003).

Figure 6.25 presents the spectral dependencies of the real (a) and imaginary (b) parts of $\dot{\varepsilon}_{N\alpha}(\lambda)$ for three values of bubble shell thickness. For comparison, we present on this diagram similar spectral dependencies of the effective dielectric permeability of a medium $\dot{\varepsilon}_C(\lambda)$ calculated by Odelevskii's formula (6.6). Formula (6.6) reflects the traditional method of electrodynamic description of a foam structure as a two-phase, statistically homogeneous mixture of water and air (static approximation), which certainly does not correspond with physical reality, as already shown in Section 6.5.1. The dashed region in Figure 6.25 is confined by two $\dot{\varepsilon}_C(\lambda)$ curves, which correspond to the values $C_0 = 0.01$ and $0.20$. Such values of relative water content in a mixture almost completely overlap the possible range of water content in foam structure layers (Figure 6.8).

As seen from Figure 6.25, the Lorentz–Lorenz and Odelevskii models give principally different results, just as should be expected. The region of dispersion of the dependence of an imaginary part of $\dot{\varepsilon}_C(\lambda)$ for a homogeneous mixture turns out to be rather narrow and basically occupies the millimeter range. At the same time, the spectral dependence of an imaginary part of $\dot{\varepsilon}_{N\alpha}(\lambda)$ possesses a very broad maximum and even propagates onto decimeter waves. In this case the imaginary part of $\dot{\varepsilon}_{N\alpha}(\lambda)$ has a maximum in the range of $\lambda = 0.8$–$2$ cm, which is caused by the corresponding spectral dependence of an imaginary part of the complex dielectric permeability of water. However, the width of this zone and the maximum value are related to the macroscopic and diffraction properties of a medium. Note that the maximum of $\dot{\varepsilon}_{N\alpha}(\lambda)$ essentially exceeds (by five times and greater) the absorption of the homogeneous mixture considered even at a high water concentration of 20%.

The dispersion relation in the Hulst model—formula (2) in Table 6.7—allows us to analyze the effect of multipole moments in the induced radiation of particles—that is, of the highest terms in the expansion of a complex function of "forward"

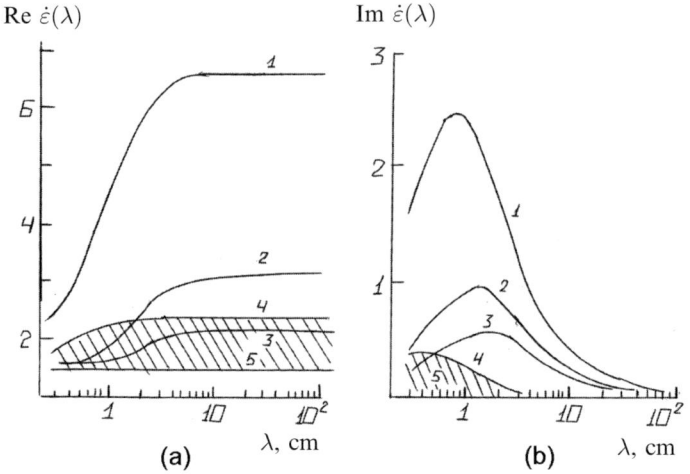

**Figure 6.25.** Theoretical spectral dependence of effective dielectric permittivity (real and imaginary parts) for the dipole interaction model $\dot{\varepsilon}_{N\alpha}(\lambda)$ (equation 1 in Table 6.7) and the homogeneous mixture $\dot{\varepsilon}_C(\lambda)$ (equation (6.6)) of a polydispersive system. Temperature $T_0 = 293$ K; water salinity $S = 0‰$. Estimation of $\dot{\varepsilon}_{N\alpha}(\lambda)$: 1—$\delta = 10^{-2}$ cm; 2—$\delta = 5 \cdot 10^{-3}$ cm; 3—$\delta = 10^{-3}$ cm. Estimation of $\dot{\varepsilon}_C(\lambda)$ for the volume water content in the system: 4—$C_0 = 0.20$; 5—$C_0 = 0.01$ (shaded region).

scattering $S(0, x)$ with respect to the diffraction parameter $x = 2\pi a/\lambda$—on the effective characteristics of a medium. Figure 6.26 presents the theoretical dependencies of real and imaginary parts of $\dot{\varepsilon}_{NS}(a)$ for a monodisperse system consisting of bubbles of various sizes. The concentration of particles was specified in the form $N(a) = k/(3/4)\pi a^3$, where $k = 0.74$ (compact-hexagonal packing). In this case the complex function of "forward" scattering was calculated by the accurate Mie formulas. The character of $\dot{\varepsilon}_{NS}(a)$ curves is nontrivial: an increase in bubble radius results in the appearance of resonance dependencies of an imaginary part $\text{Im}[\dot{\varepsilon}_{NS}(a)]$ as a function of $a$. These maxima, however, have nothing in common with Mie's resonance effects, because the natural frequencies of oscillations of a hollow spherical particle are determined by its size. Nevertheless, the appearance of maxima is caused by multipole moments. We can easily be convinced of this, if we take into account not one, but two terms in the scattering function expansion in parameter $x$ (Hulst, 1981):

$$S(0, x) = A(a)x^3 + B(a)x^5 + \cdots \tag{6.20}$$

where $A(a)$, $B(a)$ are complex coefficients in the expansion (6.20). Substituting (6.20) into the relation of Hulst's model (Table 6.7) and having in mind that $N \sim a^{-3}$, we obtain

$$\dot{\varepsilon}_{NS}(a) = 1 + j33k[A(a) + B(a)x^2 + \cdots] \tag{6.21}$$

From this we can immediately see that the real and imaginary parts can assume, in

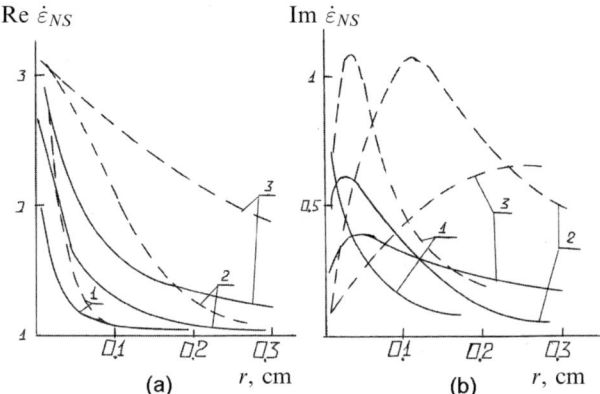

**Figure 6.26.** Theoretical spectral dependence of the effective dielectric permittivity (real part) for the Hulst model as a function of hollow particle size. Estimation: temperature $T_0 = 293$ K; water salinity $S = 0$‰. 1—Wavelength 0.26 cm; 2—0.86 cm; 3—2.08 cm. The solid curve corresponds to the thickness of an envelope $10^{-3}$ cm; the dashed curve $5 \cdot 10^{-3}$ cm.

principle, extreme values for $A, B \neq$ const. (we omit further deductions because of their length). In the opposite case, where both $A$ and $B =$ const. (i.e., for $a =$ const.), resonances will already be determined by Mie effects (scattering function oscillations).

Thus, to account for the dipole interaction and multipole moment of hollow spherical particles in the model results in increasing both the real part and—especially—the imaginary part of the effective dielectric permeability of a medium. In other words, the total electromagnetic losses in a medium due to the diffraction mechanism essentially grow. It is also characteristic that these conditions are realized within a rather wide range of wavelengths, including decimeter ones. The latter condition cannot be achieved using formulas for heterogeneous mixtures (i.e., neither statistically homogeneous nor matrix mixtures), which exclude from consideration the boundary effects associated with the geometry of phase components.

### 6.6.6 Layered, inhomogeneous diffraction model

The disperse system will be modeled by a planar, layered, inhomogeneous structure using some law of complex dielectric permeability variation in depth $\dot{\varepsilon}_{EF}(z)$. To calculate the emissive characteristics we shall make use of a numerical method based on quantization of an inhomogeneous half-space into the system of plane-parallel and dielectrically homogeneous layers, not necessarily of equal thickness (see Section 6.6.2). We shall consider two radiothermal models, corresponding to disperse systems of emulsion and polyhedral types, as our working models. These models are shown schematically in Figure 6.27. Their geometrical parameters are chosen on the basis of empirical estimations made from microphotographs of Figures 6.1 and 6.8, and the dielectric parameters correspond to calculated values of $\dot{\varepsilon}_{EF}(z)$.

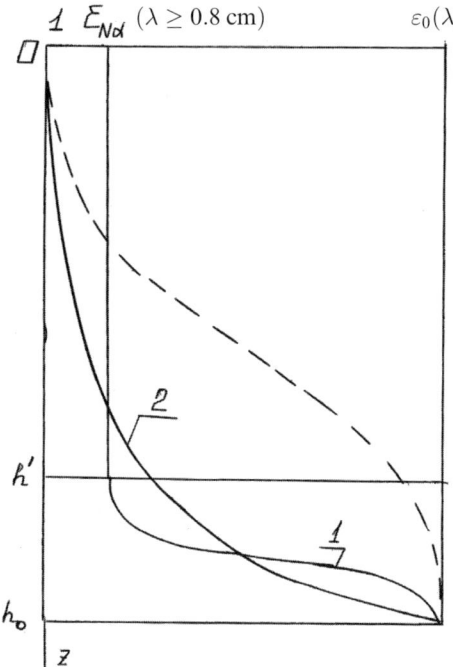

**Figure 6.27.** Schematic models of vertical profiles for effective dielectric permittivity (real part) of dispersive structures: 1—the emulsion structure with $h_0 = 0.013$ cm and $h' = 0.10$ cm; 2—the polyhedral structure with $h_0 = 1$ cm. The dashed curve corresponds to a fair transition of the real part of dielectric permittivity $\varepsilon_C(z)$ using equation (6.8) for a homogeneous mixture model.

A two-layer structure with a blurred lower boundary can be accepted as an emulsion system model (curve 1 in Figure 6.27). The thin transition layer, introduced at the underlying medium boundary by formula (6.8), describes the conforming effect of small-scale roughness or water surface "blurring" as a result of floating bubbles. These geometrical features are well demonstrated by the microphotograph in Figure 6.1.

In the case of a polyhedral system (curve 2 in Figure 6.27) the model provides vertical distribution of bubbles in the disperse layer. It is typical of a thick foam layer for particle size to vary within very wide limits: from fractions of a millimeter near the water surface up to a centimeter in the region bordering the atmosphere. As height increases, the bubbles lose their sphericity and turn into multi-sided cells fastened to each other by liquid films (Figure 6.1). Choosing, for example, the dependence of bubble radius $a$ on coordinate $z$ to be linear and using the Hulst approximation, we obtain the continuous profile (curve 2 in Figure 6.27). The same figure shows the smooth transition (the dotted line) from the dielectric properties of water to air for the homogeneous mixture model.

Table 6.8 lists the current radiothermal models of foam cover, and their adequacy for laboratory experiment is demonstrated (Section 6.5.1) (Raizer and Sharkov,

**Table 6.8.** Comparison of radiothermal models for dispersive systems.

| Model name and calculation method | Reference | Agreement with laboratory experiments | | | |
|---|---|---|---|---|---|
| | | The emulsion monolayer | | The polyhedral structure | |
| | | $\kappa(\lambda)$ | $\kappa(\theta)$ | $\kappa(\lambda)$ | $\kappa(\theta)$ |
| 1. The homogeneous dielectric layer with sharp boundaries. The matrix mixture of water and air | Matveyev (1971) | $--$ | $+-$ | $--$ | $+-$ |
| 2. The discrete flat-layer structure in water films and air | Stogrin (1972) | $--$ | $--$ | $--$ | $--$ |
| 3. The inhomogeneous dielectric layer with the statistical mixture of water and air | Raizer and Sharkov (1981) | $+-$ | $+-$ | $+-$ | $+-$ |
| 4. The pure passage layer with smoothly changing parameters | Raizer and Sharkov (1981) | $+-$ | $+-$ | $+-$ | $+-$ |
| 5. Discrete scattering media in hollow water spheres: | | | | | |
| (a) the scalar radiative transfer theory excluding polarization properties; Mie theory | Dombrovskiy (1979) | $+-$ | $--$ | $+-$ | $--$ |
| (b) the scalar radiative transfer theory including polarization properties; Mie theory | Raizer and Sharkov (1981) | $+-$ | $--$ | $+-$ | $--$ |
| 6. Continuous layered inhomogeneous media including effective dielectric properties: | | | | | |
| (a) the homogeneous dielectric layer including scattering of hollow water spheres and the non-sharp boundary between "water–layer"; the Lorentz–Lorenz formula | Raizer and Sharkov (1981) | $++$ | $++$ | | |
| (b) the inhomogeneous dielectric layer including the height distribution for hollow water spheres; the Hulst approximation | Raizer and Sharkov (1981) | $++$ | | $++$ | $++$ |

*Note*: the sign $+-$ denotes that the laboratory experiment data agree with theory qualitatively but not quantitatively; the sign $++$ denotes the agreement qualitatively and quantitatively; the sign $--$ denotes the absence of agreement qualitatively and quantitatively. $\kappa(\lambda)$ and $\kappa(\theta)$ denote the spectral and polarization properties of system emissivities.

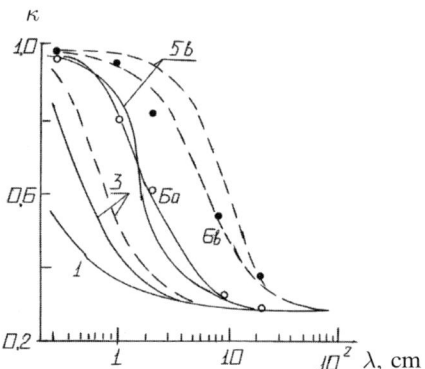

**Figure 6.28.** Spectral dependences of foam structure emissivity for various models (Table 6.8). Temperature $T_0 = 293$ K; water salinity $S = 0$‰; incidence angle $\theta = 35°$; horizontal polarization; $\delta = 5 \cdot 10^{-3}$ cm. 1—A smooth water surface. Solid curve is the estimation of the emulsion structure; dashed curve corresponds to the polyhedral structure. Experimental data: circles correspond to the emulsion layer, dark circles to the polyhedral structure. The numbers near curves correspond to a model's number in Table 6.8.

1981). Figures 6.28 and 6.15 compare theoretical and experimental data on the spectral $\kappa(\lambda)$ and polarization $\kappa_{VH}(\theta)$ dependencies of emissivity for two types of systems (see Table 6.8).

As was noted above, models of the homogeneous or inhomogeneous dielectric layer that have parameters corresponding to a heterogeneous mixture of water and air do not provide quantitative agreement with the experiment. The same conclusion relates to models of a discrete layered structure and to models of smooth transition layers as well (see Section 6.6.2 and Table 6.8).

Models of a discrete scattering medium take into account the diffraction properties of bubbles, which allows us to obtain essentially better agreement with the experiment (Table 6.8). "Optical" models are based on solution of the classical radiation transfer equation—that is, they suggest the complete incoherence of scattering and absorption, which, generally speaking, does not satisfy the case of a roughly disperse system with closely packed particles. In addition, the optical model does not qualitatively describe the polarization characteristics of radiation $\kappa_{VH}(\theta)$ observed experimentally (Figure 6.15). Agreement with spectral dependencies $\kappa(\lambda)$ is violated as the optical thickness of a scattering layer increases.

Macroscopic description of a system by introducing the dipole (multipole) moment into the effective parameters most completely reflects the properties of a roughly disperse medium in the microwave range. In this case the dipole approximation takes into account the cooperative effects associated with the close packing of particles, which is essential for the centimeter range. Dielectric models, whose profiles are presented in Figure 6.27, give both qualitative and quantitative agreement with the data of laboratory measurements of the spectral and polarization characteristics of radio emission for disperse systems of the types considered.

## 6.7  CONCLUSIONS

**1**  On the basis of detailed theoretical and experimental investigation of the interaction of microwave-range electromagnetic waves with polydisperse closely packed structures, which form in the ocean–atmosphere system under gravity wave breaking conditions, the basic physical mechanism of the natural radiation of disperse structures is established. It is related to diffraction absorption of electromagnetic waves by bubbles of an emulsion medium and polyhedral cells in the cellular structure of foam layers. In the systems considered, the cooperative effects associated with close packing of bubbles in the emulsion layer and of tracery cells in the cellular foam layer are not determined. The high emissivity of foam systems on a disturbed sea surface is not a specific effect of the close packing of structures, but the consequence of relatively weak scattering and, simultaneously, of the high absorption of electromagnetic waves by individual hollow spheres and hollow tracery cells in foam systems. The presentation of roughly disperse foam media in the form of heterogeneous water–air mixtures is not adequate for the microwave range. Macroscopic description of the electromagnetic properties of disperse systems by introducing the dipole (multipole) moment of a hollow spherical particle—a bubble—into the effective dielectric permeability with regard to the dielectric inhomogeneity of a medium makes it possible to obtain the best agreement with experimental data.

**2**  In conclusion, we note that model approaches described by the Lorentz–Lorenz and Hulst approximations for a polydisperse medium can be generalized to a wider class of natural objects. The latter can include (Sharkov, 2003) sea ice, fresh snow, slush, water–oil emulsions, humidified soils and ground, and other geophysical systems, whose electrodynamic properties essentially depend on the structural characteristics of a disperse medium, and on the configuration and packing of its particles.

# 7

# Electrodynamics of concentrated drop flows

The present chapter is devoted to studying the electrodynamics of the drop-spray phase at gravity wave breaking as a high-velocity flux in a highly concentrated drop medium. Principally, optical models for rarefied media in the radiation transfer theory and their restrictions are considered. The basic notions for the quantitative characteristics of absorption and scattering of both separated (secluded) particles, and disperse media in the form of a cloud of non-interacting (independent) randomly located scatterers, as well as in those scatterers included in high-velocity air flows, are introduced. The basic statements of the Mie theory of scattering and the approximations used in practice (mainly, in remote sensing) are presented. The results of specialized experiments, carried out under the guidance of the author of the present book, on studying the electromagnetic properties of concentrated drop flows and the possibility of their use in forming electrodynamic models of the drop-spray phase of breaking waves are analyzed in detail.

## 7.1 ELECTROMAGNETIC PROPERTIES OF SECLUDED PARTICLES

The destruction of substance in the process of breaking of gravity oceanic waves and drop-spray phase capture in the powerful airflux of a near-surface wind, as well as intensive near-surface thermal convection, causes the appearance of an aerodisperse high-velocity medium over the sea surface, the parameters of the medium highly fluctuating both in particle concentration and spectrum, and in its turbulence (see Chapter 5 and Plate 2 in the Prelims of this book). As the practice of remote sensing of the ocean–atmosphere system under high sea disturbance conditions has shown, the most effective methods of studying such complicated media can be passive and active microwave remote systems (including radar systems with synthesized and real aperture and Doppler scatterometers) (Krasiuk and Rosenberg, 1970; Meischner, 1990; Sharkov, 1998, 2003; Bulatov *et al.*, 2003; Mouche *et al.*, 2006).

From the viewpoint of possible ways of using the results of the standard radiation transfer theory, of main interest to us are the characteristics of propagation and scattering of an electromagnetic wave of the microwave range in the presence of a cloud of randomly located, electromagnetically non-interacting scatterers, as well as of a flux of a concentrated drop medium. We shall analyze this problem in this chapter in three stages. First, we shall consider a secluded particle and study its scattering and absorption characteristics. At the second stage we shall take into account the contributions of a great number of independent particles and deduce the general relationships for wave propagating in a cloud of randomly distributed particles and in a high-velocity airflows. At the third stage we shall analyze the results of specialized experiments on studying the electromagnetic properties of concentrated drop flows. In this section the first of these stages—analysis of characteristics of an individual particle—is described. This issue was investigated in many works (Stratton, 1941; Hulst, 1981; Deirmendjian, 1969; Ishimaru, 1978; Ivazyn, 1991; Born and Wolf, 1999; Sharkov, 2003). This is the reason we shall present here only the basic physical approaches to this complicated problem as applied to the tasks of microwave sensing of a disturbed sea surface in the presence of drop-spray clouds.

### 7.1.1   The scattering cross-section and scattering amplitude

When a secluded and solitary particle is irradiated by an electromagnetic wave, some part of the incident power is scattered and irrevocably leaves the particle, and the other part is absorbed and transforms eventually into heat. These two basic phenomena—scattering and absorption—can be described most conveniently by supposing the particle to be illuminated by a planar incident wave.

Consider a linearly polarized planar electromagnetic wave propagating in a medium with relative dielectric and magnetic permittivities equal to unity. The electric field of such a wave has the form (Ishimaru, 1978; Hulst, 1981; Sharkov, 2003):

$$\mathbf{E}_i(\mathbf{r}, \mathbf{\Omega}') = E_0 \mathbf{e}_i \exp[jk(\mathbf{\Omega}'\mathbf{r})] \tag{7.1}$$

where $E_0$ is the field amplitude; $k = 2\pi/\lambda$ is the wavenumber; $\lambda$ is the wavelength of a medium (external with respect to a particle); $\mathbf{\Omega}'$ is the unit vector in the direction of propagation of an external field; and $\mathbf{e}_i$ is the unit vector specifying the direction of external field polarization.

This wave falls on particles that have relative dielectric permittivity $\varepsilon_p$, which is generally speaking complex but depends on the coordinates, since the particle can be absorbing and inhomogeneous. The field at distance $R$, measured from some point inside a particle in the direction of unit vector $\mathbf{\Omega}$, is equal to the sum of field $\mathbf{E}_i$ of an incident wave and a field $\mathbf{E}_S$ of a wave scattered on a particle. At distances $R > D^2/\lambda$ ($D$ is the characteristic size of a particle—for example, its diameter) owing to the interference of waves coming from the various points of the particle, the amplitude and phase of field $\mathbf{E}_S$ vary in a very complicated manner (so-called near field mode). In this case the observation point $\mathbf{r}$ is said to be in the near zone of a particle. For

$R > D^2/\lambda$ the scattered field $\mathbf{E}_S$ behaves like a spherical wave and can be presented in the form

$$\mathbf{E}_S(R) = \dot{\mathbf{f}}(\bar{\mathbf{\Omega}}', \mathbf{\Omega}) \frac{e^{jkR}}{R} \qquad R > \frac{D^2}{\lambda} \qquad (7.2)$$

where the scattering amplitude $\mathbf{f}(\mathbf{\Omega}, \mathbf{\Omega}')$ describes the amplitude, phase, and polarization of a scattered wave in the far zone in the observation direction $\mathbf{\Omega}$ provided that the planar wave, propagating in the irradiation direction $\mathbf{\Omega}'$, falls on a particle. It should be noted that—even in the case of linear polarization of an incident wave—the scattered wave of a particle of complicated shape would generally speaking possess elliptical polarization. The scattering amplitude is very important, because its value contains information on the internal dielectric properties, geometrical shape, and size of a particle (Ishimaru, 1978; Hulst, 1981; Sharkov, 2003).

Consider now the density $\mathbf{\Pi}_S$ of a flux of power, scattered in the directions of wave $\mathbf{\Omega}$ at distance $R$ from a particle, when a wave with power flux density $\mathbf{\Pi}_0$ falls on a particle from direction $\mathbf{\Omega}'$. Here $\mathbf{\Pi}_0$ and $\mathbf{\Pi}_S$ are the vectors of the power flux density of incident and scattered waves in corresponding directions:

$$\mathbf{\Pi}_0 = \frac{|\mathbf{E}_i|^2}{2Z_0} \mathbf{\Omega}' \qquad \mathbf{\Pi}_S = \frac{|\mathbf{E}_S|^2}{2Z_0} \mathbf{\Omega} \qquad (7.3)$$

where $Z_0$ is the characteristic impedance of a medium. The total power $P$ (measured in Watts), which will be scattered by a particle into the ambient space, can be determined as

$$P_S(\mathbf{\Omega}') = \iint_{4\pi} |\mathbf{\Pi}_S(\mathbf{\Omega}', \mathbf{\Omega})| \, d\Omega \qquad (7.4)$$

and then the ratio

$$\sigma_S(\mathbf{\Omega}') = \frac{P_S(\mathbf{\Omega}')}{|\mathbf{\Pi}_0|} \qquad (7.5)$$

is called the integral scattering cross-section of a particle. It can easily be seen that this value has the dimension of $m^2$. The physical sense of the quantity introduced consists in the fact that it indicates the difference in losses as a result of power scattering by a particle with respect to its geometrical cross-section (or its geometrical shadow). If the particle has a complicated shape, then the total scattering cross-section depends on the direction from which the external radiation comes.

Now we introduce another important definition characterizing the power and spatial-angular scattering of incident external radiation falling on a solitary particle by this particle. The differential scattering cross-section of a particle is determined as follows:

$$\sigma_d(\mathbf{\Omega}', \mathbf{\Omega}) = \lim_{R \to \infty} \left[ R^2 \frac{|\mathbf{\Pi}_S|}{|\mathbf{\Pi}_0|} \right] = |\dot{f}(\mathbf{\Omega}', \mathbf{\Omega})|^2 \qquad (7.6)$$

It follows from expression (7.6) that $\sigma_d(\mathbf{\Omega}', \mathbf{\Omega})$ has the dimension of an area divided by a solid angle. Note that the differential scattering cross-section has an unambiguous physical sense only when the distances considered from a particle exceed the size of the far zone. In the opposite case (or in the presence of some other particle near the one under investigation) the physical unambiguity of the definition introduced is lost.

In radar and scatterometric applications a bistatical radar-scattering cross-section $\sigma_B$ and backscattering cross-section $\sigma_{BS}$ are often used. They are related to $\sigma_d(\mathbf{\Omega'}, \mathbf{\Omega})$ by the equations

$$\sigma_B(\mathbf{\Omega'}, \mathbf{\Omega}) = 4\pi\sigma_d(\mathbf{\Omega'}, \mathbf{\Omega}) \qquad \sigma_{BS} = 4\pi\sigma_d(\mathbf{\Omega'}, -\mathbf{\Omega}) \qquad (7.7)$$

Quantity $\sigma_{BS}$ is also called the radar-scattering cross-section. The physical sense of these definitions can be explained as follows. Suppose that within the limits of a total solid angle of $4\pi$ the power flux density is constant and equals the value of density for direction $\mathbf{\Omega}$. Then the cross-section of a plate, from which such a power is scattered, is equal to the value of $\sigma_d$ for direction $\mathbf{\Omega}$ multiplied by $4\pi$. Note that other definitions of a backscattering cross-section are sometimes used as well (Skolnik, 1980).

### 7.1.2 The absorption cross-section

We shall now consider that part of incident flux energy falling on a particle that will be completely absorbed by the particle and then will transfer into heat. Certainly, if a particle is inhomogeneous in its electromagnetic properties, then all diffraction phenomena arising inside a particle should be taken into account in calculating the absorption. To get some unification in the description of scattering and absorption processes the following definition is introduced. By absorption cross-section $\sigma_A(\mathbf{\Omega'})$ is meant the ratio of total power $P_A$ that was absorbed in a particle's volume, to the density of flux power that falls on a particle from direction $\mathbf{\Omega'}$,

$$\sigma_A(\mathbf{\Omega'}) = \frac{P_A}{|\mathbf{\Pi}_0(\mathbf{\Omega'})|} \qquad (7.8)$$

The dimension of the absorption cross-section is expressed in $m^2$. If a particle is inhomogeneous in its composition, then the absorption cross-section will depend on the direction of incident external radiation. Since the question is about the absorption of electromagnetic energy, this quantity can have no direct relation to the geometry of a particle.

### 7.1.3 The extinction cross-section

We shall now consider the following important point. Since we have noted that the energy scattered by a particle is considered in the far zone of the particle, then this part of energy therefore "leaves" the particle irrevocably. Thus, no statistical bond exists between the power absorbed by a particle and the power scattered by the same particle. Only under this condition can we introduce a definition describing the total losses (or extinction) of the particle in the form of a sum of losses for scattering and absorption:

$$\sigma_E(\mathbf{\Omega'}) = \sigma_S(\mathbf{\Omega'}) + \sigma_A(\mathbf{\Omega'}) \qquad (7.9)$$

The quantity $\sigma_E(\boldsymbol{\Omega}')$ used to be called the extinction cross-section (or the total cross-section).

### 7.1.4  Single-scattering albedo

The relation between the absorption and scattering processes that occur when a particle is irradiated by a flux of electromagnetic radiation is, doubtlessly, a very important factor in studying the total energy balance in transforming (or extracting) the energy of a basic external flux by the particle. The ratio of the extinction scattering cross-section to the total cross-section used to be called the single-scattering albedo of a solitary particle:

$$\omega(\boldsymbol{\Omega}') = \frac{\sigma_S(\boldsymbol{\Omega}')}{\sigma_S(\boldsymbol{\Omega}') + \sigma_A(\boldsymbol{\Omega}')} \tag{7.10}$$

For natural media the albedo value varies within very wide limits. So, for optically transparent media in the terrestrial atmosphere (drops of water) the albedo value is close to unity (0.95–0.99). In the microwave band the albedo of water particles lies within the limits of 0.01–0.8 (Oguchi, 1983), whereas for particles close in their electromagnetic properties to a blackbody (such as hollow water spheres), the albedo is virtually zero (Raizer and Sharkov, 1981 and Chapter 6).

Note that the earlier introduced (see Chapter 6) albedo of a unit of the medium's volume can essentially differ from the albedo of a solitary particle, since the first of these definitions depends on the polydisperse composition of a medium or, in other words, on the relationship between the working wavelength and the range in particle size (see Section 7.4).

### 7.1.5  The scattering indicatrix

It is obvious from physical considerations that any particle of complicated shape will scatter the incident radiation in space in an inhomogeneous manner. To describe the character of spatial-angular scattering on a particle a special dimensionless function $p(\boldsymbol{\Omega}', \boldsymbol{\Omega})$ is introduced, which is called the scattering indicatrix and takes the following form:

$$p(\boldsymbol{\Omega}', \boldsymbol{\Omega}) = 4\pi \frac{\sigma_d(\boldsymbol{\Omega}', \boldsymbol{\Omega})}{\sigma_E(\boldsymbol{\Omega}')} \tag{7.11}$$

The dimensionless quantity $p(\boldsymbol{\Omega}', \boldsymbol{\Omega})$ is sometimes called the phase function and is widely used in the radiative transfer theory (in the optical band, especially). Note that this name has purely historical roots. Physically, the phase function describes scattered power and has no relation to the phase of an incident wave—see equation (7.2). The name "phase function" arose as a result of its astronomical background and is related to the phases of the Moon (Ishimaru, 1978).

Using relations (7.6), (7.10), and (7.11), we obtain the equations that associate all the electromagnetic parameters of the particle introduced above:

$$\sigma_S(\mathbf{\Omega}') = \iint_{4\pi} \sigma_d(\mathbf{\Omega}', \mathbf{\Omega})\, d\Omega = \iint_{4\pi} |\dot{f}(\mathbf{\Omega}', \mathbf{\Omega})|^2\, d\Omega = \frac{\sigma_E(\mathbf{\Omega}')}{4\pi} \iint_{4\pi} p(\mathbf{\Omega}', \mathbf{\Omega})\, d\Omega \qquad (7.12)$$

$$\omega(\mathbf{\Omega}') = \frac{\sigma_S}{\sigma_E} = \frac{1}{\sigma_E} \iint_{4\pi} |\dot{f}(\mathbf{\Omega}', \mathbf{\Omega})|^2\, d\Omega = \frac{1}{4\pi} \iint_{4\pi} p(\mathbf{\Omega}', \mathbf{\Omega})\, d\Omega \qquad (7.13)$$

These relations clear up the physical sense of the parameter introduced—the scattering indicatrix. Suppose that a particle will scatter uniformly within the total solid angle $4\pi$ surrounding it—that is, $p(\mathbf{\Omega}', \mathbf{\Omega}) = 1$. Then the particle's albedo will be equal to unity, and the total cross-section of the particle will only be determined by its scattering cross-section. In such a case the particle is called purely scattering.

Note that this approach to forming the scattering indicatrix of a particle is not unique. There are other approaches to definition of a scattering indicatrix (Skolnik, 1980), in which case relations (7.12) and (7.13) will have other numerical coefficients.

## 7.2.  BASIC CONCEPTS OF THE MIE THEORY

The important problem in electromagnetic radiation scattering by material particles consists in finding the relationship between the properties (i.e., the size, shape, dielectric characteristics) of particles with the angular distribution of scattered radiation and with the external radiation absorption by particles. Such a problem arises in many fields of science and technology (such as astrophysics, biochemistry, radio physics, optical oceanography, microwave remote sensing). This is the reason numerous theoretical and experimental investigations have been carried out in the study of electromagnetic wave scattering. Historically, such investigations were first carried out in the optical band and then were spread to the IR and radio wavelength bands.

One of the first researchers, J. Rayleigh, proceeding from purely dimensional considerations, obtained the famous asymptotic approximate solution for radiation scattering by spherical particles, whose size is small compared with the wavelength of incident radiation falling on a particle. This work was followed by the general theory of radiation absorption and scattering by homogeneous particles having a simple geometrical shape, such as a sphere or a circular cylinder. This theory was formulated by G. Mie in 1908. In the Mie theory, based on solution of fundamental Maxwell equations, an idealized situation was considered—namely, a simple spherical particle made of a homogeneous, isotropic material and placed in a homogeneous, isotropic, dielectric, boundless medium and irradiated by planar waves propagating in a particular direction. A purely dielectric spherical particle does not absorb radiation, whereas an electrically conducting spherical particle partially absorbs, partially scatters, and partially transmits incident radiation. Derivation of Mie's solution, as well as the mathematical and physical aspects of his theory, including the features of numerical calculation algorithms, are contained in a number of books (Stratton, 1941; Hulst, 1981; Born and Wolf, 1999; Deirmendjian, 1969; Ozisik, 1973; Ishimaru,

1978; Bohren and Hoffman, 1983; Ivazyn, 1991). Solutions to the amplitude of a scattered wave on a sphere take the form of a complicated series containing Riccati–Bessel functions and Riccati–Hankel functions of higher orders. The results of Mie's solution are most useful for determining absorption and scattering coefficients, as well as the scattering indicatrix for spherical particles suspended in a dielectric medium, provided that the particles are spaced at a great distance from each other. Some special experiments were carried out for determining the minimum distance between spherical particles to ensure independent scattering. It was found that—for some optical scatterers—mutual interference can be neglected, if the distance between the centers of spherical particles is greater than three diameters. In the majority of applied problems (studies of cloud systems, snowfall, aerosols) the particles are separated by greater distances from each other. Note, however, that in the Mie theory an idealized case is considered—namely, a secluded spherical particle that acts like an independent point-like scatterer in a boundless medium, whereas the scatterers met in the majority of practical applications have an arbitrary geometrical shape. At present, great efforts are ongoing in the study of electromagnetic radiation scattering by particles of arbitrary shape and orientation and complicated structure (such as multilayer particles, spheroids). Nevertheless, we shall consider below the results of Mie's theory, since this is a unique fundamental theory and its results are useful in many idealized cases.

### 7.2.1   Parameters of the Mie theory

A series of dimensionless parameters is introduced in the theory that are widely used in practice.

The ratio of cross-section values introduced above to the geometrical cross-section is called the efficiency factor and designated by $Q_i$, where $i$ is equal to $A$, $S$, or $E$ (which means absorption, scattering or extinction, respectively). Thus, one can write

$$Q_i = \frac{\sigma_i}{\pi a^2} \qquad (7.14)$$

where $a$ is the radius of a sphere. As follows from (7.9), the efficiency factors satisfy the relation

$$Q_E = Q_S + Q_A \qquad (7.15)$$

By size parameter is meant the ratio of the length of circumference of a studied sphere to its working wavelength $x(0 < x < \infty)$

$$x = \frac{2\pi a}{\lambda} = \frac{\pi D}{\lambda} \qquad (7.16)$$

where $D$ is the diameter of a sphere.

The complex parameter $m$ of refraction of a sphere's substance relative to the dielectric properties of an ambient boundless space is

$$\dot{m} = \frac{\dot{n}_{SP}}{\dot{n}_S} = n + j\chi \qquad (7.17)$$

where $n_{SP}$ is the index of refraction of a sphere's substance; $n_S$ is the similar characteristic of the ambient space. If the ambient space is not a vacuum, but a medium with a high value of $n_S$, then parameter $|m|$ can be less than unity. For example, such a situation takes place when studying the propagation and scattering of electromagnetic waves of the optical band in a marine medium in the presence of air bubbles.

Since a sphere is a symmetrical particle, scattering does not depend on an azimuthal angle, but is a function of scattering angle $\theta_0$ between the directions of incident and scattered beams. Thus, we introduce another parameter—scattering angle. Here it is necessary to keep in mind that—if the incident flux possesses a strictly linear polarization—then the (secondary) radiation scattered by a sphere acquires the character of elliptically polarized radiation (Stratton, 1941), and its description requires the introduction of the azimuthal angle. If, however, the primary field is non-polarized (the case of natural thermal radiation), then the secondary radiation is weakly polarized. This makes it possible to present the scattering indicatrix in the form of a series of Legendre polynomials

$$p(\cos \theta_0) = 1 + \sum_{j=1}^{\infty} A_j P_j(\cos \theta_0) \tag{7.18}$$

where $\theta_0$ is scattering angle; $P_j(\cos \theta_0)$ Legendre polynomials; and $A_j$ expansion coefficients that are functions only of parameter $x$ and the parameter of refraction.

### 7.2.2   The main results of the Mie theory

To get an idea about results of the Mie theory, we shall write the expressions for the efficiency factors of extinction and of scattering, which can be presented in the form of an infinite series:

$$Q_E = \frac{2}{x^2} \sum_{n=1}^{\infty} (2n + 1)\{\mathrm{Re}(\dot{a}_n + \dot{b}_n)\} \tag{7.19}$$

$$Q_S = \frac{2}{x^2} \sum_{n=1}^{\infty} (2n + 1)\{|\dot{a}_n|^2 + |\dot{b}_n|^2\} \tag{7.20}$$

where Re is the real part of a sum. If the particle does not absorb the incident radiation (i.e., the index of refraction is a real number and the particle is a pure scatterer), then expressions (7.19) and (7.20) lead to identical results. If the particle absorbs the incident radiation, then the index of refraction is complex, and the efficiency factor of absorption $Q_A$ is obtained from definition of $Q_E$ (7.15) in the form of

$$Q_A = Q_E - Q_S \tag{7.21}$$

The efficiency factor for the backscattering cross-section $Q_{BS}$ can be presented as follows (see Section 7.5 for details):

$$Q_{BS} = \frac{\sigma_{BS}}{\pi a^2} = \frac{1}{x^2} \left| \sum_{n=1}^{\infty} (2n + 1)(-1)^n (\dot{a}_n - \dot{b}_n) \right| \tag{7.22}$$

In radar technology this parameter used to be called the effective scattering area (ESA) of a target (Skolnik, 1980). In this case the diagram showing the dependence of ESA on the angle of wave incidence on a scatterer is called the ESA diagram (this is the scattering indicatrix in its basic form).

The complex $\dot{a}_n$ and $\dot{b}_n$ coefficients in formulas (7.19), (7.20), and (7.22) are called the Mie coefficients. They represent complicated functions, expressed in terms of Riccati–Bessel functions, and are written in the form:

$$\dot{a}_n = \frac{\Psi_n(x)[\Psi'_n(y)/\Psi_n(y)] - \dot{m}\Psi'_n(x)}{\xi_n(x)[\Psi'_n(y)/\Psi_n(y)] - \dot{m}\xi'_n(x)} \tag{7.23}$$

$$\dot{b}_n = \frac{\dot{m}\Psi_n(x)[\Psi'_n(y)/\Psi_n(y)] - \Psi'_n(x)}{\dot{m}\xi_n(x)[\Psi'_n(y)/\Psi_n(y)] - \xi'_n(x)} \tag{7.24}$$

where the primes denote differentiation with respect to the argument under consideration. The Riccati–Bessel functions $\Psi_n(z)$ and $\xi_n(z)$ are associated with the Bessel function of non-integer order by the relations:

$$\Psi_n(z) = \left(\frac{\pi z}{2}\right)^{1/2} J_{n+1/2}(z) \tag{7.25}$$

$$\xi_n(z) = \left(\frac{\pi z}{2}\right)^{1/2} J_{n+1/2}(z) + (-1)^n j J_{-n-1/2}(z) \tag{7.26}$$

where $z = x$ or $y$, and the complex argument $y$ is determined as follows $y = \dot{m}x$.

The physical sense of the Mie coefficients is as follows. A primary (external) electromagnetic wave excites some particular forced oscillations inside the substance of a sphere and on its surface. These forced oscillations can be sub-divided into the electric and magnetic modes of oscillations on the basis of existence of a corresponding radial component in a scattered (forced) field. So, if the electric vector of a scattered field has a radial component that is caused by electric charges distributed over the surface, then such a mode of oscillations is called an oscillation of electrical type. The amplitudes of oscillations of such a type are expressed in terms of $b_n$ coefficients. If the scattered field is excited by means of $a_n$ coefficients only, then the structure of the field will be such that would be produced by variable magnetic charges disposed on the surface of a sphere. Such a field is called a field of magnetic type. Thus, it can be considered (Stratton, 1941) that $a_n$ coefficients represent the amplitudes of oscillations of magnetic type, and $b_n$ coefficients those of electrical type. If the frequency of an impressed (external) field approaches any characteristic frequency of the natural electromagnetic oscillations of a system, then a resonance phenomenon arises. This is just the condition in which the denominators in expressions (7.23) and (7.24) tend to zero. But, since absorption is always present in a system (inside the sphere), the denominators of the Mie coefficients can be reduced to their minimum values, but they cannot be made equal to zero. Thus, the mathematical catastrophe—the advent of infinite amplitudes—does not occur.

Though the Mie solution is strictly applicable to the whole range of $m$–$x$ values, it was quickly found that the numerical calculations of a scattering indicatrix and the efficiency factors for arbitrary $m$ and $x$ values are laborious. For example, the convergence of series determining the Mie coefficients becomes very slow when the relative size of a sphere increases compared with the incident radiation wavelength. Another difficulty consists in the irregularity of the values of the $a_n$ and $b_n$ coefficients. On the one hand, this makes the interpolation procedures unreliable (Deirmendjian, 1969), and, on the other hand, in performing detailed numerical calculations a lot of resonance modes arises, some of which can be "false" (Conwell *et al.*, 1984). Fortunately, for many practically important tasks (including remote sensing) there is no need to perform calculations by the Mie theory throughout the range of $m$–$x$ values. We can restrict calculations to the limiting values of the Mie solution, which in turn can be determined by simplified techniques. So, for example, for high values of parameter $x$ (i.e., for a large spherical particle as compared with wavelength) the convergence of the accurate Mie solution becomes very poor. However, in such cases geometric optics laws are applicable for determining the scattering indicatrix and efficiency factors, and the final expressions become rather simple. For very small $x$ values the accurate Mie formula is essentially simplified if we apply the power series expansions of spherical Bessel functions with respect to Mie's $a_n$ and $b_n$ coefficients. However, the procedure involved in the expansion of efficiency factors in power series with respect to small $x$ values and the physical interpretation of expansion terms turns out to be complicated.

### 7.2.3   The three regions of Mie scattering

Detailed investigations of the mathematical features of the expressions for efficiency factors, undertaken for a wide frequency band of electromagnetic waves and for the dielectric properties of substances encountered in natural media, have shown that three regions can be found where the scattering on particles has some peculiarity.

The first region—the Rayleigh scattering region—is characterized by the following conditions: first, particle size is small compared with the wavelength of an external field—that is, $a \ll \lambda$ ($x \ll 1$); and, second, $|\dot{m}|x \ll 1$. The first condition implies that we are in a quasi-static approximation and can make use of the laws of electrostatics. The second condition requires the absence of electromagnetic resonances inside a particle. As usual, the traditional value of $a = 0.05\lambda$ ($x < 0.3$) is accepted as the upper limit of a particle's radius for this approximation. But, in such a case the second condition should certainly be satisfied as well.

The second region—the resonance (Mie's) scattering region—is characterized by the presence of a great number of resonance features and very complicated scattering indicatrices. These are the reasons this region turned out to be most complicated for investigations. The values of $x$ are usually found within the limits from 0.25–0.5 to 50.

The third region—the high-frequency region, or the geometry optics region—is characterized by the presence of a geometrical shadow behind the particle. This results in a situation in which the extinction cross-section will tend toward the doubled geometrical cross-section of a particle (of arbitrary shape, it should be added). This phenomenon used to be called the extinction paradox and has several (different) physical explanations.

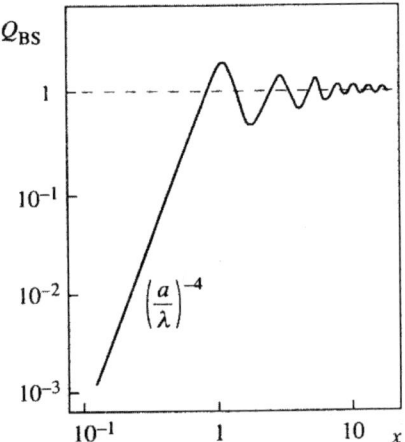

**Figure 7.1.** The backscattering efficiency factor of a metal sphere as a function of the size parameter (at microwave bands).

As an indicative example, we shall consider the dependence of the efficiency factor of backscattering for a metal sphere (Figure 7.1). Such an object is often used as an experimental standard for calibrating microwave antenna systems and complicated receiving radio-engineering complexes (radar early detection stations, for instance).

From analysis of the plot (Figure 7.1), presented in the bi-logarithmic scale, it can easily be seen that the whole region of size parameter values can really be sub-divided into three characteristic sub-regions: the Rayleigh scattering region, where $Q_{BS}$ decreases as $1/\lambda^4$; the resonance Mie region, where resonance dependencies are explicitly exhibited; and the geometrical optics region, where the $Q_{BS}$ value is equal to the geometrical cross-section value of a large particle. It is interesting to note that—to get a dimension parameter value equal to unity—the first and strongest Mie resonance, at which the backscattering cross-section exceeds nearly three-fold the size of the geometrical shadow, needs to take place. Physically, this is due to the fact that a sphere intensively scatters "backwards" like a resonance half-wave vibrator (i.e., $\pi a = \lambda/2$).

## 7.3  SCATTERING PROPERTIES OF AQUEOUS PARTICLES

An important class of scatterers in the ocean–atmosphere system is aqueous drops, which are present in quite various physical media, such as spray sheets over a stormy sea surface (see Plate 2 in the Prelims and Chapters 5 and 6), sub-surface fogs of various classes, and precipitations and cloudy systems of various types.

As noted earlier, the dielectric characteristics of water possess prominent frequency features in the microwave band (Sharkov, 2003). For these reasons the general picture of scattering of aqueous particles will essentially change depending on

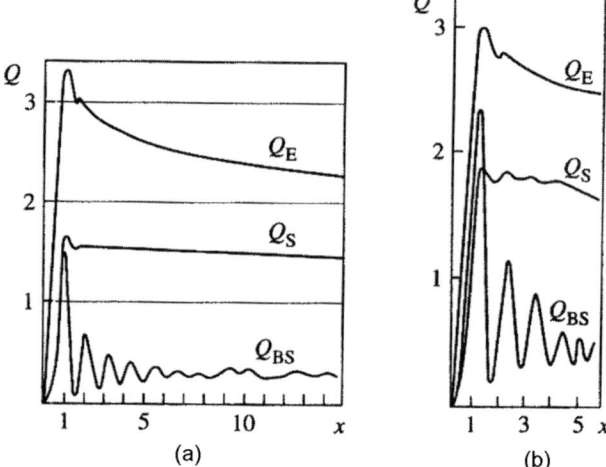

**Figure 7.2.** The efficiency factors of scattering ($Q_S$), backscattering ($Q_{BS}$), and extinction ($Q_E$) as a function of the size parameter for aqueous spheres: (a) $\lambda = 0.2\,\text{cm}$ and $t = 20°\text{C}$; (b) $\lambda = 0.8\,\text{cm}$ and $t = 20°\text{C}$.

the relationship between the physical size of particles and the working wavelength band, and in each particular case of experimental investigation the detailed calculation of scattering parameters is required. It is necessary to create the available calculation tables for the microwave band with subsequent interpolation procedures (Krasiuk and Rosenberg, 1970; Skolnik, 1980; Oguchi, 1983; Lhermitte, 1988; Ivazyn, 1991). Nevertheless, we shall demonstrate in a series of examples some general properties of electromagnetic wave scattering by aqueous spheres (see Figures 6.21 and 7.2).

Figure 7.2 presents calculated values of the efficiency factors of extinction, scattering, and backscattering for aqueous spheres at wavelengths of 0.8 and 0.2 cm depending on the size parameter in the range of its values up to 15. Considering the plots of the region of values for $x < 1$, it can easily be seen that the behavior of efficiency factors corresponds to features of the Rayleigh region (see Section 7.2.3). A prominent maximum is observed in the Mie region for all efficiency factors at $x = 1$. However, as the size parameter increases, extinction and scattering decrease very slowly to values equal to two and unity, respectively, but they do not exhibit any prominent resonance properties in this case. Unlike extinction and total scattering, backscattering possesses sharp and strong resonance properties up to values of $x = 10$. In the geometrical approximation the efficiency factor of extinction becomes equal to two—that is, it twice exceeds the geometrical diameter of a sphere ("the extinction paradox"). In this case the rate of tending toward their limiting values is essentially different for extinction and for scattering; as a result, for large drops the losses for scattering exceed the losses for absorption. This is well illustrated by the calculated plot (in the bi-logarithmic scale) of a single-scattering albedo that depends on the size parameter of spheres (Figure 7.3). Data were calculated in a wide range of

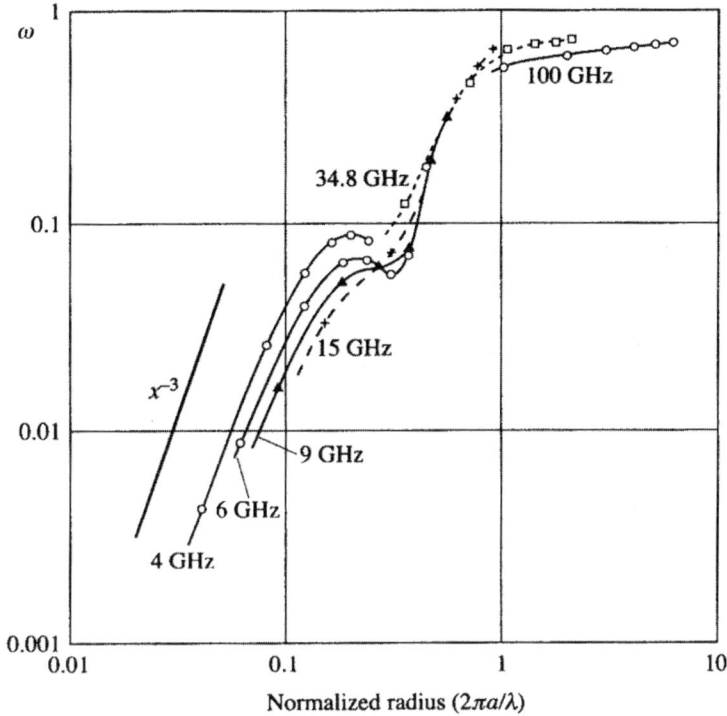

**Figure 7.3.** Single-scattering albedo of aqueous spheres as a function of size parameter (normalized radius) at 4, 6.9, 15, 34.8, and 100 GHz. The points on each curve correspond—from the left—to drop radii 0.5, 1.0, 1.5, 2.0, 2.5, and 3.0 mm, respectively (Oguchi, 1983).

frequencies—from 4 GHz (wavelength of 7.5 cm) up to 100 GHz (wavelength of 3 mm)—and of drop radii (0.5–3.0 mm). Virtually irrespective of the wavelength band, for $x < 0.5$ the albedo is lower than 0.1, and the scattering contribution to total losses is very small. Note that in this case a decrease in albedo for small drops has a prominent character of exponential dependence as $x^3$, as should be expected for the Rayleigh region (see Section 7.2.3). For $x > 1$ the contribution of scattering to the total losses of large drops sharply increases, reaching 60–70% of total losses (in other words, of the extinction).

Figure 7.4 presents the frequency dependencies (in the bi-logarithmic scale) of the efficiency factor of extinction for aqueous spheres with fixed radii in a wide frequency band—from 5 GHz (6 cm) to 300 GHz (1 mm). Analysis of this figure shows all the characteristic regions of scattering—for large wavelengths an exponential drop as $\lambda^{-2}$ is observed for a sphere of fixed radius. This dependence characterizes the beginning of the Rayleigh scattering region. For short wavelengths the efficiency factor of extinction tends toward the value of 2. At intermediate wavelengths a resonance Mie maximum is observed (it is smeared in this coordinate system). It is interesting to note that the plot clearly demonstrates only the transition region from the first Mie

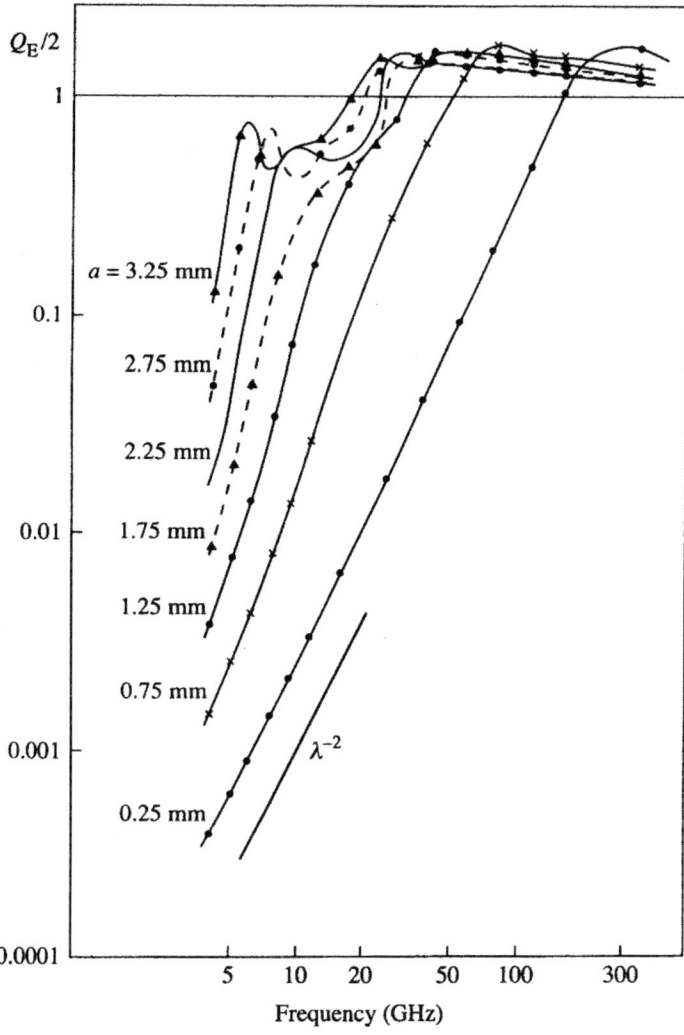

**Figure 7.4.** The extinction factor of spherical raindrops as a function of frequency ($t = 20°C$) (Oguchi, 1983).

maximum to the Rayleigh region and the very beginning of the Rayleigh region. So, as the fixed radius decreases, the exponent of the power-law drop also decreases and finally reaches a value equal to two, but only in a purely Rayleigh region.

As already noted (see Chapter 6), numerical calculations of the electrodynamic characteristics of scattering, even for water spheres in the microwave range, are rather rare in the scientific literature. This is the reason we shall present below (Figure 7.5) the numerical calculations, according to the full Mie scattering scheme, of spectral dependencies within a wide range of wavelengths (from 0.2 to 10 cm) of the efficiency factors of absorption and scattering for spherical drops of various radius: $a = 0.5$, 1,

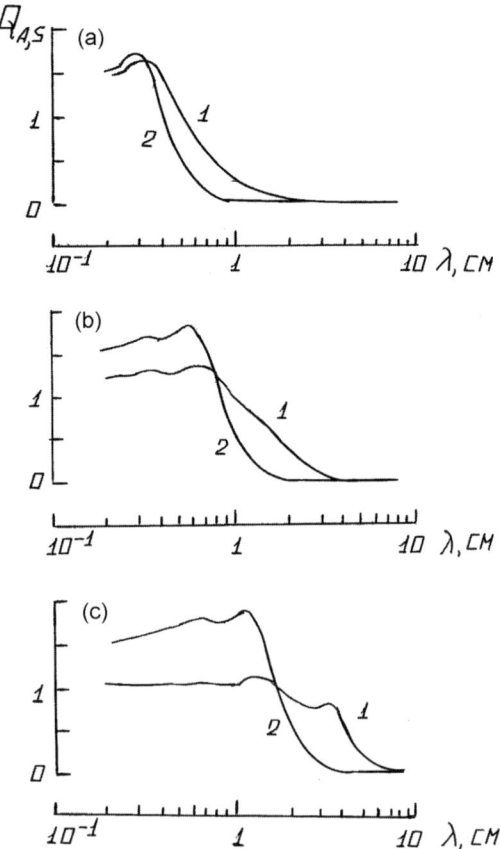

**Figure 7.5.** The spectral dependences of the efficiency factors of absorption (1) and scattering (2) for spherical salt drops of different radii: $r = 0.05$ cm (a), 0.1 (b), 0.2 cm (c) ($t = 20°$C and $S = 35‰$).

and 2 mm (Dombrovskiy and Raizer, 1992). Since there exists a relaxation maximum of water absorption in the centimeter range (Sharkov, 2003), and, hence, high variations in the spectral properties of the dielectric parameters of water are observed in this range, the spectral character of the absorption and scattering properties for drops of various sizes can take on a peculiar form, as just demonstrated by the plots in Figure 7.5. In the long-wave limit, the absorption factors for drops of any size essentially exceed the values of scattering factors, which, as a matter of fact, follows from the Rayleigh approximation properties. With decreasing wavelength and, accordingly, with approaching the first Mie maximum, the general picture of the relationship between the absorption and scattering properties for water drops drastically changes, as should be expected. The value of scattering sharply grows and exceeds the value of absorption, and in this case—for drops of all dimensions— the spectral dependence of the scattering factor passes through an extremum at the

Mie parameter value of $x = 1$, after which weakly expressed variations are observed for the next Mie maxima. The general tendency of curves is decreasing the scattering of value in the high-frequency range of Mie scattering and limiting tending of the scattering factor to unity. The first maximum of the absorption factor is slightly shifted—with respect to the scattering maximum—into the long-wave region (Figure 7.5a) for small-sized drops ($a = 0.5$ mm); in this case the single albedo of a particle equals the value of 0.5. For large-sized drops ($a = 1$ and 2 mm) the absorption maximum value considerably decreases with respect to the scattering value, and in this case an additional maximum arises in the long-wave region (Figure 7.5b–c). The appearance of this maximum is most likely related to the sharp increase in absorption in the drop substance in this particular wavelength range alone. The centimeter wavelength range corresponds to the long-wave branch of the relaxation maximum of water absorption (Sharkov, 2003). In this case the single albedo of a particle equals the value of 0.67 for the first Mie maximum and then slowly drops down to the value of 0.5 with decreasing wavelength. Note that—in all three cases considered—in the high-frequency limit ($x \gg 1$) the extinction paradox is satisfied (i.e., the extinction factor tends to the value of 2—Ishimaru, 1978).

## 7.4  ELECTROMAGNETIC PROPERTIES OF POLYDISPERSE MEDIA

It is common knowledge (Stratton, 1941; Born and Wolf, 1999) that the structure of substance in the Maxwell theory is specified by introducing the phenomenological dielectric and magnetic parameters for continua. In the radiative transfer theory (the macroscopic version) the structure of substance is presented in a different manner—that is, in the form of a cloud of randomly distributed particles in a continuum (e.g., in the atmosphere or in the sea) with the parameters of attenuation and scattering in the medium calculated per unit of a beam path in the medium. There are many different physical structures in the atmosphere and the sea that can be attributed to such kinds of media. Virtually all have the character of polydisperse media (i.e., media with particles of different size). Physically, this is related to the fact that—because all polydisperse media are open physical systems—a certain dynamical equilibrium can only be established in the dynamical process of a particle's birth and death. This stipulates the principal presence of polydisperse-composed particles in media. In this section we shall consider the basic mechanical characteristics that are used to describe disperse mixtures, and the procedure of transition from the characteristics of scattering of individual particles to the electromagnetic parameters of a unit volume.

Physical systems consisting of some combination of substances (which are in themselves in different phase states) are usually sub-divided into two large classes: heterogeneous mixtures and homogeneous mixtures (Nigmatulin, 1978). By heterogeneous mixtures we mean systems that contain the macroscopic (with respect to molecular scales) and chemically non-interacting inhomogeneities (or admixtures). From the huge amount of heterogeneous mixtures that exist in nature, some comparatively regular structures called *disperse mixtures* can be distinguished. Such systems, which usually consist of two phases, include, for example, the aqueous drop clouds in air or air bubbles in sea water (so-called aerated layer) and the hexagonal structures found in sea foam. In this case the particles are called a disperse phase, and their carrier medium the dispersed phase. By homogeneous mixtures we mean those systems in which substances are intermixed at the molecular level. So-called *colloidal mixtures* occupy an intermediate position.

We shall now consider some mechanical characteristics determining the dispersity of disperse systems.

### 7.4.1 The density function

The most important characteristic of the microstructure (dispersity) of disperse systems is the differential density function of particles in size, designated by $n(r, x, y, z, t)$, where $r$ is the radius of particles (instead of radius, sometimes the diameter, surface, volume, or mass of particles are used); $x, y, z$ are spatial coordinates; and $t$ is time. Proceeding from the definition, quantity $n(r, x, y, z, t)\, dr\, dx\, dy\, dz$ is the number of particles having a radius from $r$ to $r + dr$ in the volume of $dx\, dy\, dz$ in the vicinity of point $(x, y, z)$ at time instant $t$. Naturally, in practice it is impossible to determine the value of $n(r, x, y, z, t)$ at all points of the space studied simultaneously. This is the reason the microstructure of a disperse system is often characterized by the *size spectra* of particles $n(r)$, averaged over time and space, or by related integral distribution parameters, which are proportional to the distribution moments of any order. For example, the important characteristics of a disperse medium's microstructure are the density of particles, the total mass of water (or water content), and radar reflectivity.

Theoretically, the density function plays a fundamental part, since it determines the physical features of a system and its possible evolution.

It follows from physical considerations that for $r \to 0$ and for $r \to \infty$ the density of a number of particles must tend to zero (within the framework of the given physical system). The dimension of this parameter, as can easily be seen from its definition, equals $\mathrm{cm}^{-4}$.

### 7.4.2 The volume density of particles

An important integral parameter is the volume density of particles (or the number density) $N$ ($\mathrm{cm}^{-3}$), defined by the following integral transformation from the spectrum of particles:

$$N(\mathrm{cm}^{-3}) = \int_0^\infty n(r)\, dr = \int_0^\infty n(D)\, dD \qquad (7.27)$$

where $D$ is the diameter of particles. This characteristic determines the absolute number of particles in a unit volume. It follows from this relation that $n(r) = 2n(D)$.

### 7.4.3    The integral distribution function

In experimental practice it is often convenient to present observational results in the form of the volume density of particles with a lower variable limit $N(r)$ (cm$^{-3}$)—that is, in the form:

$$N(r) = \int_r^\infty n(r)\,dr \qquad (7.28)$$

This characteristic determines the absolute number of particles in a unit volume beginning with some particular (fixed) value of size. A lot of measuring devices that record the size of particles operate with this particular characteristic, and in order to transfer to the density function it is necessary to perform numerical (or graphical) differentiation of the results obtained.

### 7.4.4    The relative density function

In theoretical analysis, as well as in processing and comparing experimental results of various types it is expedient to use the relative density distribution function in the form:

$$f(r) = \frac{n(r)}{N} \qquad \int_0^\infty f(r)\,dr = 1 \qquad (7.29)$$

As seen from the definition, the dimension of this parameter is cm$^{-1}$.

### 7.4.5    The density sampling probability

Experimentally, the dispersity characteristics of a system are usually found by detecting and estimating the density sampling probability, or, in other words, by forming and constructing experimental histograms. This procedure is rather complicated, in general, and requires a researcher to have both experience and skills in solving such tasks and a clear understanding of the basic physical problem. We shall briefly describe this procedure, albeit predominantly qualitatively.

Let the experimental data on particles be obtained in the radii range from $a$ to $b$. The total number of recorded particles is $N$. We divide the range of radii by $j$—the number of sampling intervals. In each of these intervals $N_j$ particles will be recorded. The size of a sampling interval equals $\Delta r_j$. Then by

$$P_j = \frac{N_j}{N} \qquad j = 1, \ldots, k \qquad (7.30)$$

will be meant a sampling probability of the presence of particles in the given sampling interval. Here $k$ is the total number of sampling intervals such that

$$N = \sum_{j=1}^k N_j \qquad (7.31)$$

The sampling probability $P_j(r_j)$ presented in graphical form just represents the experimental histogram.

By the density function of sampling probability we mean the following quantity:

$$f_j(r_j) = \frac{N_j}{N\left(\dfrac{b-a}{k}\right)} \tag{7.32}$$

From normalization conditions it follows that

$$\sum_{j=1}^{k} f_j(r_j)\Delta r_j = 1 \tag{7.33}$$

If the sizes of sampling intervals are the same ($\Delta r$), then the relationship between the density function of sampling probability and the sampling probability (the experimental histogram data) can be presented as follows:

$$f_j(r_j) = \frac{P_j}{\Delta r}\,(\mathrm{cm}^{-1}) \tag{7.34}$$

It can easily be concluded from the obtained relation that $f(r_j)$ is the finite-difference analog of the relative distribution function (10.46).

### 7.4.6   The total mass and the relative volume concentration of water

In some meteorological problems, as well as in microwave sensing of the atmosphere, it is necessary to know the total mass of a substance (e.g., water) in a unit of volume of a disperse medium (e.g., a cloud). If the disperse medium consists of regular spheres of various diameters, then, by definition, the total mass of a substance in a unit volume (known conventionally as water weight content) $W$ (g cm$^{-3}$) can be obtained from the following relation:

$$W = \tfrac{4}{3}\pi\rho \int_0^\infty r^3 n(r)\,dr = \tfrac{4}{3}\pi\rho N \int_0^\infty f(r) r^3\,dr \tag{7.35}$$

where $\rho$ is the density of substance of spheres.

There are many application problems where it is necessary to know the relative volume concentration $C$ (a dimensionless quantity) of a substance (or volume concentration); this can be obtained from the following relation:

$$C = \frac{W}{\rho} = \tfrac{4}{3}\pi \int_0^\infty r^3 n(r)\,dr \tag{7.36}$$

The characteristics considered are proportional to the third moment of the size distribution of particles. However, some remote investigations, such as radar studies of the structure of cloud systems, require knowledge of moments of much higher order.

### 7.4.7  Radar reflectivity

By $Z$ (cm$^3$) we mean the following quantity:

$$Z = \int_0^\infty n(r)r^6\, dr \tag{7.37}$$

The physical sense of $Z$ can easily be understood using the expression for back-scattering of an individual particle in the Rayleigh approximation in calculating the backscattering of a unit volume $\sigma_0$ with the density function of reflective spheres $n(r)$:

$$\sigma_0 = \int_0^\infty \sigma_{BS} n(r)\, dr = \frac{64}{\lambda^4}\pi^5 \left|\frac{\dot{\varepsilon}-1}{\dot{\varepsilon}+2}\right|^2 \int_0^\infty n(r)r^6\, dr \tag{7.38}$$

This relation indicates that a signal, scattered back from a cloud structure, is proportional to the sixth moment of the density function of drops in a cloud mass. It can easily be seen from this result that drops of large and super-large size (having $r > 100$ microns) play very essential parts in the process of backscattering electro-magnetic radiation from cloud systems. Moreover, these drops are the main carriers of radar remote information about drop cloud systems and weather hazards (Doviak and Zrnic', 1984; Doviak and Lee, 1985). However, in radiothermal investigations the situation is principally different: thermal radiation depends on the total mass of water in a drop cloud and, thus, the signal is proportional to the third moment of the density function of drops in a cloud mass. All these features are important when interpreting observational data.

### 7.4.8  Rainfall rates

If disperse systems possess prominent dynamical properties (e.g., drop-spray clouds, precipitation of various phase types), then—for their remote analysis—a parameter characterizing the quantity of substance precipitated on a unit area per unit of time is important. Such a characteristic, which is widely used in meteorological and remote investigations, is called the rainfall rate $R$ (cm s$^{-1}$). It is determined by the following expression:

$$R = \int_0^\infty n(r)v(r)\tfrac{4}{3}\pi r^3\, dr \tag{7.39}$$

where $v(r)$ is the velocity of the motion of drops that have a corresponding radius. To get data on rainfall drops in the atmosphere a series of empirical relations was established between the precipitation rate and the radius of a drop. They include, in particular, the linear relation $v(\text{m s}^{-1}) = 75r(\text{cm})$ for rainfall drops of sufficiently small size, and for larger drops the precipitation rate becomes constant and does not depend on size (Kollias *et al.*, 1999). This indicates that rainfall rates will be proportional to the fourth moment (or to the third one—depending on the diameter of drops) of the density function of drops in a cloud mass of precipitating drops. In meteorological practice the dimension of $R$ is usually reduced to the value of mm hr$^{-1}$.

### 7.4.9  Analytical forms of the density function

Over the last 50 years a large amount of experimental work has been carried out, devoted to searching for the most acceptable analytical form of a density function for disperse systems of various physical natures. Theoretically, efforts were directed to solution of the complicated problems of the kinetics of physicochemical media using, for example, solutions provided by a system of Fokker–Planck–Smolukhovsky equations. The following analytical expression for the density function, known as the gamma-distribution, is considered as the most theoretically substantiated:

$$n(r) = ar^{\mu} \exp(-br^{\gamma}) \qquad (7.40)$$

where $a$, $b$, $\mu$, $\gamma$ are the parameters determining all the characteristic features of the distribution. Almost all empirical distributions, formed earlier from the experimental data, can be obtained from the given distribution.

Using expressions (7.36) and (7.40), we obtain the following formulas for relative density:

$$C = \frac{4\pi}{3} \int n(r) r^3 \, dr = \frac{4}{3} \pi a \gamma^{-1} b^{-\frac{\mu+4}{\gamma}} \Gamma\left(\frac{\mu+4}{\gamma}\right) \qquad (7.41)$$

Letting $\gamma = 1$, we obtain the expression for volume density:

$$N = \int_0^{\infty} n(r) \, dr = ab^{1-\mu} \Gamma(\mu+1) \qquad (7.42)$$

Here, as in (7.41), $\Gamma x$ denotes the gamma function (Gradshteyn and Ryzhik, 2000).

Using (7.29) and (7.40), we can obtain the expression for the relative density function:

$$f(r) = b^{\mu-1} \frac{r^{\mu}}{\Gamma(\mu+1)} \exp(-br) \qquad (7.43)$$

This distribution is now characterized by two parameters only: $-b$ and $\mu$. This expression is often written in a slightly different (but equivalent) form:

$$f(r) = \frac{1}{\Gamma(\mu+1)} \mu^{\mu+1} \frac{r^{\mu}}{r_m^{\mu+1}} \exp\left\{ -\mu \frac{r}{r_m} \right\} \qquad (7.44)$$

where parameter $\mu$ characterizes the distribution halfwidth; and parameter $r_m$ determines the so-called modal (most probable) distribution radius. Serious efforts have recently been undertaken to determine these parameters for natural disperse media by means of laboratory experiments. So, for fogs and clouds final values of parameter $\mu$ are given within the limits of 1–10, and those of the modal radius within the limits of 0.1–10 microns.

It is interesting to note that as long ago as 1948 J. Marshall and W. Palmer suggested a simple empirical relation for the density function of rainfall in the form:

$$n(r) = N_0 e^{-\Lambda r} \qquad (7.45)$$

where

$$N_0 = 1.6 \cdot 10^4 \ (\text{m}^{-3} \, \text{mm}^{-1}) \qquad \Lambda = 8.2 R^{-0.21} \ (\text{mm}^{-1}) \qquad (7.46)$$

Here radius is expressed in mm and $R$ in $\text{mm} \, \text{hr}^{-1}$.

Such a distribution was found to successfully describe averaged experimental data for various types of rain: drizzle, widespread, convective, and thunderstorm (albeit, with significant modification of the numerical values of $N_0$ and $\Lambda$). The Marshall–Palmer drop-size distribution, as well as distributions close to it (such as the Laws–Parsons relation), are widely used now as well (Oguchi, 1983). We also managed to obtain a rather simple empirical relation between precipitation intensity and water content (the mass of substance in a unit volume) in a medium, namely:

$$W = 0.06R^{0.88} \tag{7.47}$$

where water content takes the dimension of $g\,m^{-3}$, and precipitation intensity that of $mm\,hr^{-1}$. There also exist other numerical versions of the given formula.

Theoretical and numerical investigations of physicochemical kinetics problems—including the processes of condensation of water vapor, and coalescence between drops and drop breakup—have shown that, generally, the theoretical spectra of drops are qualitatively close to the exponential Marshall–Palmer distribution, though there are some features that are not described by the empiricism of the given distribution. This mainly relates to the multimodal character of theoretical distributions and to considerably greater density in the small drop-size range ($r < 0.1$ mm) than in the case of exponential approximation. All these features result in noticeable variations in the electromagnetic properties of a medium (Jameson, 1991).

### 7.4.10   Parameters of attenuation and scattering of a polydisperse medium

In accordance with the basic concept of radiative transfer theory—namely, the electromagnetic rarefaction of a medium—the incident radiation falling on an investigated volume from outside completely "illuminates" all the particles that are present in the unit volume. Therefore, when the medium contains a cloud of spherical particles of the same composition, but of different size, the spectral coefficients of extinction and of scattering can be calculated by formulas

$$\gamma(\text{cm}^{-1}) = \int_0^\infty Q_E \pi r^2 n(r)\, dr \tag{7.48}$$

$$\sigma(\text{cm}^{-1}) = \int_0^\infty Q_S \pi r^2 n(r)\, dr \tag{7.49}$$

When a radiation beam propagates through a medium that contains $N$ spherical particles of the same composition and the same size (each having radius $R$) in the unit volume, the cross-sections of absorption and scattering (or the efficiency factors of extinction $Q_E$ and of scattering $Q_S$) can be related to the spectral coefficients of total attenuation (extinction) and scattering by simpler relations:

$$\gamma(\text{cm}^{-1}) = Q_E \pi r^2 N \tag{7.50}$$

and

$$\sigma(\text{cm}^{-1}) = Q_S \pi r^2 N \tag{7.51}$$

If particles are grouped together in size into intervals that have radius $r_j$ ($j = 1, 2, \ldots, M$), then the integrals presented above can be replaced by sums. If the integrals cannot be obtained analytically in expressions (7.48)–(7.49), then numerical integration is carried out and tables are composed (Krasiuk and Rosenberg, 1970; Skolnik, 1980; Oguchi, 1983; Lhermitte, 1988; Ivazyn, 1991). It should be emphasized once again that all these expressions were obtained under important physical limitations: the electromagnetic rarefaction of a medium and the absence of interactions between particles.

Let us first consider the Rayleigh approximation. Since the absorbing properties of particles prevail in this approximation, the spectral absorption coefficient can be presented as:

$$\gamma = k_1(\lambda) \int_0^\infty r^3 n(r)\, dr \qquad (7.52)$$

where $k_1(\lambda)$ is the numerical coefficient. On the other hand, the total mass of a medium's substance $W$ (in a unit volume) will be equal to

$$W = \frac{4}{3} \pi \rho \int_0^\infty r^3 n(r)\, dr \qquad (7.53)$$

Comparison of these expressions indicates that the spectral extinction coefficient for a medium with particles in the Rayleigh approximation is proportional to the total mass of substance in a unit volume:

$$\gamma = k_2(\lambda) W \qquad (7.54)$$

and, what is very important, it does not depend on the form of the density function. Thus, during remote investigations in the Rayleigh region, information on the form of the density function of a disperse medium cannot be obtained (at least directly).

Since water possesses prominent spectral properties in the centimeter and millimeter bands, for the band of 0.5–10 cm and for liquid-drops (fresh water) clouds ($t = 18°C$) the following simple approximation can be established:

$$\frac{\gamma}{W} = \frac{0.43}{\lambda^2} \qquad (7.55)$$

where $W$ is expressed in $g\,m^{-3}$ and absorption in $dB\,km^{-1}$. Physically, this is related to the fact that the parameter in expression (7.54), which depends on the dielectric properties of water, has an approximation of type $1/\lambda$ on the long-wavelength branch of the Debye relaxation maximum (see Sharkov, 2003). However, in the case of crystalline clouds (hailstones, snow flakes) the extinction decreases by two to three orders of magnitude (all other things being equal). Thus, wavelength dependence can be accepted to be $1/\lambda$ (because the explicit wavelength dependence of the real part of a dielectric constant is absent for ice). There also exist some other experimental data approximations in the Rayleigh region. However, all have a frequency character close to (7.55).

Consideration of a wider range of particle sizes and wavelengths already requires numerical operations with (7.48) and (7.49). Special calculations of the coefficient of extinction per unit of path—that is, the specific attenuation (for a 1-km propagation

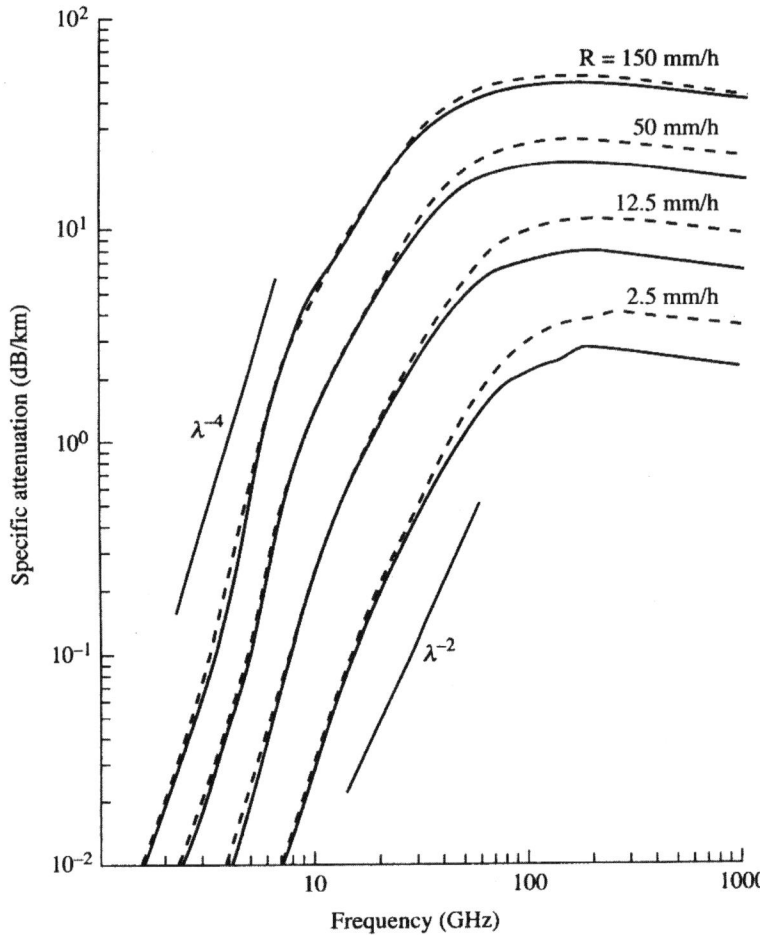

**Figure 7.6.** Frequency characteristics of rain attenuation at rain temperature of 20°C, for Laws–Parsons (solid curves) and Marshall–Palmer (dashed curves) drop-size distributions. Parameters are rain rates (mm/h) (Oguchi, 1983).

path) carried out in the 1–1000 GHz range for various drop-size distributions and precipitation intensities (Figure 7.6)—have shown that the frequency dependencies are characteristic in form: a rapidly growing rise from the side of large wavelengths, a weak maximum in the frequency range of about 100 GHz, and a slow drop to the side of higher frequencies. As should be expected, these dependencies do not reveal any sharp maxima that are specific for the Mie region of an individual particle. Moreover, growing regions can be characterized as transition regions because of the "smeared" Mie maximum to the Rayleigh region. In this case the frequency approximation of extinction for high-intensity precipitation that possess large-sized drops and, accordingly, great scattering, is closer to the $1/\lambda^4$ dependence. Whereas for weak precipitation (with a small dispersity of drops and, accordingly, with very weak scattering and strong absorption), this approximation is closer to that of the Rayleigh region: $1/\lambda^2$ (see relation (7.55) and Figure 7.6). It should also be noted that the distinctions in extinction values for various distributions are the greater, the higher the working frequency and the lower the precipitation intensity. We have already said that the Rayleigh region is not sensitive to the form of distribution of drops—see expression (7.54). As for strong precipitations—well, in this case the intensive scattering of large-sized drops in some sense "blocks" the contribution from absorption of small-sized drops to the total extinction of a disperse medium. Special experiments (Oguchi, 1983) have really shown that in the millimeter frequency band for rainfall in the terrestrial atmosphere the sensitivity of the degree of extinction in a medium to the type of distribution is very high, and, therefore, this band is efficient for remote investigation of the fine features of disperse media.

It is instructive to compare the frequency properties of the value of extinction (per unit of path) of the various disperse media that are typical of the terrestrial atmosphere (Figure 7.7). Certainly, in this case the question concerns the qualitative picture of the phenomenon, and presented data are not intended for quantitative interpretation of particular experiments. The solid lines in Figure 7.7 give the wavelength dependencies of the extinction coefficient for rainfall with intensities of 0.25 (curve 1), 1.0 (curve 2), 4 (curve 3), and $16\,\mathrm{mm\,hr^{-1}}$ (curve 4). According to the

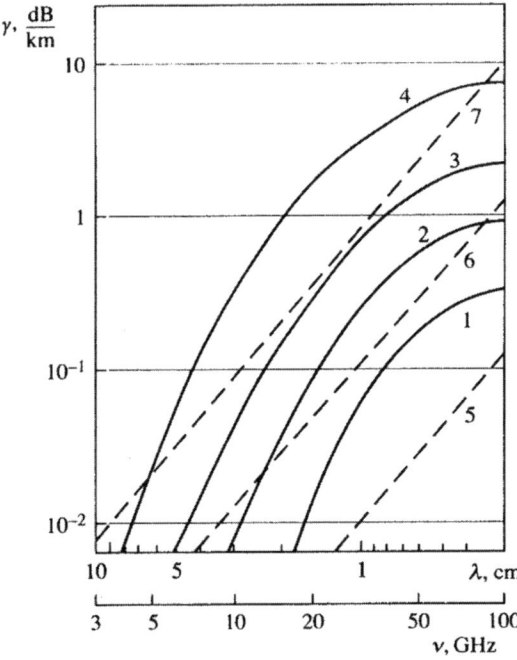

**Figure 7.7.** Frequency characteristics of attenuation for natural disperse media: rain (solid curves) and fog (dashed curves). The notation is explained in the text.

current meteorological classification, these intensities correspond to drizzle, light, moderate, and heavy rainfall. The dashed curves in Figure 7.7 show extinction in clouds and fogs calculated by formula (7.55) for water content of 0.032 (curve 5), 0.32 (curve 6), and 2.3 g m$^{-3}$ (curve 7). These fogs correspond to visual ranges (in the optical band) of about 600, 120, and 30 m. As follows from analysis of these data, the picture is in general ambiguous. So, extinction in a thick sea fog (curve 7) exceeds that in moderate rainfall (curve 3) in the millimeter and centimeter wavelength bands. And in the long-wave centimeter band, extinction in a thick fog even exceeds that in heavy precipitation. This seems paradoxical at first sight. However, physically, this is associated with the different relationship between contributions to extinction from large-sized (scattering) and small-sized (absorption) drops in various disperse media. So, for intensive rainfall the frequency dependence of extinction is proportional to $1/\lambda^4$ in the centimeter band, and for fog to $1/\lambda^2$—which highlights an apparent paradox at long centimeter waves.

Note that high-frequency variations in the dielectric parameters of salt water in the centimeter and decimeter ranges (Sharkov, 2003) can essentially change the general spectral picture of the electrodynamic properties of a drop-spray phase arising over the sea surface. Figure 7.8 presents numerical calculations of extinction per unit length in a model drop-spray medium in the Rayleigh approximation, which takes due account of the dielectric properties of water: in one case pure fresh water (with a salinity of 0‰) was taken, in the other case salt water with a salinity of 35‰ (i.e., sea water). It follows from analysis of the figure data that for fresh water the $1/\lambda^2$ approximation in the spectral dependence of extinction per unit length is valid within a wide range of wavelengths: beginning with 0.3 up to 100 cm. However, for sea water the frequency dependence of extinction per unit length behaves in a completely different manner: the $1/\lambda^2$ approximation in the spectral dependence of extinction per unit length is valid within a very narrow wavelength range: beginning with 0.3 up to 2–3 cm. In the long-wave range the extinction per unit length essentially exceeds (by more than an order of value) the values corresponding to fresh water. This is related to a drastic increase in the imaginary part of the dielectric constant of salt as a highly absorbing electrolyte in the decimeter range (Sharkov, 2003). Certainly, this will stipulate peculiar properties of electromagnetic wave propagation and of natural radiation of such a medium with respect to the similar properties of atmospheric cloud systems.

## 7.5  BACKSCATTERING BY NATURAL POLYDISPERSE VOLUME TARGETS

Active methods of remote sensing in the mode of reception of backscattered signals are widely and successfully used when studying the internal dynamics and phase structure of cloud systems in the terrestrial atmosphere (Doviak and Zrnic, 1984; Oguchi, 1983; Sharkov, 1998, 2003). It is undoubtedly important to make use of the experience gained in active remote investigation of atmospheric cloud systems and to apply active methodology and instrumental means for detailed study of the dynamic

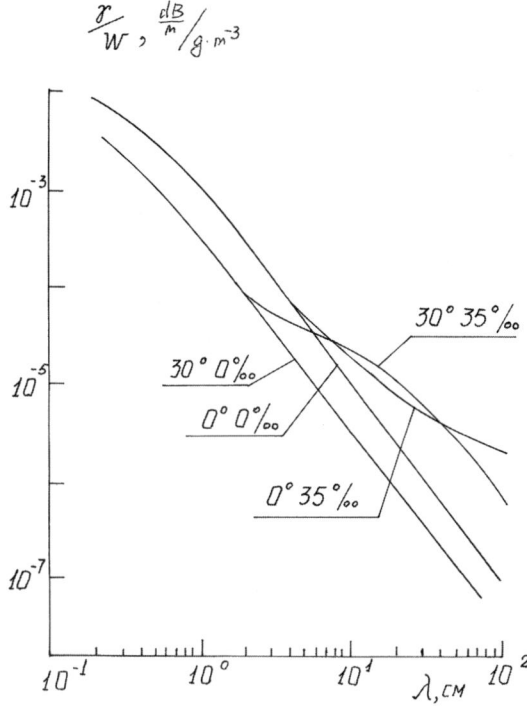

**Figure 7.8.** Frequency characteristics of attenuation per length unit with respect to the water content of drop-spray clouds involving fresh and salt ($S = 35‰$) water ($t = 0°$C and $30°$C). Calculations were made within the Rayleigh limit.

and phase structure of drop-spray systems that form over the sea surface in the gravity wave breaking process. We shall consider in this section some of the basic methodological approaches being developed in radiometeorology for remote sensing of atmospheric cloud systems.

Modeling the radar cross-section of volume targets requires the application of electromagnetic scattering theories. First, backscattering by a single particle is calculated and, second, the resultant reflection by an ensemble of particles is obtained by summation of the contributions of all individual particles. Scattering by homogeneous dielectric spheres was analyzed by Mie. The Mie expression for the radar cross-section (RCS) of a sphere is given as (Skolnik, 1980; Ivazyn, 1991; see also Eq. (7.22)):

$$\sigma = \frac{\pi D^2}{x^2} \left| \sum_{n=1}^{\infty} (-1)^n (2n - 1)(\dot{a}_n - \dot{b}_n) \right|^2 \tag{7.56}$$

where $D$ is the particle diameter; $x = (\pi D)/\lambda$, with $\lambda$ as the radar wavelength; parameters $a_n$ and $b_n$ are coefficients that involve Bessel and Hankel functions with arguments that depend on $x$; and $\dot{\varepsilon}$ is the relative complex permittivity of the sphere. Deirmendjian (1969) developed iterative procedures to calculate the radar cross-section. The Mie theory is valid for a wide range of frequencies. However, implementation in computer software is tedious, because of time-consuming computation iterations and easily occurring instabilities. When the particles are small compared with the wavelength ($D \ll \lambda$), equation (7.22) for parameter $\sigma$ reduces to the *Rayleigh expression*

$$\sigma = \frac{\pi^5}{\lambda^4} \left| \frac{(\dot{\varepsilon} - 1)^2}{(\dot{\varepsilon} + 2)} \right|^2 D^6 \tag{7.57}$$

abbreviated to

$$\sigma = \frac{\pi^5 |k|^2}{\lambda^4} D^6$$

When $\sigma$ is normalized to the geometric cross-section, it becomes

$$\sigma_0 = \frac{\sigma}{\pi D^2} = |k|^2 x^4 \tag{7.58}$$

The accuracy of the Rayleigh approximation depends on permittivity and particle size. For $|\sqrt{\varepsilon}|x < 2$, the Rayleigh approximation gives results within 3% of the exact Mie solution. The maximum value of $|\sqrt{\varepsilon}|$ at radar wavelengths used for weather forecasting is approximately 9 for water (Sharkov, 2003) and the minimum approximates 1 in the case of very loose snow. At a radar wavelength of 10 cm, the Rayleigh approximation holds well for particles with diameters smaller than 7 mm. In most types of rain this requirement is met. Snowflakes, however, can be larger than 7 mm, but their fortuitous smaller permittivity still allows use of the Rayleigh approximation. Equation (7.56) clearly expresses the fact that the radar cross-section depends on hydrometeor type—that is, it depends on the material and on the size of the hydrometeors involved.

In the strict sense, natural particles of complicated forms (sea drops, snowflakes, hale–rain particles, hollow spheres) cannot be described by Mie theory and necessitates using up-to-date electromagnetic theories (including multifractal approaches) to calculate their backscattering microwave characteristics.

The RCS of particle clouds can be presented in integral form (for non-interacting particles in polydispersed media) as (Ishimaru, 1978)

$$\sigma_{\text{tot}} = \int_0^\infty \sigma_0 \pi r^2 n(r)\, dr \tag{7.59}$$

where $n(r)$ is a particle's spectrum, and the total number of particles (in volume unit) is

$$N(\text{m}^{-3}) = \int_0^\infty n(r)\, dr \tag{7.60}$$

For a limited number of particles the radar cross-section of a volume $V$ filled with $N$ hydrometeors is—when interaction between the particles is ignored—given by the sum of the contributions of all individual particles:

$$\sigma_{\text{tot}} = \sum_{i=1}^{N} \sigma_i = \frac{\pi^5 |K|^2}{\lambda^4} \sum_{i=1}^{N} D_i^6 \tag{7.61}$$

The radar cross-section is frequency-dependent. To enable comparison of measurements of backscattering from different volumes and at different frequencies, the *reflectivity factor Z* is defined:

$$Z = \frac{\lambda^4}{\pi^5 |K|^2 V} \sigma_{\text{tot}} = \frac{1}{V} \sum_{i=1}^{N} D_i^6 \tag{7.62}$$

Usually, the unit of $Z$ is set to $\text{mm}^6\,\text{m}^{-3}$. When $Z$ is expressed in decibels, the unit dBZ is used, with $0\,\text{dBZ}$ corresponding to $1\,\text{mm}^6\,\text{m}^{-3}$. It may not be known what type of particle clouds causes radar reflection or if the conditions for the Rayleigh approximation are fulfilled. In that case an *equivalent reflectivity factor $Z_e$* is introduced:

$$Z_e = \frac{\lambda^4}{\pi^5 |K_r|^2 V} \sigma_{\text{tot}} \tag{7.63}$$

in which $K_r$ is the effective value of $K$.

This equivalent reflectivity factor (commonly called just "reflectivity" for real weather) spans many orders of magnitude. The characteristic qualitative values for $Z$ are (Lhermitte, 1988; Meischner, 1990):

for non-precipitating clouds $\sim -20$–$10\,\text{dBZ}$
moderate rain                 $\sim 30\,\text{dBZ}$
heavy rain                    $\sim 45\,\text{dBZ}$
hail–rain mixtures            $\sim 50$–$60\,\text{dBZ}$
large hail                    $> 60\,\text{dBZ}$

The results of backscattering and absorption calculations carried out for non-precipitating clouds over many years were collected by Oguchi (1983) and Ivazyn (1991).

The wide sets of experimental works performed on precipitation study in the mid-latitudes and for slow and moderate rains (see, for instance, Atlas *et al.*, 1984) facilitated obtention of important parametrizations among a measured parameter (radar reflectivity) and precipitation parameters (rainfall rate, median volume diameter, size distribution parameter, liquid water content, and optical extinction). This parametrization is shown in graphical display of Figure 6.10 (Atlas *et al.*, 1984). It should be stressed that there is not a one-to-one (one-factor) correspondence between radar reflectivity and rainfall rate.

Nevertheless, for any time interval of the experiment, the researchers, taking into account enquiries of actual practice, pursued a large body of efforts to discover the unique relationship between these radar measurements and the rainfall parameter of experimental interest. For a number of meteo-situations (e.g., widespread homogeneous precipitation in the mid-latitudes) a relation of the form $Z = aR^b$ was found by standard regression methods. The values of parameters $a$ and $b$ were found trustworthy for a number of radar wavelengths and for various ("calm") meteo-situations (Oguchi, 1983; Atlas *et al.*, 1984).

For strong convective and hurricane precipitation the situation reverses. So, Pasqualucci (1984) proposed and developed a non-linear parametrization model on the basis of his Doppler radar (operating at 35 GHz) data for storm precipitation (in South Africa). Radar-measured drop-size distributions are parameterized using an exponential distribution of the form:

$$N = N_0 e^{-\Lambda D} \tag{7.64}$$

where $N$ is the concentration of particles; $D$ is their diameter; and $\Lambda$ and $N_0$ are parameters of the Marshall–Palmer distribution. In this parametrization, $N_0$ and $\Lambda$ will be varied so that the water content $W_E$, the radar reflectivity factor $Z_E$, and the rainfall rate $R_E$ of the exponential distribution are least-squares-fitted to $W$, $Z$, and $R$, which are calculated from the radar-measured distributions. The expressions are the following:

$$W_E = N_0 \frac{\pi}{\Lambda^4} \tag{7.65}$$

$$Z_E = N_0 \frac{720}{G^7} \tag{7.66}$$

$$R_E = N_0 \left[ \frac{0.1181}{\Lambda^4} - \frac{0.1260}{(\Lambda + 0.6)^4} \right] \tag{7.67}$$

The units used in these equations are $W_E$ in $\text{mm}^3 \, \text{m}^{-3}$, $Z_E$ in $\text{mm}^6 \, \text{m}^{-3}$, $R_E$ in $\text{mm} \, \text{h}^{-1}$, $N_0$ in $\text{m}^{-3} \, \text{mm}^{-1}$, and $\Lambda$ in $\text{mm}^{-1}$. It should also be noted that the equation for $R_E$ is for a fall-speed law of the form

$$V = 10.44 - 11.5e^{-0.6D} \tag{7.68}$$

where $V$ is in $\text{m s}^{-1}$ and $D$ in mm.

The resulting $N_0$ and $\Lambda$ parameters were plotted versus the liquid water content $W$ measured, and the results showed that, for $W \geq 1.2 \, \text{g m}^{-3}$, $N_0$ increases with increasing $W$ while $\Lambda$ becomes constant and independent of values of $W$.

Drop-size distributions measured in a severe squall line show a large abundance of small drops ($< 1.5$ mm in diameter) in regions of wind shear near the edges of the main downdraft core. A possible explanation of this large concentration of smaller drops is the sorting of hydrometeors by shear in the horizontal wind that is generated by divergence of the downdraft at the ground.

In any event, there is a need to develop a more refined $Z$–$R$ parametrization for tropical rainfall and drop-spray clouds in pre-hurricane and hurricane situations.

Recovery of the spatiotemporal characteristics of rainfall areas leads to a number of complexities that reflect the complicated multifractal structures of these natural formations (Lovejoy and Schertzer, 1985; Lovejoy and Mandelbrot, 1985).

## 7.6 FEATURES OF RADIATIVE TRANSFER IN DENSE MEDIA

In connection with the intensive development in microwave diagnostics of composite natural media in the ocean–atmosphere system, it is interesting to study the features of electromagnetic wave transmission and scattering in randomly inhomogeneous media with densely disturbed, discrete, highly absorbing scatterers, where the size of particles, the distance between particles $d$, and the electromagnetic radiation wavelength $\lambda$ are quantities of the same order. Such important microwave remote-sensing tasks include the study of electromagnetic wave scattering and radiation in a cloudy atmosphere with considerable volume densities (more than 0.1%) of hydrometeors (Oguchi, 1983), in the drop-spray phase of gravitation wave breaking (Cherny and Sharkov, 1988), in snow–water disperse media, in foam-type disperse systems (Raizer and Sharkov, 1981; and Chapter 6) and in other similar natural media.

When a disperse medium has such parameters, the physical conditions for applicability of the radiative transfer theory are obviously violated (Sharkov, 2003). However, the desirability of using the numerous results that the radiative transfer theory makes available requires, doubtlessly, solution of the question on the limits of effectiveness of the theory itself. Certainly, this complicated problem cannot be solved within the framework of the radiative transfer theory itself. Its solution is possible either within a multiple scattering framework, or experimentally. Theoretical analysis of this problem is far beyond the scope of the present book. Here we shall only describe the results of laboratory experiments that are closest to the subject matter of this book—namely, microwave sensing of dense disperse media. These experiments were carried out between 1976 and 1986 under the scientific guidance of the author of the present book (Cherny and Sharkov, 1988, 1991a, b).

Though there have been many experiments on studying the electromagnetic properties of tenuous discrete systems with $d \sim (10-10^4)\lambda$ and volume density $C \sim (10^{-2}-10^{-4})\%$—see the review by Oguchi (1983)—no results from studying the electromagnetic properties in the radio-frequency band of dense dynamical media with absorbing scatterers can be found in the literature. The principal methodical problem in a statement of theory for such experiments lies in the experimental difficulties of producing dynamical, dense drop structures with strictly controlled parameters. However, such a statement of theory is now extremely topical both from general theoretic and practical points of view. Principally, these experiments are necessary to discover the limiting values of densities at which the mechanisms of electromagnetic interaction of solitary absorbing scatterers are "switched on". Experiments carried out in the optical band for transparent media and for semi-transparent particles (i.e., "soft particles") have shown (Varadan et al., 1983) that the essential contribution of multiple scattering falls on the range of densities exceeding 1 %—this boundary value being strongly dependent on the particle size parameter. These numbers certainly cannot be directly applied to discrete media with highly absorbing scatterers and can only serve as a quantitative landmark. Cherny and Sharkov (1991a, b) contain the results of experimental investigations on the characteristics of transmission, backscattering, and thermal radiation of millimeter-band electromagnetic waves in a disperse discrete medium with the volume density of spherical scatterers ranging from 0.05 to 4.5%. In this case the average distance ($d$) between the centers of particles varied within limits from 2.3 to 0.9$\lambda$.

### 7.6.1 A disperse medium and its characteristics

As already noted, fulfillment of the necessary radiophysical experiments encounters difficulties in producing, in free-fall mode, aqueous particles of a quasi-monochromatic (in size spectrum and in magnitude of velocities) flow with a high density of spherical-shaped particles. It is necessary, on the one hand, to avoid the dynamical deformation and decay of particles of rather large diameters (of the order of 2–3 mm) and having high velocities (5–10 m s$^{-1}$). On the other hand, gravitational and turbulent coalescence between drops should not be allowed. The natural cascade processes result in a very wide spectrum of particles under natural conditions (in cloud systems and precipitation, for instance). This, in its turn, essentially hampers interpretation of radiophysical experiments. The processes of deformation and decay of drops in a flow are controlled by two dimensionless numbers: the Rayleigh number (for a sphere) Re = $2aV\rho\mu^{-1}$ and the Weber number $W = a^2 V\rho(2\sigma)^{-1}$. Here $V$ is the steady velocity of a drop; $a$ is the drop radius; $\rho$ and $\mu$ are the density and viscosity of air; and $\sigma$ is the surface tension of water. The laminar regime of airflow around drops (the Stokes regime) is kept at Re $\approx$ 300, and the critical value of $W$ for ensuring dynamical stability of drops equals 10. Analysis of the various methods of forming dense media led the authors to realize the necessity of using the forced regime with a particular flow velocity rather than the free-flow regime. A disperse, highly dense medium was produced by a spray system, which entailed a special injector with a removable grid. The grid could be tilted at any given profile and had a number of orifices. The number and diameter of orifices determined the density and size of drops, whereas the profile determined the value of flow divergence, which also influenced density. The sphericity of drops was specially controlled: the eccentricity of drop ellipses did not exceed 0.3 (for high densities) and 0.1 (for low densities). Under the conditions of this experiment the Re number was 200 to 300 (for various flow velocities) and $W = 0.03$. Thus, under these experimental conditions both the laminar regime of airflow around a drop (the Stokes condition) was ensured, and the processes of decay and arising of a wide (decay) spectrum of scatterers were not allowed. If the injector was directed downwards, then the drops—accelerated under the force of gravity—produced a uniform density variation down the flow. Thus, for a single grid it was possible to obtain a wide range of variation in density with the same dispersity (Figure 7.9a–c). The control and measurement of particle density were carried out by the stereoscopic photography method using two synchronized mirror cameras with telescopic lenses and a special light flash system (with a flash duration of $10^{-6}$ s). Moreover, the velocity of drops was measured by the track method (the reflecting blinks on photo-images—Figure 7.9a). The operator analyzed the stereo pairs obtained using a stereoscope and, comparing them with a test object, determined the number of drops and their disperse properties (at a fixed time instant). Radiophysical measurements were carried out for the two types of disperse media whose histograms are presented in Figure 7.10. The form of particle distribution functions $n(r)$ $(\text{dm}^{-3}\,\text{mm}^{-1})$ was approximated by the gamma distribution:

$$\left.\begin{aligned} n_1(r) &= 0.38 N_1 r^9 \exp(-0.73 r^3) \\ n_2(r) &= 73.5 N_2 r^8 \exp(-3.66 r^3) \end{aligned}\right\} \qquad (7.69)$$

The values of $N_1$ and $N_2$ are proportional to the volume density of particles for medium 1 and medium 2. The average value of radius for medium 1 equals 0.15 cm (and, accordingly, the size parameter $x_1 = 1.18$), and for medium 2 it equals 0.09 cm ($x_2 = 0.7$) (Figure 7.10a, b). A special statistical estimation of fluctuations in countable particle flux density showed that the root-mean-square deviation of density was less than 2% of the average value of $N$. In this case, samplings—spaced in time from 1 to 3 h—relate to the same general set. It was clearly seen from analysis of histograms that—by forming a dense flux—we managed to avoid decay and coalescence processes, and the spectrum of particles could be considered to be close to being monochromatic. For these types of media the authors calculated the extinction, scattering, and absorption coefficients in accordance with (7.48) and (7.49). In addition, the single-scattering albedo was calculated for the unit volume of a polydisperse medium using a function of the size distribution of particles obtained from the experiment. Calculations were carried out for the working radiation wavelength $\lambda = 8$ mm and the complex index of refraction of water $m = 5.39 - j2.81$, which corresponded to a water temperature $t = 22°\text{C}$ and a salinity $S = 0‰$. By virtue of the fact that a tenuous medium approximation with a near-monochromatic spectrum was used here, the scattering albedo for the unit volume of a medium corresponded to the value of albedo of a solitary particle (so, for medium 1 $\omega = 0.63$, and for medium 2 $\omega = 0.43$), but did not depend on the medium's density.

### 7.6.2  Experimental technique and instruments

The purpose of the experiment was to measure the radiophysical characteristics of a disperse dynamical medium by strictly controlling its parameters. Measurements were carried out in three modes: bistatic (radiation transmission through the medium within line-of-sight limits), scatterometric (backscattering investigation), and radiometric. Extinction of a medium was measured in the first version, the backscattering cross-section in the second version, and the thermal radiation of a disperse medium in the third version. The fluctuations of scattered radiation intensity were measured in addition to its average values. The extinction, absorption, and scattering coefficients; the scattering and backscattering albedo (in the "cold" layer approximation); and the thermal radiation of a disperse medium with a spherical scattering indicatrix were calculated using an analytical solution of the equation for a plane-parallel layer (in the "pure" absorption approximation).

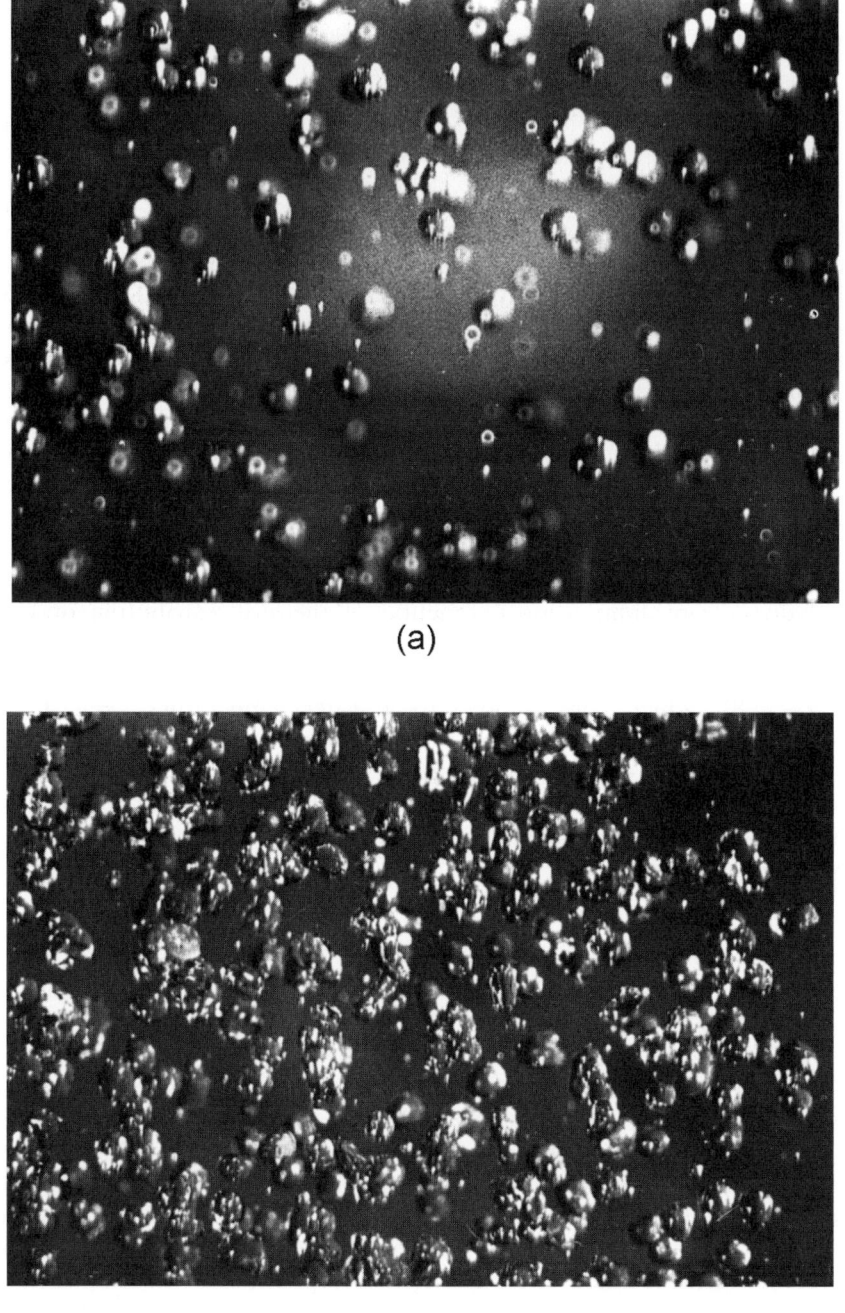

(a)

(b)

**Figure 7.9.** Photographs of a disperse water drop medium with relative volume concentration: (a) 0.28%; (b) 1.5%.

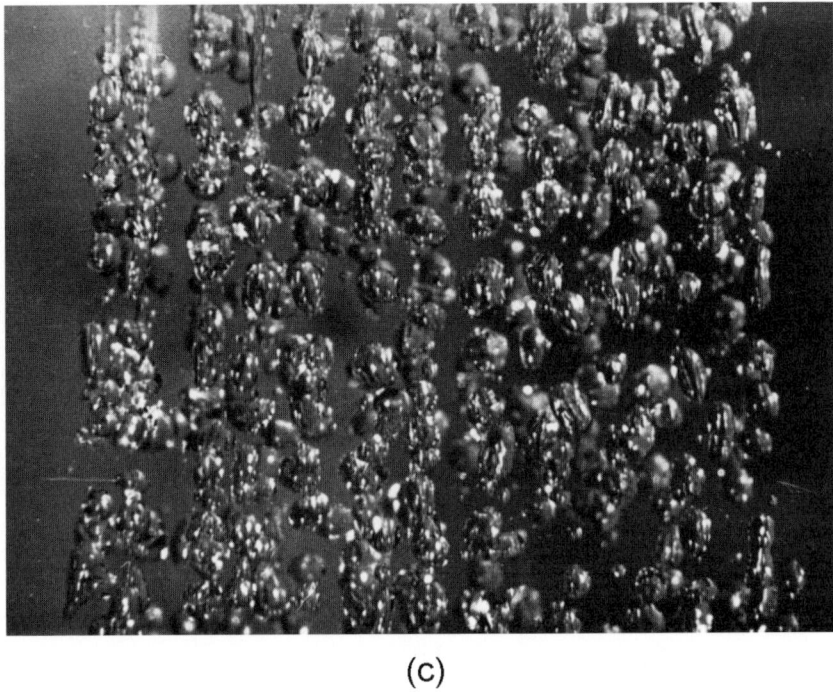

(c)

**Figure 7.9** (*cont.*). (c) 4.50%.

### 7.6.3  Average values of electrodynamical characteristics

Comparing the experimental and theoretical values of the extinction and thermal radiation of disperse medium 1 (the average diameter of particles was 0.3 cm), we can see a distinction, which becomes apparent as the density of particles inceases (Figure 7.11). We can distinguish the region of low deviation of experimental data from theoretical ones and the region of stronger deviation. The boundary that separates these regions corresponds to a value of particle volume density approximately equal to 0.8%, this boundary being the same both for extinction and for thermal radiation. By considering the results of investigation of the extinction value for disperse medium 2 (where the average diameter of particles was 0.2 cm), we can see that the afore-mentioned boundary lies in the region of particle volume density values of 0.15%. Now we shall analyze the dependence of extinction values for a disperse medium on the number $N$ of particles in a unit volume (the countable density) rather than on the particle volume density $C$. It is seen from Figure 7.11 that in this case these boundaries lie in the range of $N_0 = 500$–$550 \, \text{dm}^{-3}$ for both types of disperse medium (i.e., they virtually coincide). In its turn, quantity $N_0$ determines the average distance between particles as $d \sim N_0^{1/3}$. Therefore, we can now characterize these boundaries by the distance between particles (i.e., by $d \sim 1.5\lambda$).

Thus, from analysis of the experimental data and from theoretical calculations it follows that the radiative transfer theory in a tenuous medium approximation

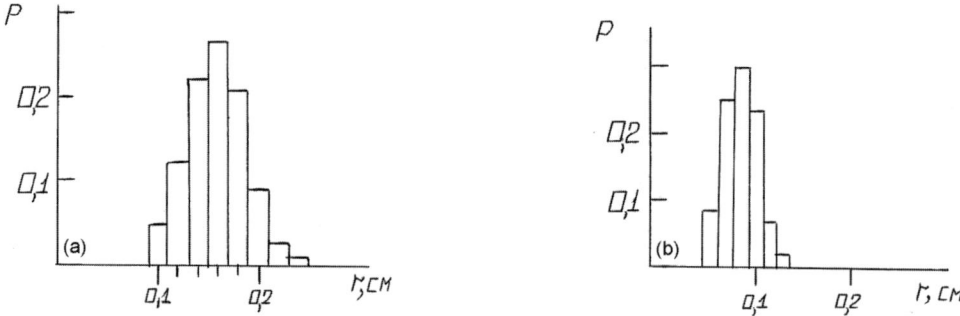

**Figure 7.10.** Experimental histograms of droplet radii for medium $N_1$ (a) and medium $N_2$ (b).

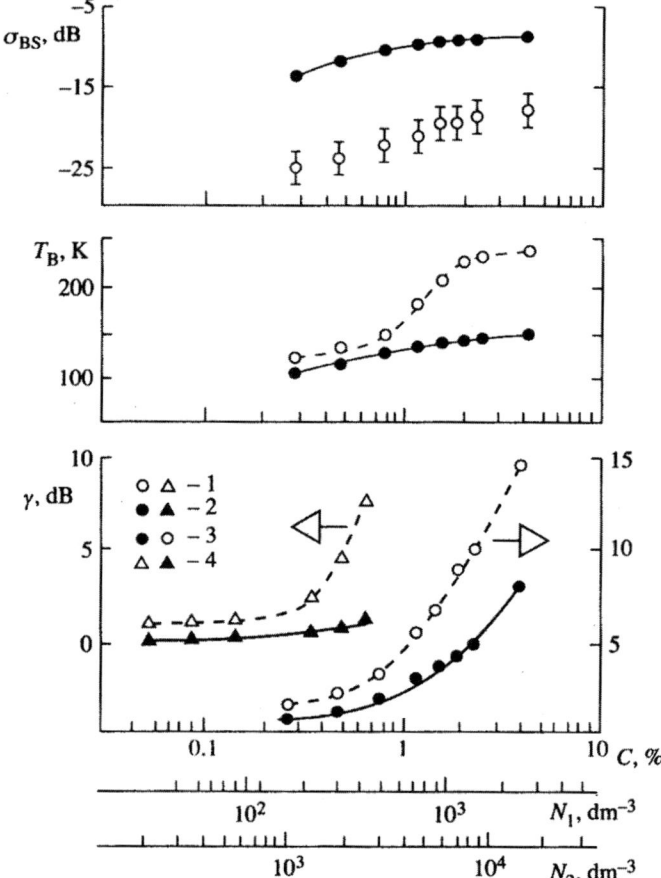

**Figure 7.11.** The extinction coefficient ($\gamma$), radiobrightness temperature ($T_B$), and back-scattering cross-section ($\sigma_{BS}$) of disperse water drop media as functions of concentration ($C$) and number density ($N$): 1—experimental data; 2—theoretical results; 3—data for medium $N_1$; 4—data for medium $N_2$.

satisfactorily describes the electromagnetic properties (average values) of a discrete disperse medium with absorbing scatterers, as long as the distance between particles is $d > 1.5\lambda$. In the case where $d < 1.5\lambda$, experimental data principally differ from calculated ones. For example, for a particle volume density $C = 4.5\%$ ($d = 0.9\lambda$) the distinction in radio-brightness temperature equals 86 K, and for the extinction value $-6.5\,$dB. As far as backscattering is concerned, here the experimental and theoretical data (in the cold layer approximation) essentially differ (by about 10 dB) throughout the range of particle densities. Using experimental data, we shall now estimate the disperse medium parameters for the particle volume density value $C = 4.5\%$ ($d = 0.9\lambda$).

Electrodynamical parameters were estimated by means of a specially developed technique that combines (albeit in a complex way) the data of active and passive measurements for the same investigated medium (Cherny and Sharkov, 1991a). In this case, expressions for radio-brightness temperature were obtained in the "pure" absorption approximation. It is important to note that inclusion of the integral term in the transfer equation, which describes the "internal re-scattering" in a layer, does not essentially change the spectral characteristics of a medium in the considered case of absorbing scatterers. This follows from comparison of the calculations with the results of solving a similar problem by the double spherical harmonics method and by the Monte-Carlo method. Of importance is the fact that—for such values of the density of particles in a medium—the electrodynamical parameters of a disperse medium have essentially changed compared with those calculated (for a tenuous medium) obtained in the single-scattering approximation. So, the scattering albedo of a unit volume of disperse medium $N_1$ decreased three-fold (from the value of 0.63 down to 0.22). The extinction and absorption coefficients, on the contrary, increased about 1.5 times (from the value of 0.63 up to 0.94 cm$^{-1}$) and three times (from the value of 0.23 up to 0.73 cm$^{-1}$), respectively. The scattering coefficient value decreased twice in this case (from the value of 0.40 down to 0.21 cm$^{-1}$). This result indicates that—for a disperse dense medium with absorbing scatterers—the interaction between particles primarily results in a growth in the absorption of a medium and, therefore, in increasing its thermal radiation and, what is more, in decreasing its scattering properties.

### 7.6.4  Fluctuation mode of extinction

It is important to note that Cherny and Sharkov (1991a) demonstrated experimentally the principal change in character of the fluctuation mode of extinction in a dense medium. This effect is visually illustrated in Figure 7.12, which presents the registrograms of an external harmonic signal transmitted through the medium, this signal being considered at an intermediate frequency. Fluctuations in the intensity of the radiation transmitted are observed in the mirror-symmetrical amplitude modulation of a signal. Measurements were carried out with extinction recording by exposure to microwave radiation. It can easily be seen that the statistical characteristics of a signal sharply change in the case of two different densities. A possible physical cause—explaining fluctuations in radiation transmitted through a

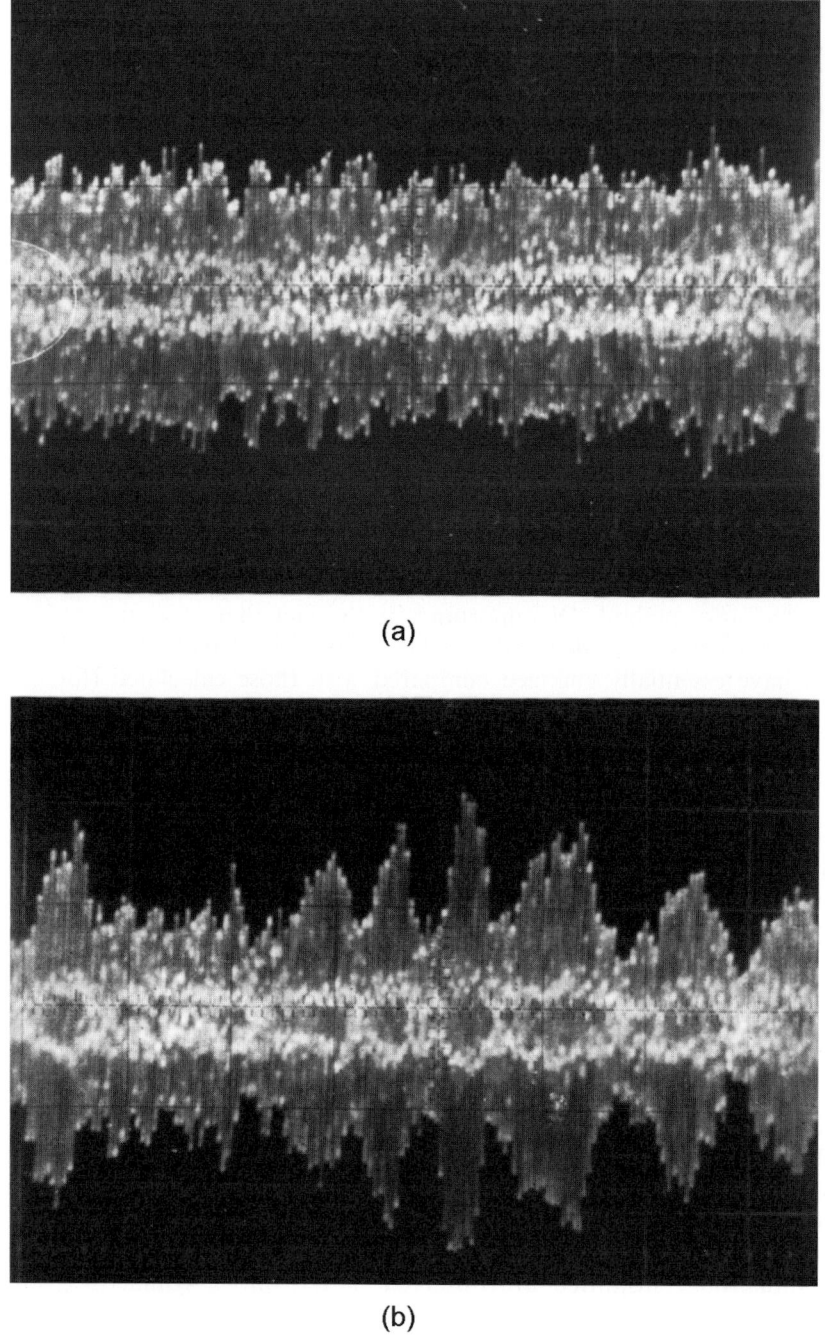

(a)

(b)

**Figure 7.12.** Photographic registration of a signal (at intermediate frequency) transmitted through a water drop medium with volume concentrations 0.28% (a) and 4.5% (b).

medium—could be associated with changing the countable number of particles in the volume under study. However, we shall show that this is not the case.

So, let us consider—in accordance with the Bouguer law—the ratio of intensities of external radiation, weakened by a medium, for different time instants

$$(I_1/I_2) = \exp(\tau_2 - \tau_1) \tag{7.70}$$

or

$$\tau_2 - \tau_1 = \ln(I_1/I_2) \tag{7.71}$$

where $\tau_1$ and $\tau_2$ are the opacities of an investigated disperse medium at different time instants. Since—in the single-scattering approximation for a medium with a monochromatic spectrum of particles—we have $\tau = Q_E \pi r^2 Ns$ (here $s$ is the linear size of a medium), we can write the following finite-difference relation:

$$(\Delta N/N) = (\Delta \tau/\tau) = (1/\tau)\ln(I_1/I_2) \tag{7.72}$$

Proceeding from this relation, we shall estimate the maximum value of $(\Delta N/N)$ for the volume density of particles of a disperse medium $C = 4.5\%$ in which $I_{max}/I_{min} = 61$ and the quantity $\tau = \gamma/4.34 = 3.34$ represents the average value of opacity. Substituting this value into (7.72), we find $(\Delta N/N) = 54\%$; but, this is impossible, since the particle density fluctuations in a disperse flow do not exceed 2% with a probability of 0.95. Thus, the sharp growth in the variance of fluctuations of a medium's extinction are not determined by fluctuations in the number of particles in a flow, but has a different physical cause.

Let us pay attention to the principal point which consists in the fact that as the density of particles increases, so the character of fluctuations also changes. Thus, the probabilistic distribution of the intensity of a signal, transmitted through the investigated medium at $C = 0.28\%$, has a prominent normal character, whereas at $C = 4.5\%$ intensity fluctuations are distributed according to the normal logarithmic law (Figure 7.13). This is clearly observed on records as well (Figure 7.12). For these volume density values and, accordingly, for $(d/\lambda) \sim 1.5$, the so-called scintillation index sharply increases (Figure 7.13b). The latter characteristic is often used in optical observations, which gave rise to its name. However, these characteristics do not provide a detailed picture of fluctuations in distribution over the scales of interactions. Let us consider the behavior of a structural function expressed in terms of spatial coordinates. The transition from spectral–temporal coordinates $t$ and $f$ to a spatial-frequency presentation of $R$ and $k$ (the spatial wavenumber) can be accomplished based on the hypothesis of "freezing" inhomogeneities in a moving flow:

$$R = Vt \qquad k = \frac{2\pi f}{V} \tag{7.73}$$

where $V$ is the particle flow velocity in the direction perpendicular to radiation transmission. Figure 7.14 presents the structural function of intensity fluctuations in three values of volume density of a disperse medium. We shall now analyze the behavior of the structural function that represents the mean square of the magnitude of an increase in the fluctuation component $I(R)$ of intensity

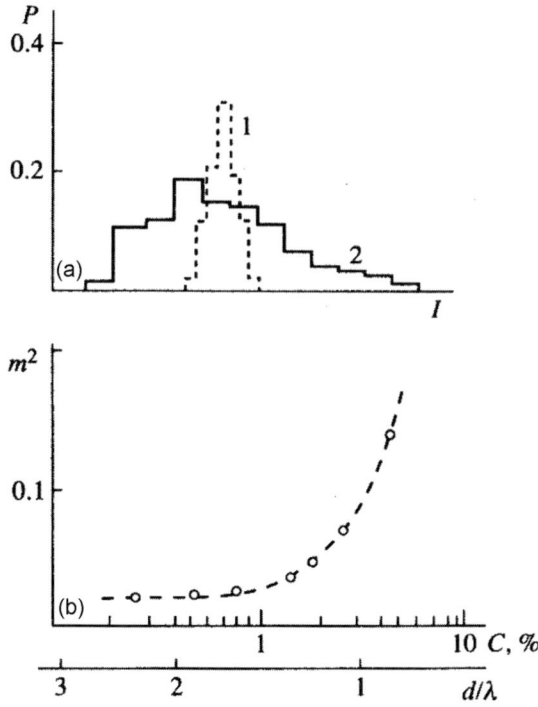

**Figure 7.13.** Statistical characteristics of the radiation intensity of a signal transmitted through a water drop medium: (a) density function (experimental histograms) with two volume concentrations: (1) 0.28%, (2) 4.5%. (b) The scintillation index $m^2$ as a function of volume concentration and of distance between particles.

(Rytov et al., 1978):

$$D(R_1, R_2) = \langle |I(R_1) - I(R_2)|^2 \rangle \tag{7.74}$$

If the spatial field studied takes on the character of a locally homogeneous one (i.e., depending only on the difference in the scales of interactions $R = R_1 - R_2$), then the form of a structural function can be essentially simplified:

$$D(R) = 2[B(0) - B(R)] \tag{7.75}$$

where $B(R)$ is the spatial correlation function (Rytov et al., 1978). The important property of a structural function consists in the fact that it excludes from consideration large-scale inhomogeneities $L_0$. In our case the latter represent the characteristic size of a particle flux. The correlation function takes into account fluctuations of any scale in equal measure. This is the reason that use of a structural function turns out to be physically justified in those cases where we are interested in fluctuations on scales much smaller than $L_0$.

For $C = 0.28\%$ and $C = 1.5\%$ the rapid saturation of a structural function takes place on scales of the order of $R = 3$cm. For $C = 4.5\%$ the form of a structural

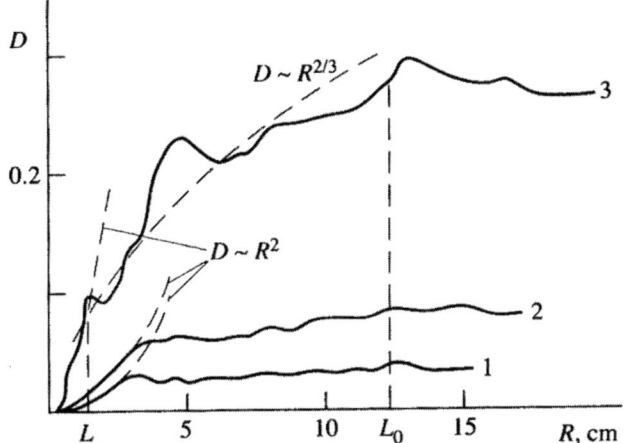

**Figure 7.14.** The structure function of intensity fluctuations for a signal transmitted through a water drop medium with three volume concentrations: 1—0.28%, 2—1.5%, 3—4.5%.

function principally differs from previous cases. Here both the internal ($L = 1.5$ cm) and external ($L_0 = 12$ cm) scale of inhomogeneities is clearly exhibited (Figure 7.14), and in the interval $L < R < L_0$ the structural function grows as $D \sim R^{2/3}$. The limiting value of a structural function (in the saturation region) is equal to double the value of variance in fluctuations.

Thus, analysis shows that—for a volume density of particles $C = 4.5\%$ ($d = 0.9\lambda$)—the scattering of electromagnetic radiation in a medium occurs in spatial inhomogeneities whose scales lie in the interval between $L = 1.5$ cm and $L_0 = 12$ cm, which turns out to be much greater than particle size (diameter $= 0.3$ cm) and the distance between them ($d = 0.7$ cm). This fact, in its turn, confirms the existence of collective effects in scattering. The fact that intensity fluctuations are distributed according to the normal logarithmic law and that the spectrum of fluctuations and a structural function can be described by the well-known exponential laws of "$-5/3$" and "$2/3$", respectively, is indicative of the turbulent-vortex character of fluctuations with quasi-vortex inhomogeneities. Therefore, a discrete disperse medium for $d < \lambda$ can be considered to acquire the properties of a continuous, randomly inhomogeneous medium, in which spatial fluctuations in dielectric permittivity take place (Rytov et al., 1978).

Let us now consider the results obtained using a Doppler scatterometer in the mode of observation of microwave radiation backscattering by the same disperse medium (Cherny and Sharkov, 1991b). Figure 7.15 presents the Doppler spectra of a scatterometric signal backscattered by a disperse medium. The measurements were carried out in such a way that the moving flux of particles had a velocity component in the direction of the instrument. As a result, the power of radiation scattered by particles lies in the spectrum of a scattered signal at the Doppler frequency $f_D$ determined by the velocity component in the direction of the instrument

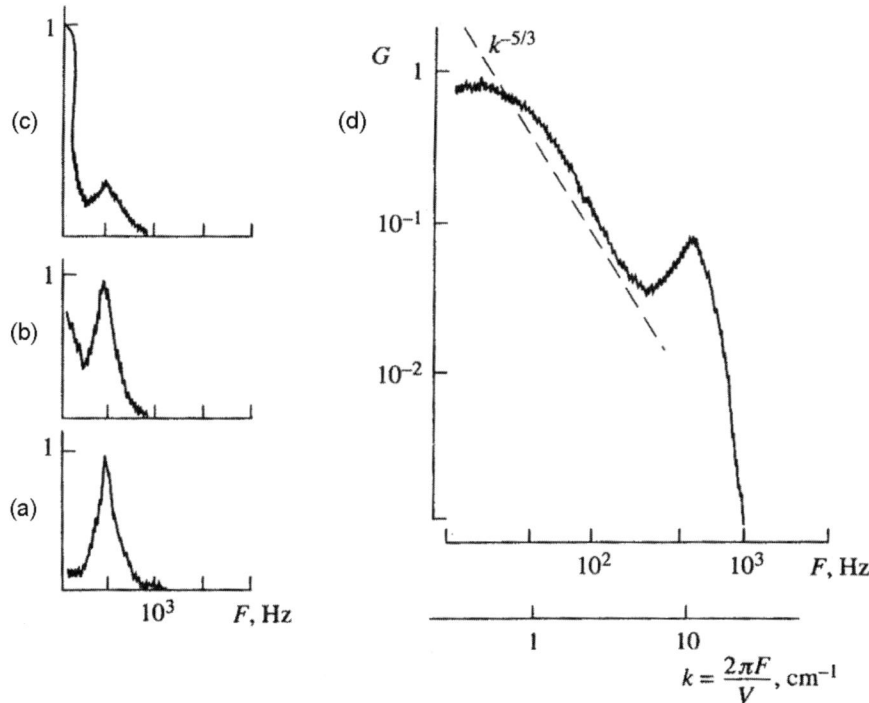

**Figure 7.15.** Normalized Doppler spectra of the backscattering signal from a disperse medium with three volume concentrations: (a) 0.28%, (b) 1.5%, (c) 4.5%, (d) 4.5% in bi-logarithmic coordinates. $V_D = 1.7\,\mathrm{m\,s^{-1}}$; $V = 2\,\mathrm{m\,s^{-1}}$.

$f_D = 2V_D/\lambda$, which is clearly seen for particle volume density $C = 0.28\%$. However, as particle density grows (Figure 7.15b, c), the form of spectrum principally changes and, along with the Doppler components, the additional component appears in a spectrum that is concentrated near "zero" frequencies (for $C = 4.5\%$). The appearance of "zero" frequencies in a spectrum in the case of moving scatterers can be physically related to the loss of temporal coherence of a scattered signal. This makes it impossible to detect the phase of a signal with the purpose of obtaining information on the object velocity based on the Doppler effect.

The presentation of results in the bi-logarithmic coordinate system (Figure 7.15c) reveals an interesting point: the spectrum of an "incoherent" component obeys the exponential law of "$-5/3$" in the frequency band of 20–200 Hz. Moreover, the range of spatial frequencies $k$, where the spectrum obeys the "$-5/3$" law in backscattering mode, is exactly the same as in the case of radiation transmission within line-of-sight limits (by exposure to microwave radiation) (see Figure 7.14). It can be supposed that—in both bistatic and scatterometric modes of measurements—fluctuations in intensity are identical in nature. The exponential law in the spectrum of scattering, as well as the gamma distribution of intensity amplitudes, can be considered the result of scattering from a fractal, geometrically bound structure (or from a turbulent-vortex space) in a volume body of discrete flow (Lakhtakia *et al.*, 1987; Varadan *et al.*, 1983).

Analysis of calculated and experimental data indicates that there exists a quite certain (critical) value in the distance between absorbing scatterers ($d/\lambda \leq 1.5$), which stipulates principal re-arrangement of both the average values of electrodynamical parameters and the fluctuation mode.

It is interesting to compare experimental results with the electrodynamic, closely dense medium condition, obtained theoretically in Rytov *et al.* (1978):

$$n\alpha \geq 1 \tag{7.76}$$

where $n$ is the average density of scatterers; and $\alpha$ is the polarizability of particles. In the Rayleigh approximation this is equal to

$$\alpha = r^3 |(\dot{\varepsilon} - 1)/(\dot{\varepsilon} + 2)| \tag{7.77}$$

where $\varepsilon$ is the complex dielectric constant of a scatterer's material. Condition (7.76) physically describes the contribution of induced dipoles, closest to the original particle, to the effective field. So, taking into account the critical value of $d = 1.5\lambda$, $n \approx 2.6\,\mathrm{cm}^3$, we have

$$n|a| = 1.2 \cdot 10^3 \ll 1$$

Thus, long before satisfying condition (7.76) a dense discrete medium principally changes its properties and becomes similar, in a certain sense, to a continuous medium that has fluctuating parameters.

It is interesting to note, however, that seemingly similar physical structures (a set of hollow aqueous spheres) manifest themselves in a quite opposite manner: even the compact, dense packing of scatterers of such a type does not make any noticeable contribution to the electrodynamics of a system owing to the very weak effect of single-structure interaction. Each hollow aqueous sphere represents an almost black-body emitter that does not possess any noticeable scattering properties and does not interact with the surrounding components of a system (Raizer and Sharkov, 1981; and Chapter 6).

## 7.7   CONCLUSIONS

**1**   The results of specialized experiments, carried out under the guidance of the author of the present book, on studying the electromagnetic properties of concentrated drop fluxes and the possibility of using them to form electrodynamic models of the drop-spray phase of breaking waves, make it possible to establish the boundaries of applicability of the transfer theory to describing the electrodynamic parameters of discrete aerodisperse media that have a high density of absorbing scatterers. It is found experimentally that—in the electromagnetic interaction of solitary scatterers—the aerodisperse discrete system acquires the properties of a continuous turbulent medium with spatial fluctuations in its dielectric characteristics.

2  A qualitatively new electrodynamic model of a discrete medium that takes due account of the interaction of scatterers is proposed.

# 8

# Field optical–microwave remote sensing of the air–sea transition zone in the atmosphere–ocean system

This chapter presents the results of experimental remote field investigations directed at revealing the physical features and electromagnetic properties of dispersed random structures generated during the process of gravity wave breaking and drop-spray phase capture in the powerful airflux of a near-surface wind. In such a task it is vitally important to achieve instrumental controllability of hydrometeorological conditions: the spatiotemporal characteristics of an oceanic wave breaking field, the wind field turbulent regime, the thermal stratification state, and the atmospheric pressure field. Measurement of the spatial characteristics of these parameters by remote techniques represents a complicated experimental problem in itself. The best approach to revealing the physical features of random dispersed media in the ocean–atmosphere system seems to lie primarily in facilitating complex (radio-optics) and multifrequency (radiospectroscopic) active–passive sensing of the ocean–atmosphere system, and in improving the air–space information processing techniques and radiophysical models of the ocean–atmosphere system.

## 8.1 SPACE OCEANOGRAPHY PROBLEMS

A topical problem encountered when developing air–space oceanography nowadays is the study of the interactive mechanisms between electromagnetic radiation and a rough sea surface. In spite of the fact that investigations over the last 25–30 years have demonstrated, in general, the validity of the two-scale radiophysical scattering model for signals with vertical polarization (in transmission–reception mode) (Moor and Fung, 1979; Alpers and Hasselmann, 1982), the results of many works indicate the urgency of organizing special experiments with the purpose of revealing the physical features of scattering signals with horizontal polarization.

The latter is related to the fact that in this mode the powerful splashes of a scattered signal can be observed in both real and synthesized images. These splashes form a peculiar speckle structure of images and essentially restrict the possibilities of observation and spectral processing of the regular structure of heavy seas and surface manifestations of intraoceanic phenomena (up to the complete disappearance of a quasi-regular structure). As a characteristic example, we present in Figure 8.1 the radio images of the same area of oceanic surface in the northwest part of the Pacific Ocean, near the Kamchatka Peninsula. These images were obtained by air-based side-looking radar investigations within the framework of the USSR Academy of Sciences Space Research Institute's[1] complex of works on remote investigation of the sea surface structure. Under moderately rough sea conditions (Figure 8.1a) the quasi-regular structure of surface manifestations of internal waves is clearly seen. However, under highly rough sea conditions—6 on the Beaufort Scale (Figure 8.1b)—the speckle structure of a scattered signal fully suppresses the quasi-regular structure of the surface manifestations of internal waves.

It has been firmly substantiated that speckle signals are correlated with the process of breaking of large gravity waves illuminated by a sensing signal in a spatial resolution element (Kalmykov et al., 1976; Lewis and Olin, 1980). However, further analysis revealed quite different points of view on the physical nature of scattered signal formation. One widespread viewpoint (e.g., Moore and Fung, 1979; Kwoh and Lake, 1985; Bulatov et al., 2003; Kanevsky, 2004) is based on the hypothesis that the dominant contribution to backscattering is made by quasi-mirror reflection (with respect to an observer) of electromagnetic waves from overturning crests, including diffraction phenomena on sharp crests and resonance scattering on the capillary wave trains captured by a wave (the Bragg mechanism). Another viewpoint (Kalmykov et al., 1976) states that a backscattered signal forms as electromagnetic waves interact with the dispersed phase of a breaking wave. It should be noted, however, that the authors of both hypotheses do not present experimental evidence of the requisite accuracy, since the corresponding experiments were carried out either in laboratory conditions (Kwoh and Lake, 1985), or in the rather exotic condition of a wave running along the coast (Lewis and Olin, 1980) or on an underwater breakwater (Kalmykov et al., 1976). In recent years a series of radiophysical experiments was performed in a special wave pool (Sletten et al., 2003). These experiments, however, still did not throw full light on the physical mechanism of backscattering.

Of course, the results of these investigations cannot be correlated fully with the real situation in the open ocean. To clarify the physical causes of speckle structure formation of a backscattering field it is vital to undertake a detailed radiophysical experiment that involves filming and photorecording of the process of individual, large gravity waves breaking. Such an experiment should be carried out under field sea conditions, but with regard to its geometrical parameters (observation angle, resolution element, etc.) it should be similar to the situation arising at observation from movable air- and space-based carriers. The next section is devoted to description of the results of experiments of such a kind.

---

[1] IKI AN SSSR is the Russian name of this organization.

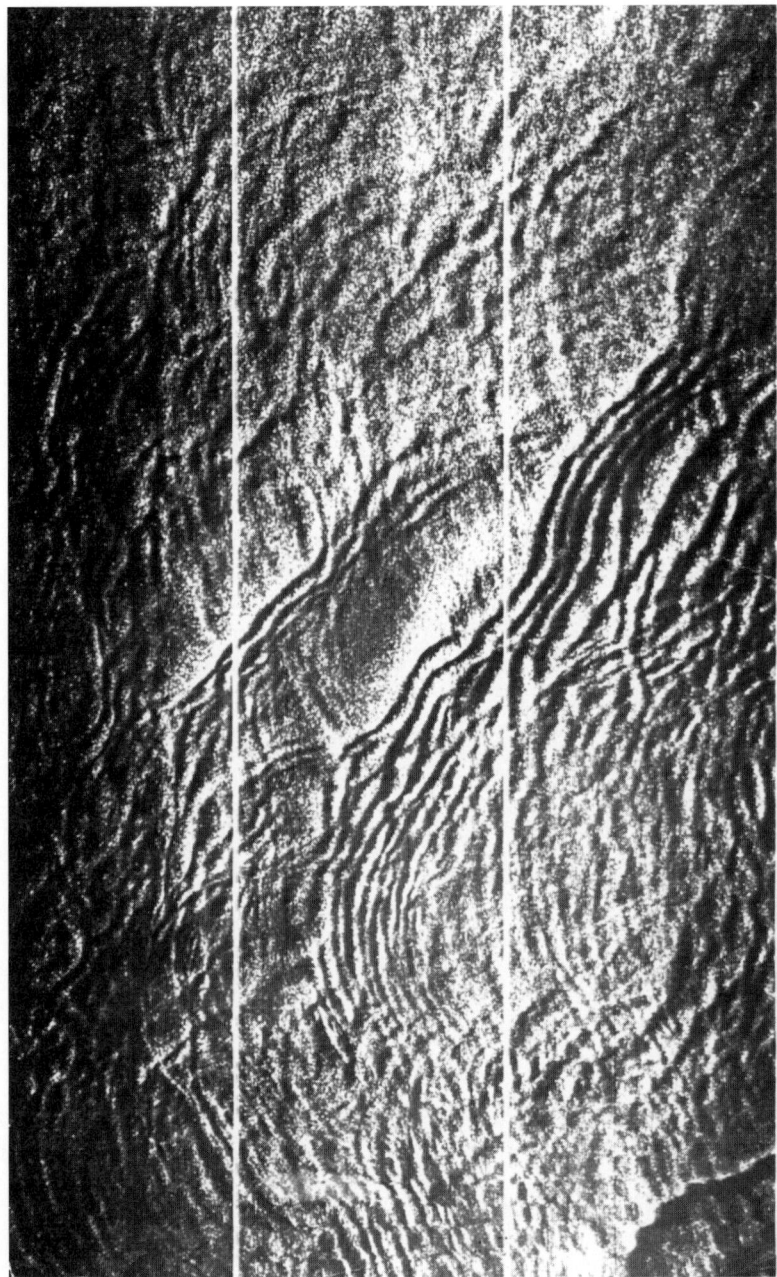

(a)

**Figure 8.1.** Side-looking radar survey experimental pictures of a disturbed sea surface in the Pacific Basin near the Kamchatka Peninsula. The radar survey was carried out by "TOROS" side-looking radar onboard the Russian airplane laboratory AN-24. (a) The radar image indicates the surface manifestations of internal waves in the form of diverging waves positioned in the middle of a sea. The space interval between the white lines is equal to 5 km.

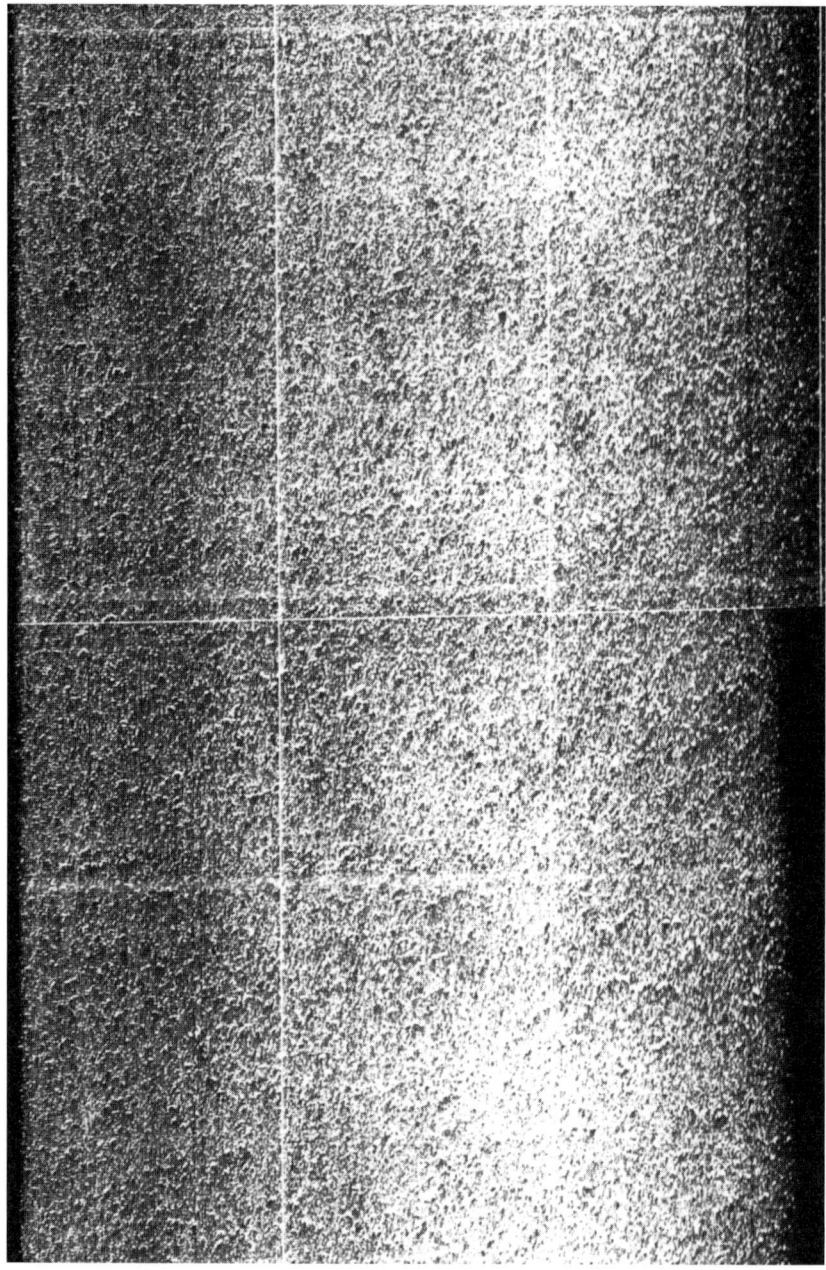

(b)

**Figure 8.1** (*cont.*). Side-looking radar pictures of a disturbed sea surface in the Pacific basin near the Kamchatka Peninsula. The radar survey was carried out by "TOROS" side-looking radar onboard the Russian airplane laboratory AN-24. (b) The radar image of the same basin indicates the special structure of the backscattering field. The space interval between the white lines is equal to 5 km.

## 8.2  OPTICAL AND RADIOPHYSICAL INVESTIGATIONS OF THE OCEANIC GRAVITY WAVE BREAKING PROCESS

This section presents the results of experimental investigations directed at finding the most adequate method of describing and forming electrodynamic models of the dispersed structures generated during the gravity wave breaking process under natural heavy sea conditions. These field investigations were carried out within the framework of the USSR Academy of Science Space Research Institute's complex of works on remote study of the sea surface structure. The experiments were carried out during the 26th trip of a research vessel—the RV *Dm. Mendeleyev*—to the Indian Ocean Basin (Cherny and Sharkov, 1988).

### 8.2.1  Instruments, techniques, and conditions for performance of the experiment

The principal feature of this experiment was the use of radiophysical remote sensing: a unique active–passive 8-mm complex made up of a radiothermal radiometer and a Doppler scatterometer combined with an analog magnetic recorder. The latter device made it possible to record the fluctuating components of signals up to 5 kHz and, then, to perform detailed Fourier-processing of the Doppler components of a scattered signal and to reveal their time evolution in detail.

The block diagram and principle of operation of the combined radiothermal radiometer and Doppler scatterometer are presented in Cherny (1982).

The basic characteristics of the radiometer–scatterometer used in the experiment were as follows:

- the fluctuation sensitivity of the radiometer was $\Delta T = 0.2\,\mathrm{K}$ at an integration time constant of 1 s;
- the radiometer's energetic and radiometric passbands were measured—using a technique developed by Sharkov (2003)—and found to be equal to 1160 and 1670 MHz, respectively;
- the passband of the low-frequency section (after the quadratic detector) was 20–20,000 Hz;
- the power in the emitting antenna (the scatterometer channel) was 40 mW;
- the dynamic range of measurements in the scatterometric channel relative to the inherent noises of the receiver was not worse than 60 dB;
- the minimally detectable effective scattering surface (ESS), measured by means of corner reflectors at a distance of 55 m, equaled $3 \cdot 10^{-4}\,\mathrm{m}^2$;
- the resolution in Doppler velocity of a measured object (the root-mean-square deviation) was equal to $0.04\,\mathrm{m\,s}^{-1}$; and
- the width of the main lobe of the horn in the antenna's directional pattern (ADP) at the 3-dB level was equal to $12°$ (the installation of antennas provided the possibility of carrying out measurements in radiation–reception mode at all polarizations (VV, HH), as well as in cross-polarization mode (HV and VH).

Remote sensing of the sea surface was carried out by means of the radiometer–scatterometer, whose high-frequency unit and antenna system were installed on the gyrostabilized platform of a rotating device on the vessel's port side at a height of 11 m from the water surface mean level. The angular stabilization system of the device compensated for the effect of roll to an accuracy of $\pm 0.20°$ and, thus, kept the observation angle of antennas constant (equal to $80°$ from the vertical) irrespective of the roll angle of the vessel. The spatial element of the scatterometric channel resolution was formed as a result of the antenna directional pattern and the peculiarities of coherent reception of a signal with FM modulation (the Bessel function envelope). This element can also be determined by the convolution integral $P = P_1 \times P_2$, where $P_1(x, y)$ is the sea surface zone illuminated by ADP ($i = 1$) and formed due to FM modulation ($i = 2$). Special calibration of the scatterometer's resolution element using the movable corner reflector showed that the zone (at the 3-dB level) of maximum sensitivity represents an ellipse with major axes of 120 and 30 m, and a 70-m slope distance to the ellipse center. The radiophysical measurements of drop-spray dispersed formations, carried out during the breaking process of large gravity waves, were accompanied by synchronous filming. The latter was performed by means of a movie camera at a rate of 48 frames per second. Synchronization was achieved by delivering a special marking pulse to a tape recorder, the temporal position of this marker characterizing the actuation instant of the movie camera. Subsequent comparison of the two types (optical and radio) of information was performed by frame attribution to registrograms of active and passive radio signals reproduced on an analog self-scriber from the tape recorder. The total time of filming a cycle of measurements as a rule did not exceed 20 s. As a result, the time synchronization error, with regard to the $\pm 1\%$ deviation from nominal of the tape recorder's tape speed, was less than $\pm 0.2$ s.

During the experiment, special attention was given (and this turned out to be the basic complication of the way in which the experiment was carried out) to distinguishing and fixing the initial wave breaking stage. By the latter we mean the stage of formation of a sharp crest (but still without a foam phase). To do this, the initial phase in a wave had to be fixed at the instant the wave was situated precisely in the maximum sensitivity zone of a resolution element. This operation—the peculiar location of a gravity wave at the instant of breaking in range—was performed by a skilled operator (with field glasses), who also actuated the recording system's markers. The breaking of surface waves was observed in a developed rough sea system. According to the data of standard hydrometeorological measurements, the sea surface state corresponded to a wind force of 5–6, the wind velocity (at an altitude of 20 m) was 12–13 m s$^{-1}$, the height of waves was 2.5 m, and the average period of the energy-carrying component of sea waves was 5 s.

### 8.2.2   Experimental investigations of natural breaking

A typical example of synchronous observations of the sea surface for such a geometry is the registration of Figure 8.2, which shows the time dependence of backscattered and radiothermal signals at identical time constants and at two types of polarizations. A characteristic feature here is the presence of a backscattered signal of prominent splashes in HH mode (up to 10 dB with respect to an average background) against

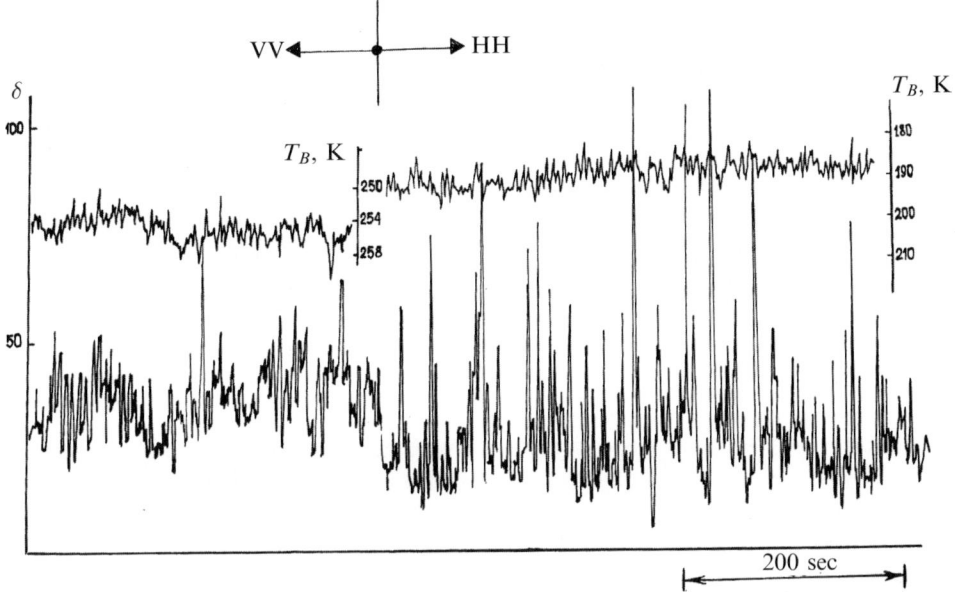

**Figure 8.2.** The temporal diagram of the scatterometer and radiothermal channels recovered during observation of sea state surface. The upper record is the thermal signal; the lower record is the scatterometric signal with HH and VV polarizations. Sea state was 3–4 on the Beaufort Scale. The time constant is 0.5 s for two channels.

a fluctuating "Bragg" background. Note that no "sharp" changes in signal ("splashes") are observed in the radiothermal channel (despite the equality of time constants for active and passive channels). The change in the general average level of natural radiation (the brightness temperature—$T_B$) in VV and HH modes is related to the polarization properties of surface emissivity, and the prominent modulation of a radiothermal signal is determined by variations in the slopes of large-scale waves passing through the instantaneous field of view of a complex system's ADP. So, spectral analysis of the time dependence of $T_B$ allowed us to determine that the average period of an energy-carrying component (making an allowance for vessel motion) was 4.7–5 s, which coincides with the results of hydrometeor measurements. However, similar analysis for a scatterometric signal cannot be performed fully because of the presence of a strong-noise spectral background (up to 150–200 Hz) from the "delta"-pulses of splashes and a fluctuating Bragg background. These scatterometric signal splashes are caused by the breaking of individual gravity waves.

The possibilities provided by using remote instruments allowed us to study in detail the features of radiothermal and scatterometric signals of the breaking of individual gravity waves. The frame sequence of the wave-breaking process is presented in Figure 8.3. Synchronous registrograms of changing brightness temperature $T_B$ and effective backscattering surface (EBS) $\sigma$ are presented in Figure 8.4, and the

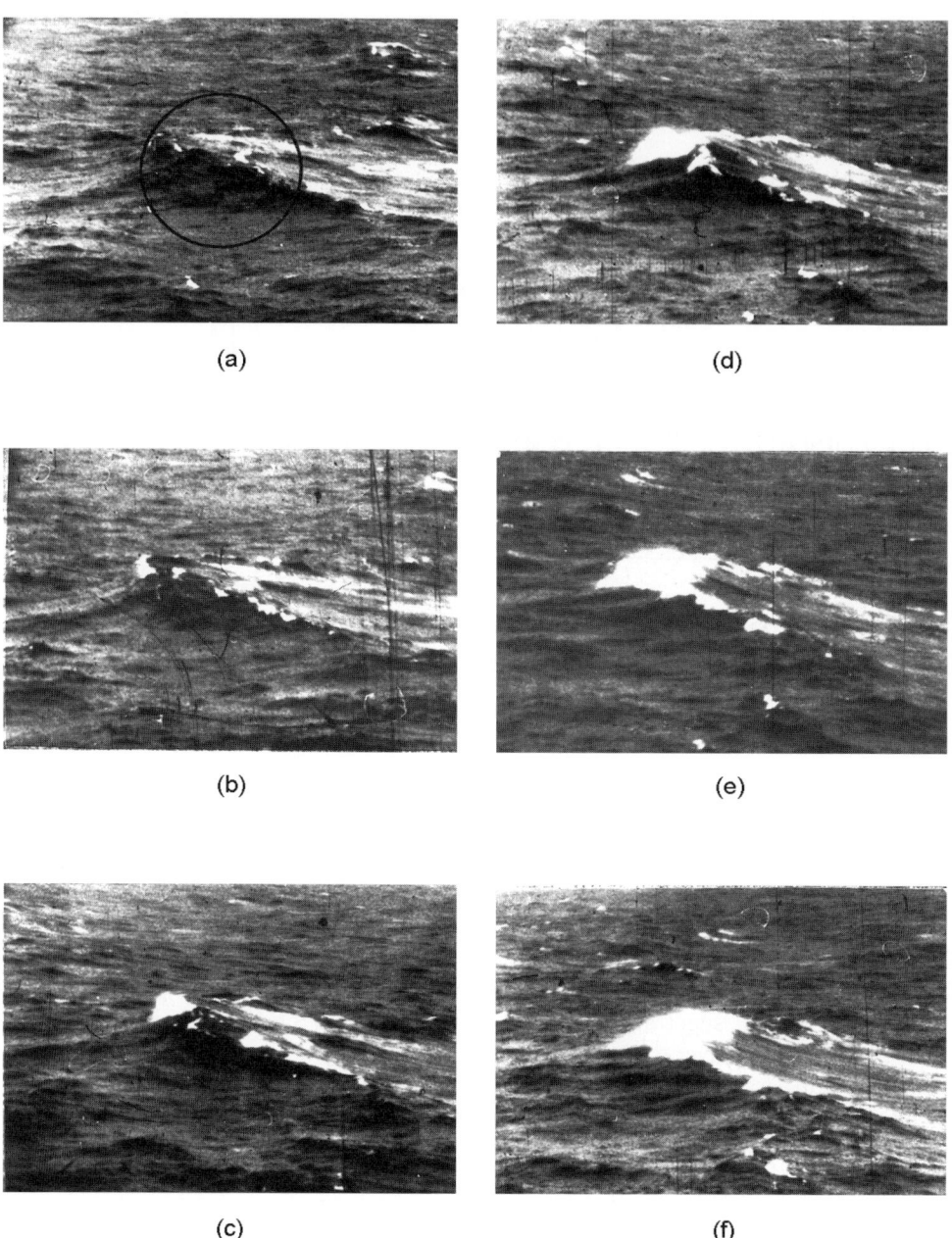

(a)  (d)

(b)  (e)

(c)  (f)

**Figure 8.3.** A motion picture sequence of the breaking process for the gravitational wave under study. The zero instant of time (0) corresponds with wave crest sharpening: (a) $t = 0$ s; (b) $t = 0.5$ s; (c) $t = 1.0$ s and so on; (d)–(f) correspond with instants of time 1.5–5.5 s). The black circle field on the image in (a) corresponds to the instant field of the antenna view in the picture plane.

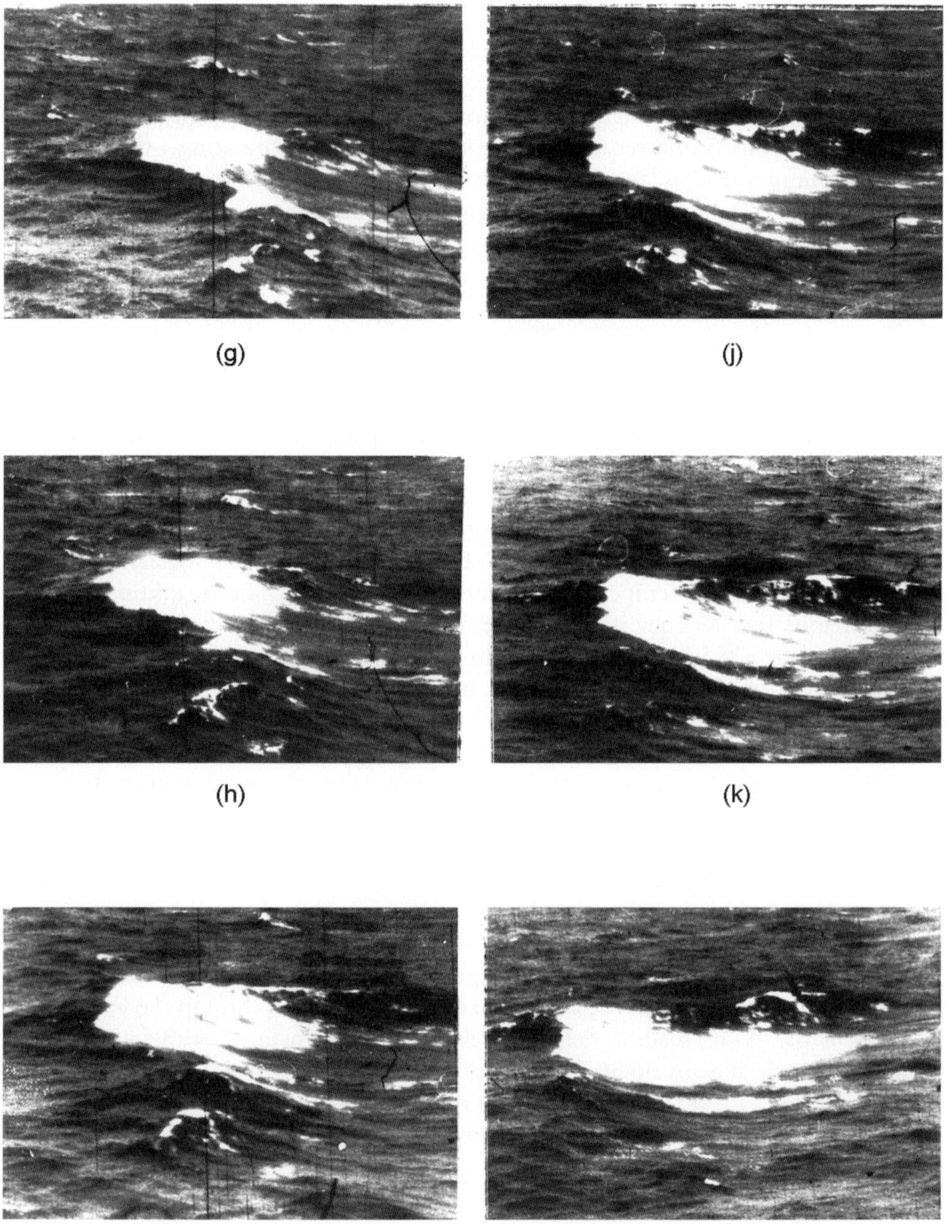

(g)

(j)

(h)

(k)

(i)

(l)

**Figure 8.3** (*cont.*).  A motion picture sequence of the breaking process for the gravitational wave under study. The zero instant of time (0) corresponds with wave crest sharpening: (g)–(l) correspond with instants of time 1.5–5.5 s.

result of spectral processing of Doppler components of a scatterometric signal in Figure 8.5. The instant of time at which counting began was strictly synchronized in all three figures.

We shall now run through the features of measurements we have carried out. It can be seen from frame sequence analysis that—before breaking—the wave configuration changes and is characterized by a sharp top (Figure 8.3a; $t = 0$ s). It is important to note that at exactly this instant no features are observed in the scatterometric signal (as opposed to statements of hypothesis on quasi-mirror reflections), and an intermediate maximum is noticed for brightness temperature variations (Figure 8.4). The appearance of a maximum is related to the circumstance that some part of the surface in the ADP's instant field of view is considered under an observation angle smaller than the remaining surface, which just causes a positive contribution to the brightness temperature value $T_B$ at horizontal polarization.

This is followed by wave breaking and formation of a "boiling" breaker (Figure 8.3b–m; $t = 0.5$–2.0 s), and the intensive drop-spray phase arises here, which is testified to by a sharp increase (of about 10 dB) in the backscattering signal. The lifetime of the intensive phase of a drop-spray structure equals about 1 s (see Chapter 4), after which the foam field is formed (Figure 8.3e; $t = 2$ sec). Analysis of frame sequence and brightness temperature variations, caused by foam formation, indicates that foam spot size reaches its maximum in 4–5 s after the start of wave breaking (Figure 8.3i–m; $t = 45.5$ sec), and then the foam field breaks up.

From visual estimations, foam spot size was 10–15 m. Brightness temperature contrast from the foam field only equaled 5–6 K. This is explained by the fact that the angular size of the foam cover section is essentially smaller than the angular size of the ADP instant field of view, and the brightness temperature is averaged over the antenna spot area. Taking this into account (according to appropriate rules— Sharkov, 2003), we can estimate the brightness contrast of the foam field. It equaled about 80 K ($\lambda = 8$ mm). This estimate correlates well with field measurements (see Chapter 6). It is interesting that the intermediate minimum of the radiothermal signal (Figure 8.4; $t = 2$ s) exactly corresponds with the instant of sharp increase in the backscattered signal at wave breaking (Figure 8.3; $t = 2$ s) and to the disappearance of a prominent slope (toward the observer) of a gravity wave, though the wave itself is observed in the ADP instant field of view up to the instant $t = 3.5$ s. The foam structure, as such, has a velocity—according to some authors' data (Lewis and Olin, 1980)—of about 11.5 m s$^{-1}$ at birth ($t = 11.5$ s), and then its motion virtually stops, which is clearly demonstrated by the frames in Figure 4.2i–m in the time interval $t = 4.0$–5.5 s.

As far as the amplitude of a splash of the scattered signal is concerned, it can easily be seen that the "true" EBS value $\sigma_0$ of a scattering object is related to its measured contrast ($\Delta\sigma$) and geometrical area ($S_0$) as $\sigma_0 = \sigma_F + \Delta\sigma(S/S_0)$, where $S$ is the area "illuminated" on the ADP surface, and $\sigma_F$ is the "background" EBS. If the linear size of a scattering object is supposed to be about 5 m, then $\sigma_0/\sigma_F = 30$–35 dB.

We shall now analyze the result of spectral processing of a scatterometric signal. A specific feature of the scatterometric instrument used in the experiment—namely, its operation in continuous emission mode with sine frequency modulation and its

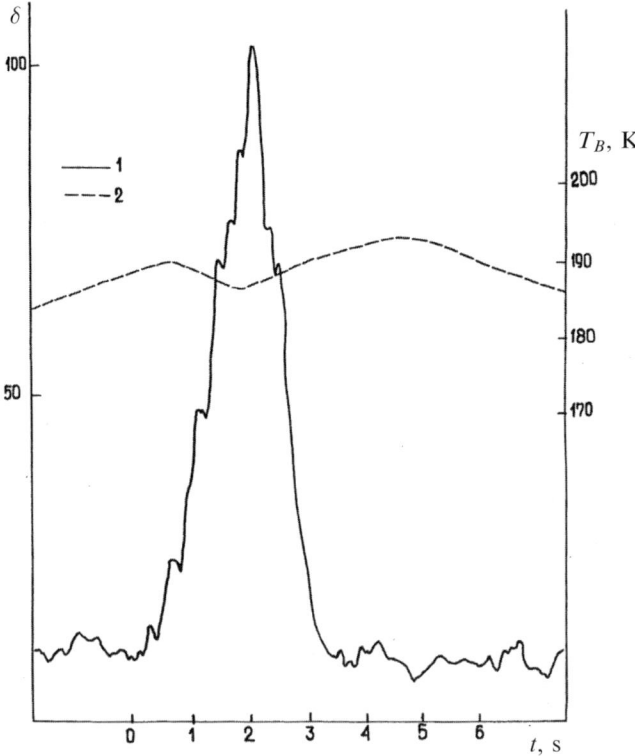

**Figure 8.4.** The temporal diagram of the scatterometer (1) and radiothermal (2) channels recovered in the gravity wave breaking process with HH polarization. The temporal coordinate is closely synchronised with the filming (Figure 8.3).

ability to separate the second harmonic of modulation frequency in the instrument's receiving section (Cherny, 1982)—made it possible to obtain information on the velocity of objects from spectral analysis of the amplitude fluctuations of a signal. Spectral processing of a scatterometric signal was performed on a specialized computer using the rapid Fourier transformation (RFT) algorithm, and the results were presented in the form of an evolution diagram of the current spectrum in the three-dimensional coordinate system "spectral density value–Doppler frequency–time" (Figure 8.5). The basic characteristics of the processing process were: sample length of 0.1 s; effective spectral resolution of 20 Hz; total band of analysis of 20 Hz–2.5 kHz; and root-mean-square relative error of 0.7 (or 2.3 dB).

The main feature of evolution of the scatterometric signal's spectrum is its multimodal dynamical structure. Along with the low-frequency (LF) region, which is rather stable during the whole breaking process and falls down as $f^n$ (where $n = 1.5$), a series of high-frequency (HF) components is observed in the form of "islets", which possess short lifetimes (of about 0.1–0.2 sec) and intensities comparable with or even slightly exceeding the LF components. The frequency

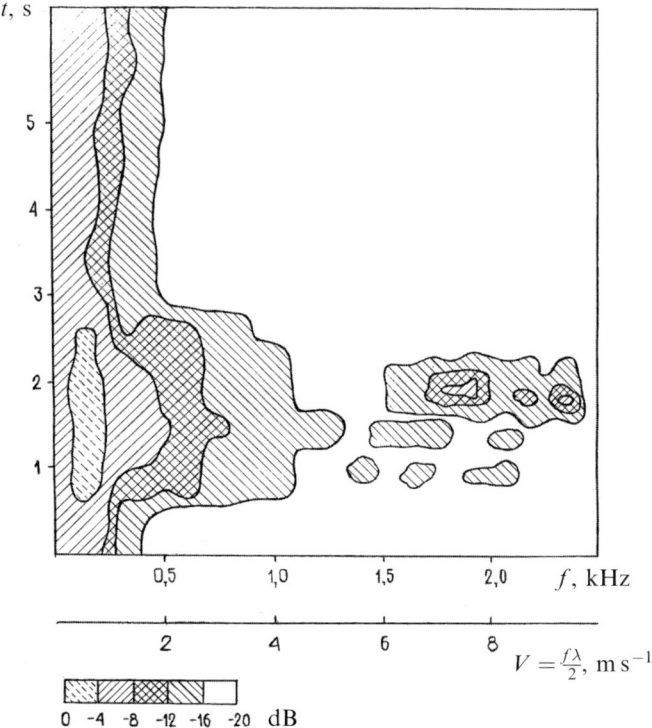

**Figure 8.5.** Time evolutiuon of Doppler spectra recovered in the gravity wave breaking process. The Doppler frequency is $f = (2V/\lambda)$, where $V$ is the vector modulus for the velocity of relative displacement in the direction of instrument-target; and $\lambda = 0.8$ cm. The temporal coordinate is closely synchronised with filming (Figure 8.3) and remote-sensing signal registrations (Figure 8.4).

approximation of "islets" can be estimated as $(f - f_0)^{-2m}$, where $m = 3$–$4$, and $f_0$ is the frequency of "islet" maximum.

With the appearance of high-frequency (HF) components the spectrum of the low-frequency (LF) region essentially broadens up to frequencies of $1.0$ kHz; however, the general character of the frequency dependence of a spectrum remains the same: $f^{-n}$ ($n = 1.5$–$2$). Analysis of synchronous filming shows that the current spectrum of a scattered signal drastically changes in the time section when the intensive drop-spray phase appears in the breaking process ($t = 1$–$2.5$ s). Taking into account the relationship between the vessel velocity vector, the general direction of sea disturbance (and the wind velocity vector) and the azimuth of breaking wave observation direction, we can estimate the horizontal components of velocities of scatterers (Figure 8.5): for the HF region of the Doppler spectrum they equal $6$–$9$ m s$^{-1}$. In its turn, estimation of the vertical components of scatterer velocities gives for the LF region values of $10$–$18$ m s$^{-1}$, and for the HF region values of $30$–$40$ m s$^{-1}$.

### 8.2.3   Experimental investigation of the breaking of ship waves

Because of the difficulty in reaching full coincidence, in time and space, of the direct breaking process and the ADP instant field of view under real, rough sea conditions, some model experiments have been performed in the same cycle of observations. In these experiments the backscattering and natural radiation of the sea surface, which arise at the breaking of ship waves excited by a vessel's motion, have been recorded. As a typical example, Figure 8.6 presents a photograph of a ship wave (obtained from onboard a moving vessel) at the instant of wave breaking and drop-spray phase arising. The antennas were pointed exactly toward that sea surface section (at a distance of 10 m from the body) where wave breaking occurred approximately every 5 seconds. The observation angle was 55° in this case; polarization was in emission–reception VV mode; the ADP instant field of view was 8 m; the natural state of rough sea corresponded to a wind force of 2 at a wind velocity of 3–5 m s$^{-1}$; and the height of breaking of a ship wave was about 1.5 m (Figure 8.6). The geometry of the experiment allowed us to get a stable and (what is very important methodically) reproducible picture of the breaking phenomenon. Certainly, from the hydro-dynamical point of view of breaking these two processes—breaking on deep water and breaking of a ship wave—are quite different hydrodynamical phenomena. But from the viewpoint of their electrodynamic properties these processes, as we shall see below, are very similar. So, analysis of Figure 8.7, which presents synchronous registrograms of backscattering $\sigma$ and brightness temperature $T_B$, indicates that the qualitative picture of the regularly repeated process of breaking of a ship wave corresponds well with the process of gravity wave breaking on deep water (Figure 8.4). The quantitative characteristics of the process, however, are determined by a new experiment geometry and by nearly complete filling of the ADP field of view with the foam mass of crests, with breaking strip foam, and with drop-spray phase volume. Detailed synchronous filming (not presented in the given text) allowed us to establish that the presence of a drop-spray phase in a breaking wave (the "boiling" breaker at instants $t = 2, 8, 12, 17$ s on the diagram of Figure 8.7) corresponds to the back-scattering maximum, and brightness temperature maxima fall on those time instants when strip foam has generated on the surface ($t = 4, 9, 14, 21$ s on the diagram of Figure 8.7). Analysis of the current spectrum of a scatterometric signal (Figure 8.8) also demonstrates its qualitative similarity to the spectrum of the backscattering value of a breaking wave. However, the former spectrum is essentially richer in details, which is to be expected if we remember the more detailed geometry of the present experiment. As in the case of individual wave breaking, this spectrum pos-sesses two prominent regions: a low-frequency (LF) one in the form of a spectrum falling as $f^{-n}$, where $n \approx 2$, and the high-frequency (HF) components that form as "splashes" ("islets") in the frequency area at the instants of appearance of a "boiling" crest with a drop-spray phase over it. At these time instants the HF components stretch up to frequencies of 2.5 kHz, which testifies to the Doppler velocities (toward an observer) of scattering particles reaching values of 8–10 m s$^{-1}$. With regard to the experiment geometry, we can estimate the vertical components of scatterer velocities as 4–7 m s$^{-1}$ (for the LF region) and 14–16 m s$^{-1}$ (for the HF region). As in the case of

**Figure 8.6.** A photograph of a ship wave breaking in a 1-point (Beaufort Scale) sea state. The picture was taken from the research vessel. The area within the black circle on the picture agrees with the field of view of the antenna main lobe in the picture plane.

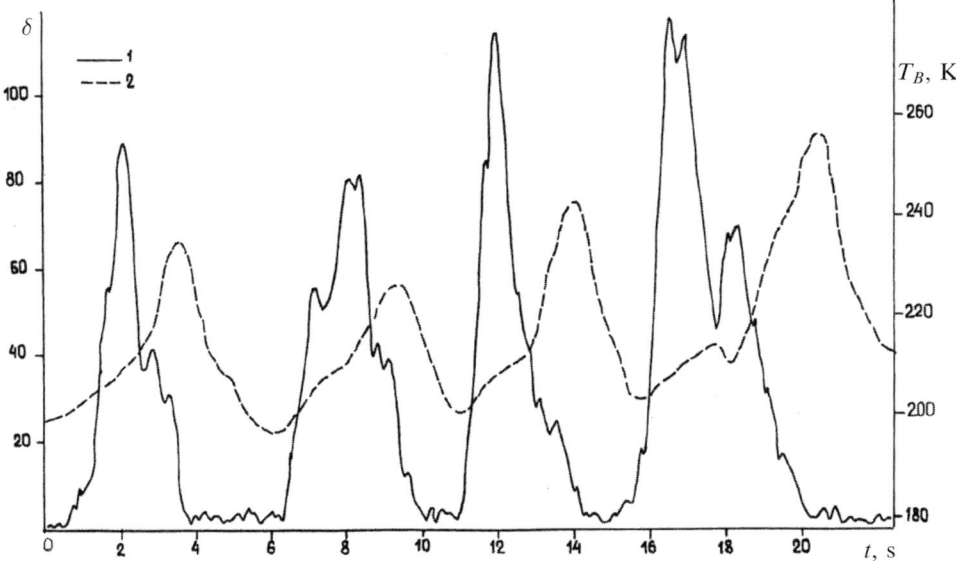

**Figure 8.7.** The temporal registration of the scatterometer (1) and radiothermal (2) channels used in four ship wave breakings with VV polarization.

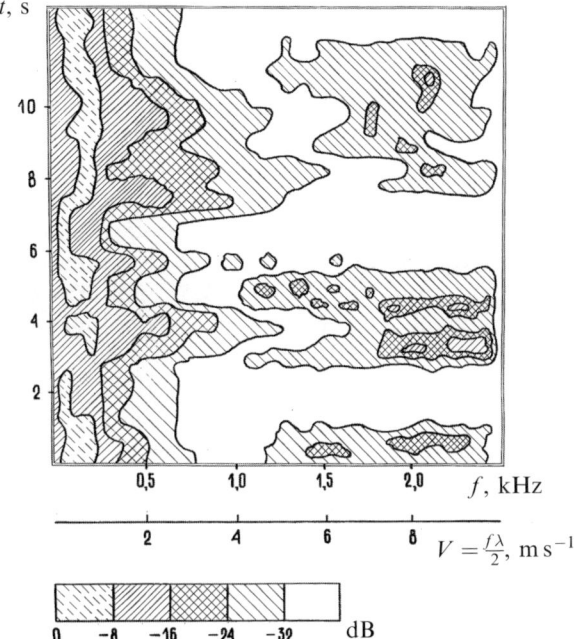

**Figure 8.8.** Evolution with time of Doppler spectra recovered in the ship wave breaking process. The Doppler frequency is $f = (2V/\lambda)$, where $V$ is the vector modulus for the velocity of relative displacement in the direction of the instrument target; and $\lambda = 0.8$ cm. The temporal coordinate is closely synchronised using remote-sensing signal registrations (Figure 8.7).

gravity wave breaking (Figure 8.4), in the presence of foam fields in the ADP field of view (at $t = 9$–11, 14–16 s) no features are observed in the scattered signal spectrum, and the total intensity of the scattered signal is also small—it remains at the level of scattering from a free (from foam) sea surface. Note also that at backscattering maximum instants the registrograms of brightness temperature values clearly show suppression of the natural signal from foam structures (as in the case of gravity wave breaking—Figure 8.4). The only physical object capable of suppressing foam structure radiation is the drop-spray cloud that forms over the foam cover exactly at this time instant.

### 8.2.4   Interpretation of results: the drop-spray model and a "radioportrait" of a breaking sea wave

The physical hypothesis, proved in this investigation, consists in the statement that it is the backscattering from water drops of a dynamical dispersed structure—the drop-spray phase (DSP)—formed at some particular phase of gravity wave breaking (and, necessarily, after crest sharpening), which makes the dominating contribution to the speckle signal. Analyzing our data (Figures 8.3–8.8), we can easily conclude that the quasi-mirror reflection at crest sharpening (as well as the diffraction effects) do not make any noticeable contribution to backscattered signal formation (at least, in the given experiment geometry). Similar results are demonstrated by the data of Lewis and Olin (1980) in another geometry: the wavevector is directed toward an observer. However, in our case the noticeable backscattered signal at the instant of crest sharpening (at $t = 0$ s, conventionally) is absent as well. A backscattering maximum is achieved only at the instant of generating the foam crest and drop-spray phase formation over it. The latter can be noticed on film frames as a weak (pale) veil above the bright image of a crest (Figure 8.3, $t = 1.5$–$2.5$ s). This phase can clearly be seen—even up to visualizing individual drops—on the photograph of the breaking of a ship wave (Figure 8.6).

Thus, the only physical scatterers with considerable vertical and horizontal translation velocities are water drops in the DSP, since the foam mass (of a crest and strip foam) possesses very weak scattering properties (this follows from laboratory investigations—Chapter 6). On the other hand, according to the data by Lewis and Olin (1980) and Bezzabotnov (1985), we can place a limit on the velocity of motion of a foam mass in the laboratory coordinate system (at least, up to $1$–$1.5 \, \mathrm{m \, s}^{-1}$). Thus, foam scatterers cannot give any contribution to LF and HF components.

Though systematic experiments on studying the dynamic characteristics of DSP and a dispersed structure at the breaking of large gravity waves (see Chapter 5) were not described in the literature; nevertheless, for comparing them with our experimental data we can use DSP parameter estimates obtained under semi-field and laboratory conditions (Egorov, 1977; Wu *et al.*, 1984; and see Chapter 5), as well as the data of numerical experiments (Vinje and Brevig, 1981).

Experiments have shown (see also Chapter 5) that a DSP is born at the instant of foam crest formation due to the breaking of bubble structures and an aerodynamic "jet breaking the water surface" (near the wave crest). In this case the horizontal velocities of water particles equal $(0.7$–$0.9)U_{10}$; the mass concentration $W$ (water content) of a DSP over a crest is estimated as $10$–$1000 \, \mathrm{g \, m}^{-3}$; the mean diameter of particles $d_m \sim 0.01$ cm; the maximum size of particles $d_m \sim 0.2$ cm; the spectrum of particles falls down as $d^{-2}$; the DSP lifetime corresponds to the breaker lifetime (before foam structure transition into strip foam); and the vertical components of velocity of so-called "jet drops" can reach values of $10$–$50 \, \mathrm{m \, s}^{-1}$.

We now turn to interpreting the dynamic parameters of the experiment under consideration. So, having assumed the logarithmic profile of the mean value of wind force to be $u(z_2) - u(z_1) = (u_*/k) \ln(z_2/z_1)$ (where $u_*$ is the dynamic velocity, $k$ is the Karman constant, $z_2 = 20$ m), at level $z_2 = 10$ m we have $u_{10} = 9.5$–$10.5 \, \mathrm{m \, s}^{-1}$ and, accordingly, we obtain the range of possible velocities of drops as $7.5$–$9.5 \, \mathrm{m \, s}^{-1}$. Comparison with the data of Doppler measurements shows a good quantitative agreement between the dynamic parameters of scatterers and the model estimates presented above.

We shall now turn to estimations of the DSP's natural radiation and backscattering. Having used the model of a homogeneous "rain cloud" over the foam surface, we get the following expression for the total specific EBS of the "cloud–surface" system (Spiridonov and Pichugin, 1984):

$$\sigma^0 = \sigma_S^0 \exp(-2\tau) + 2\pi\eta h \frac{1 - \exp(-2\tau)}{\tau} \qquad (8.1)$$

where $\sigma_S^0$ is the specific EBS of the underlying medium studied; $\eta$ is the radar reflectivity of a "cloud" volume unit; $\tau$ is its optical thickness which is equal to $\gamma h$ (where $\gamma$ is the attenuation in a cloud layer per unit length); and $h$ is the cloud height. The brightness temperature variation $\Delta T_B$, caused by the appearance of "cloud" over the surface with emissivity $\kappa = 1 - |R|^2$, is as follows (Sharkov, 2003):

$$\Delta T_B = T_0 |R|^2 (1 - e^{-2\tau}) \left[ 1 - \frac{\omega}{|R|^2} \frac{1 + |R|^2 e^{-\tau}}{1 + e^{-\tau}} \right] \qquad (8.2)$$

where $T_0$ is the thermodynamic temperature of a system; and $\omega$ is the single-scattering albedo. Having used the Rayleigh approximation of the Mie scattering on spherical drops (Chapter 7) and the aforementioned data on dispersed DSP parameters, we obtain the following estimates of the electromagnetic characteristics of a DSP with a cloud height of $h = 0.5$ m: $\eta/W \cong 10^{-3}$ m$^{-1}$(g m$^3$)$^{-1}$ for $\lambda = 0.8$ cm and $9 \cdot 10^{-4}$ for $\lambda = 0.2$ cm; $\gamma/W \cong 5 \cdot 10^{-3}$(dB m$^{-1}$)(g m$^{-3}$)$^{-1}$ and $10^{-2}$ ($\lambda = 0.2$ cm); single-scattering albedo $\omega = 0.45$ ($\lambda = 0.8$ cm) and $0.35$ ($\lambda = 0.2$ cm). Thus, for the mean value of a DSP's EBS—estimated by relation (8.1) for $W = 100$ g m$^{-3}$ and $\lambda = 0.8$ cm—$\sigma$ is of the order of $1$ m$^2$ ($0$ dB for EBS) and—for maximum values $W = 10^3$ g m$^{-3}$—$\sigma = 10$–$20$ m$^2$. For the two-millimeter range the following tracing estimates are typical: the mean $\sigma \cong 0.1$–$0.2$ m$^2$ and the maximum $\sigma \cong 2.5$–$3$ m$^2$.

Analyzing (8.2) it is immediately apparent that for high-emitting surfaces ($\kappa > 0.9$) the presence of a cloud with scattering elements causes a decrease (a negative additive) in the total brightness temperature of the "cloud–surface" system (see Sharkov, 2003 for a more detailed description of this effect). So, in our case ($\lambda = 0.8$ cm, $\omega = 0.45$, $h = 0.5$ m, $W = 100$ g m$^{-3}$) the negative contrast $\Delta T_B$ will be $-(5$–$8)$ K (see Figure 8.4 at the time instant $t = 2$ s), and for a more concentrated DSP ($W = 10^3$ g m$^{-3}$) $\Delta T_B = -(30$–$40)$ K. For a weak DSP cloud ($W = 10$–$30$ g m$^{-3}$) the influencing effect of the emissive characteristics of the total system is virtually imperceptible ($\Delta T_B = -0.1$ K). Nevertheless, at the same time, backscattering will be fully determined by the DSP "cloud", since in this case $\sigma = 1$ m$^2$ ($0$ dB), whereas a disturbed sea surface (without wave breaking) possesses scattering at the level of $-(30$–$35)$ dB (for the given geometry of the experiment and a rough sea state). The absence of "instantaneous" measurements of a DSP's dispersed structure does not allow more detailed calculations to be carried out on the present experiment. But, the estimates presented show qualitative and even, partly, quantitative agreement between the values of an EBS of experimental "splashes" and the calculated EBS of a DSP. The weak "reaction" of the radiothermal channel to the presence of a DSP cloud also becomes clear. So, in model experiments on observing the breaking of ship waves, we can clearly see the regions of "suppression" of thermal radiation of the foam structure due to the presence of a scattering cloud (see time instants $t = 7, 12, 18.5$ s in Figure 8.7).

Within the framework of our hypothesis a much fuller and more logical explanation is found for the results of observation of both scattering from breaking waves in the two-millimeter range (Lubyako and Parshin, 1986) and scattering from internal wave interaction with surface disturbance (Veselov et al., 1984). So, the "quasi-pulse" character of a scattered signal in the range of 2 mm and considerable displacement of the central frequency in the Doppler spectrum are explained by the sporadic passage of breaking waves through the instrument's field of view and by the high velocities of DSP drops. Moreover, even rough estimates of a DSP's EBS (for $\lambda = 0.2$ cm) (see above) indicate quite satisfactory agreement with experimental data. For example, according to the data by Lubyako and Parshin (1986), the mean experimental value is $\sigma \cong 0.1$–$0.8$ m$^2$, the maximum value $\sigma_m \cong 1$–$5$ m$^2$, whereas according to our estimates, $\sigma \cong 0.1$–$0.2$ m$^2$ and $\sigma_m \cong 2.5$–$3$ m$^2$.

As to observation of zones in which internal waves (IW) interact with surface waves (the interaction effect is accompanied by intensive breaking of gravity waves of various scales at the corresponding IW phase—Veselov et al., 1984), in this case the amazing "antiphase character" (almost 100% negative correlation) of scatterometric and radiothermal signals ($\lambda = 0.8$ cm) and the independence of radiothermal signal variation on the observation polarization allows us to state that these effects are more logically explained within the DSP hypothesis framework than within the framework of the hypothesis on quasi-mirror reflections that Veselov et al. (1984) insist upon.[2]

It should be noted that, in spite of the fact that the experiments on spectral analysis of backscattering signals from a disturbed sea surface have been carried out for many years (see, e.g., Lewis and Olin, 1980; Shibata et al., 1985; Lubyako and Parshin, 1986; Melnichuk and Chernikov, 1971; Bass et al., 1975), nevertheless, these authors did not manage to identify the features of the time evolution of backscattering intensity $\sigma(t)$ and its Doppler spectrum with various breaking wave phases for completely different reasons. They include the great accumulation time at analyzing the Doppler spectra (100–400 s), the full absence of a Doppler channel, a very narrow band in analysis of the Doppler spectrum, and other reasons.

The most detailed investigations into time dependence $\sigma(t)$ were carried out in Lewis and Olin (1980) using pulse radar (the Doppler channel was absent), where the time of component decorrelation by a "spike" was shown to be estimated as 3–10 ms. Obviously, this, in its turn, causes high relative velocities of scatterers in the volume studied—at least, no lower than 7–10 m s$^{-1}$. Such velocities of mutual motion can be provided only by drops in a rather small DSP volume. Lewis and Olin (1980), however, interpret the speckle structure of backscattering as a result of interaction in the foam structure of highly associated (in the electromagnetic sense) zones separated by distances lower than $\lambda/4$, which provides, in their opinion, little more than a backscattering mode with strong "splashes".

However, as was shown in Chapter 6, the foam structure of both a breaker and a strip foam can be considered, in the electromagnetic sense, as a set of small "black-bodies", which do not possess noticeable scattering properties. And, hence, the

---

[2] The investigations described in Veselov et al. (1984) and the results presented in this section were carried out by means of the same instrument set (Cherny, 1982).

hypothesis (Lewis and Olin, 1980) on the highly interacting zones in foam systems and on their contribution to the speckle structure cannot be accepted.

The hypothesis about the DSP effect on a backscattered signal substantially clears up, in our opinion, the situation with a set of probabilistic models of non-Rayleigh fluctuations of radar signals (the so-called K distribution and those similar to it—Fante, 1984; Shlaychin, 1987; Teich and Diament, 1989). The increasing number of such models raises a question on the physical reasons for such a variety. This, however, can be logically explained within the framework of our hypothesis. Since the K-distribution model describes the process of interaction of two random variables (the random vector "wandering" in the plane with a random number of steps—Teich and Diament, 1989), the following physical model can be offered: a breaking wave's DSP cloud with a random EBS value as the scatterer, and a random number of breaking waves situated in a spatial resolution element as the spatial "wandering".

The study of the spatial–statistical characteristics of a breaking field was described in Chapter 2. Here we shall only note that there exist quite certain spatial regularities of random field breaking, which strongly depend on the spatial frame of observation—from the geometrical distribution (for small frames) through the negative binomial to the Gaussian. If we assume from general physical considerations that the EBS distribution at wave breaking has a gamma distribution, then the set of these processes can certainly explain the appearance of the K distribution in the field of backscattering of a disturbed sea surface. And if we take into account here the variety of hydrometeorological conditions and the effect of intraoceanic processes on surface disturbance statistics, then the variety of laws of distribution of EBS of a disturbed sea surface's backscattering field, which depresses Fante (1984) and Shlaychin (1987), can easily be explained from the physical point of view.

Thus, synchronous measurements in radiothermal and radar-tracking modes allow us to obtain new information on the breaking process and to distinguish two time phases associated with the appearance of a drop-spray structure over a breaker as it forms and with the formation of a strip foam field. The high-frequency zone of the Doppler spectrum carries information on the field of Doppler velocities of moving scatterers in the drop-spray phase, which causes formation of the speckle structure in the field of a signal backscattered from a disturbed sea surface.

## 8.3   RADIO EMISSION OF CREST AND STRIP FOAM: FIELD SHIP INVESTIGATIONS

In August, 1978, the Space Research Institute of the USSR Academy of Sciences (IKI AN SSSR), the Moscow State Pedagogical Institute (MGPI), and the Marine Geophysical Institute of the Ukrainian Academy of Sciences carried out joint experimental works onboard the RV *Mikhail Lomonosov* in the Black Sea. These field investigations were performed within the framework of the USSR Academy of Sciences' complex of works on remote investigation of the detailed structure of a sea surface. In the course of this expedition a special experiment was accomplished on

**Figure 8.9.** Placement of the gyroscope stabilization platform with its highly sensitive radio-thermal instruments aboard the RV *Mikhail Lomonosov* (Black Sea basin, August 1978).

studying the brightness characteristics of foam structures in the centimeter wave-length range (Vorsin *et al.*, 1984).

Remote measurements were carried out by means of a two-frequency, high-sensitivity radiometric system of ranges $\lambda = 2$ and 8 cm. The normalized sensitivity was 0.03 K (for the 2.03-cm channel) and 0.1 K (for the 8-cm channel). The instrument was installed on a gyro-stabilized platform, which was rigidly fastened to a crane jib on the starboard side, near the bow of the vessel. The antennas had directional pattern widths of about 1.5 and 40° for the 2-cm and 8-cm ranges, respectively. The whole installation was situated at a height of about 7 m above sea level and was about 2 m from the vessel on a special crane jib (Figure 8.9).

External calibration of the instruments was accomplished by measuring the radio emission of a quiet water surface at various sighting angles (from horizon to nadir) and subsequently comparing these data with calculated data (see Sharkov, 2003 for more details on the external calibration of radiothermal complex sets). Both horizontal and vertical polarizations were used in measurements in "for reception" mode.

Figure 8.10 presents synchronous prescriptions (at an accumulation constant equal to 0.05 s) of variations in radio emission of a foamed sea surface. Synchronous optical surveying has shown that a caving breaker (1) and an emulsion (strip) structure (2) correspond to markers "1" and "2" on radiothermal prescriptions.

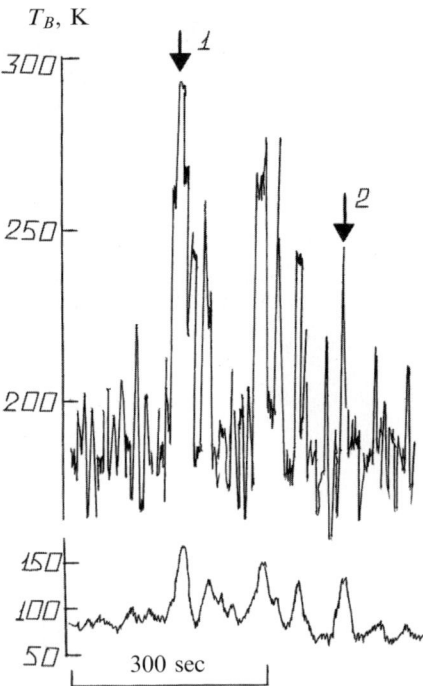

**Figure 8.10.** The simultaneous radiothermal signal registrations for the breaking and foam-generation processes at the 2.08 and 8-cm wavelengths. The angle of view is 550. The working regime is vertical polarization for the 2.08-cm channel (upper registration) and horizontal polarization for the 8-cm channel (lower registration). The sea state is 4 points on the Beaufort Scale. The arrows mark the moments of creation on the pictures (Figure 8.11) for wave crest foam (1) and residual foam (2).

The 3-dB antenna instant field of view of the sea surface was about 0.5 and 15 m for ranges $\lambda = 2$ and 8 cm (the estimation was based on the experiment geometry).

The experimental results (Figure 8.10 and Table 8.1) indicate that in the centimeter range the effect of foam structures on the radio emission of the sea can be very strong: the maximum values of brightness temperature variations for channels $\lambda = 2$ and 8 cm reach $\Delta T_B = 120$ and 80 K. They are caused by emission of a multi-structural foam of "breaker" type with a layer thickness of a few centimeters (according to visual estimations). In this case, in the 2-centimeter range the radio emission of a foam surface becomes absolutely black (marker "1"). At the same time, the presence of so-called foam strips (or spots), whose structure, probably, is close to that oof emulsion, results in brightness temperature increasing by $\Delta T_B = 70$–40 K in the wavelength range considered (marker "2"). The radio emission fluctuations caused by the dynamics of foam cover result in the increasing of output signal dispersion—up to 20 K ($\lambda = 2$) and 3–4 K ($\lambda = 8$ cm).

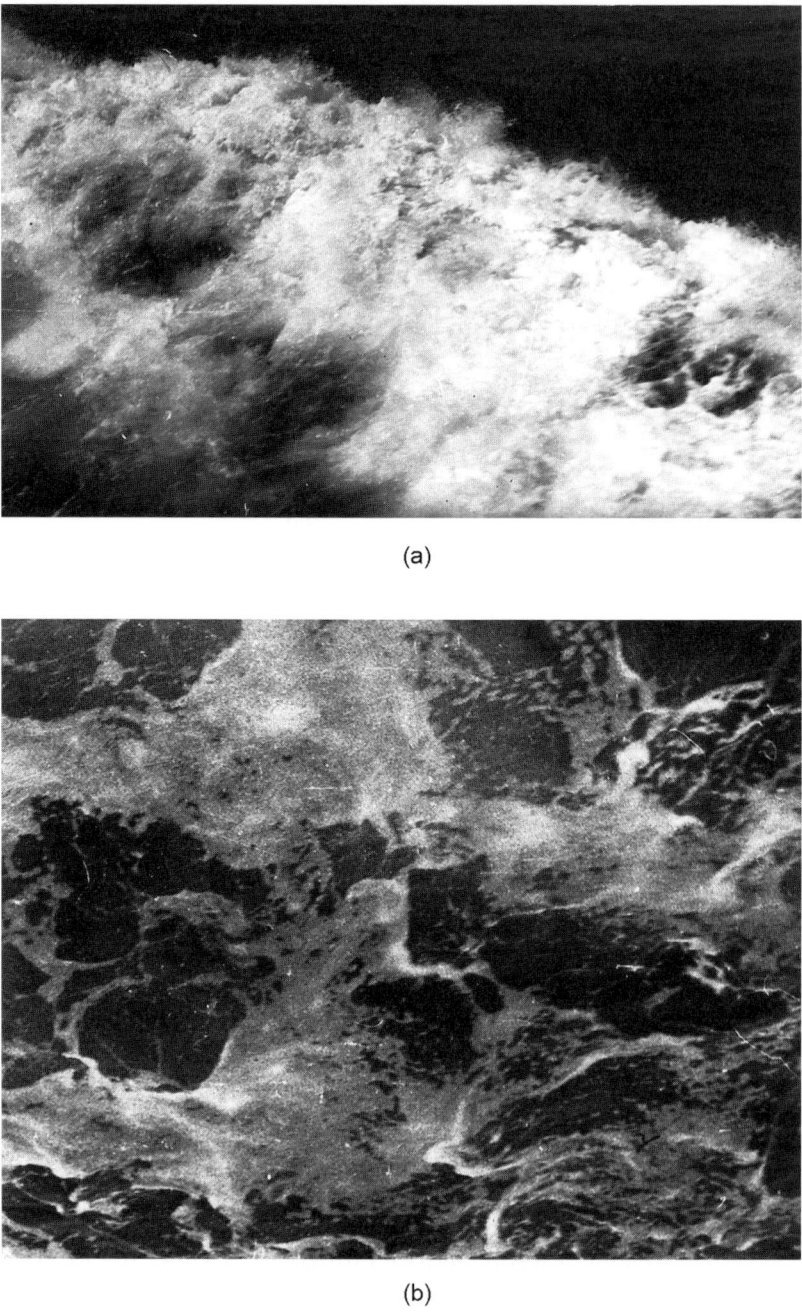

(a)

(b)

**Figure 8.11.** Photographs of sea foam surface sections. (a) The photograph of sea crest foam surface agrees with marker 1 on the registration in Figure 8.10; (b) the photograph of sea residual foam surface agrees with marker 2 on the registration in Figure 8.10 .

**Table 8.1.** Emissivities ($\kappa$) for foam structures: results of field experiments $\kappa_E$ and data of theoretical models $\kappa_T$.

| Foam structure type | Wavelength (cm) | Polarization | Model parameters | | | |
|---|---|---|---|---|---|---|
| | | | $\varepsilon_{N\alpha}$ | Layer height (cm) | $\kappa_T$ | $\kappa_E$ |
| Whitecapping | 2 | VP | $2.52-j0.88$ | 3 | 0.972 | 0.985 |
| Residual foam | 2 | VP | | 0.3 | 0.713 | 0.732 |
| Whitecapping | 8 | HP | $3.01-j0.42$ | 3 | 0.634 | 0.651 |
| Residual foam | 8 | HP | | 0.3 | 0.248 | 0.291 |

*Note*: $\varepsilon_{N\alpha}$ is the dielectric constant of the 6a model (Table 6.8).

Vorsin *et al.* (1984) were the first to show that experiments carried out under natural sea conditions by means of shipborne instruments allow, in principle, detailed investigations of the brightness characteristics of various sea foams to be carried out. It is also possible to study the dynamics of foam cover from radio emission variations. The values of emissivities $\kappa_{EXP}$ of contrasting foam formations were determined from radiothermal registrograms, and their quantitative interpretation (Table 8.1) was given with the help of the electrodynamic models described in Chapter 6. An electrodynamic model was used for plane-layered structures with effective dielectric parameters, which were calculated in the quasi-static approximation with due account of the diffraction properties of spherical bubbles of real size (Chapter 6). Semi-empirical values (estimated from photographs) of a dispersed layer's mean thickness were incorporated in the calculation as well. As seen from Table 8.1, the divergence between experimental and calculated data is about 2–3%, which indicates the reliability of the model presentations used for quantitative analysis of field measurement results.

As a conclusion to this section we shall compare the results of these field investigations with some previous field experiments, with laboratory investigations (Chapter 6), and with conclusions based on the diffraction model (Chapter 6). Figure 8.12 summarizes the basic experimental material on the emissive characteristics of foam formations in the microwave range. Despite the fact that these data were obtained under various conditions, and their dispersed identification was impeded (because of the absence of data), as a whole, they demonstrate the consistent general character of spectral dependence $\kappa(\lambda)$. A clear and satisfactory correlation between the laboratory, ship-based, and aircraft-based data is observed. At the same time, the numerical approximation presented by Stogrin (1972) turns out to be unsatisfactory (as already noted in Chapter 6).

Comparison of the data of laboratory and field radiothermal measurements allows us to suppose the existence of a quite certain electrodynamic similarity between rough dispersed systems of any type, irrespective of their origin.

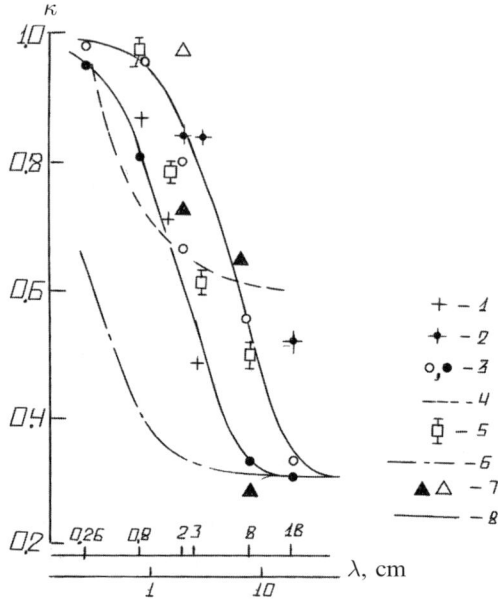

**Figure 8.12.** Spectral characteristics of the structure of foam using various experiment data: 1—field experiment during the "Bering" flight campaign (Kondratyev *et al.*, 1975), $\theta = 0°$; 2—near-shore field experiment, $\theta = 0°$ (Au *et al.*, 1974); 3—laboratory experiments, $\theta = 35°$, HP (Bordonskii *et al.*, 1978) (see Chapter 6 for details); 4—Stogryn's approximation (Stogryn, 1972), $\theta = 35°$, HP ; 5—research vessel field measurements, $\theta = 0°$ (Matveyev , 1971); 6—flat water surface emissivity, calculations using a dielectric model, $\theta = 35°$, HP ( Sharkov, 2003); 7—research vessel field measurements, $\theta = 55°$, $\lambda = 2$ cm, VP and $\lambda = 8$ cm, HP ( Vorsin *et al.*, 1984); 8—foam structure emissivity spectra for the emulsion monolayer (lower curve) and the polyhedral structure (upper curve) with $\theta = 35°$ and HP using the diffraction model presented in Chapter 6 (see also Raizer and Sharkov, 1981).

## 8.4   RADIO EMISSION OF A BREAKING WAVE FIELD: AIRPLANE INVESTIGATIONS

The complex of physical reasons stipulating a serious contribution to thermal radiation of a wave-driven sea surface are the processes of sea wave breaking and the appearance of an intensive drop-spray phase and bubble disperse phase of sea foam with their subsequent rapid and complicated spatiotemporal evolution. It should be mentioned that the experimental detection of the intensive (virtually blackbody-type) thermal radiation of foam systems by Williams (1969) under very complicated (and hazardous, we might add) hydrometeorological conditions (such as the flight of a sporting-type airplane through a tropical cyclone's "wall") was also rather unexpected for researchers. This was associated with the general supposition of the absence of noticeable absorption and scattering of microwave band electromagnetic waves on a set of rather small hollow air spheres and hollow hexagonal structures with very thin water films, which alone represented foam systems (see Chapter 6).

Nevertheless, these results gave rise to the first (and rather naive, as discovered later) ideas and determinate models, according to which the basic elements that determine thermal radiation of a wave-driven sea surface will be foam systems of various classes. And, accordingly, as the wind velocity increases (from 5–6 m s$^{-1}$ to typhoon velocities of 33–35 m s$^{-1}$) according to the power law (linear or quadratic), radiation intensity will also grow to values corresponding to blackbody radiation (Droppleman, 1970; Williams, 1971; Matveyev, 1971). Further detailed investigations of the radiophysical properties of foam systems and the spatiotemporal disperse characteristics of foam systems under full-scale conditions, carried out under the guidance of the author of this book (Sharkov, 2003), have clarified the situation in many respects and demonstrated the imperfection of early models and ideas (see Chapters 2, 3, 6 for more details). Below we shall present a brief review of remote-sensing work in this field.

The problem of detecting radioemission effects from a drop-spray zone in sea surface sensing from a low-altitude carrier (an aircraft) has some features. In essence, the question is about solitary (from the antenna field of view) and non-stationary physical objects possessing high emissivity. This imposes certain requirements on the parameters of onboard instruments, as well as on the choice of conditions for the experiment. So, the value of a radiothermal signal, recorded by a microwave instrument installed on a moving carrier, can be presented as follows (see Sharkov, 2003):

$$T_B(t) = \int_0^\infty d(x/V) h(t - x/V) \left[ \iint G(x - x'; y - y') T_{BF}(x', y', t) \, dx' \, dy' \right] \quad (8.3)$$

where $h(t)$ is the impulse response of a receiving device (with time constant $\tau_0$); $G(x, y)$ is the instantaneous field of view of the antenna directional pattern (ADP) on the surface; $(4/\tau) G^{1/2} = 2H \, \text{tg}(\theta/2)$; $H, V$ are the flight altitude and carrier velocity; $\theta$ is the angular resolution of ADP; $T_{BF}(x, y)$ is the radiothermal image of a foam structure (with geometric size $\Delta x$ and area $S$). This relation implies the great diversity of types of radiothermal signal recorded from a non-stationary, but "bright" (in the radiothermal sense) object. So, it can easily be seen that for $T_0 \ll \Delta x/V$ and $G \approx S_0$ the shape of a recorded signal will represent an isosceles triangle with base $2\Delta x/V$ and height $T_{BF} - T_{BS}$, where $T_{BS}$ is the brightness temperature of the thermal "background" of the sea surface. As $T_0$ increases up to the values of $\Delta x/V$, signal amplitude decreases, though "the appearance" of the signal does not change qualitatively. If, however, $T_0 < \Delta x/V$, but $G < S$, then the signal shape represents a "little house with a cap", and brightness contrast $\Delta T_B$ will be the same quantity $(T_{BF} - T_{BS})$. If the object has the form of a point source (i.e., $G \gg S$), then brightness contrast will have another value—$\Delta T_B = (S_0/G)(T_{BF} - T_{BS})$—and the "appearance" of the signal shape will describe that of the main lobe of the ADP.

A representative implementation of a radiothermal signal from a wave-driven sea surface (Figure 8.13) was obtained under the following conditions: Caspian Sea water area, direct track of 80 km along the section with rough sea variation from 1 to

4 points on the Beaufort Scale (according to visual estimates of a skilled meteorologist onboard); flight altitude $H = 1000$ m; and carrier velocity $V = 350$ km h$^{-1}$. The instruments consisted of a highly sensitive radiometric system in the 2-cm band with a sensitivity threshold of about $\Delta T = 0.03$ K and a time constant of 1 s, and was installed aboard the IL-18 aircraft laboratory with a mirror-reflecting parabolic antenna with a directivity pattern width of $\theta = 1.5°$. The resolution spot on the surface for nadir-looking observation was $G = 25$ m. Analysis of Figure 8.13 indicates that, along with a general increase in brightness temperature with rough sea amplification caused by small-scale rough sea components (including the azimuthal anisotropy effect), the presence of acute spikes (i.e., signals from the foam-disperse zone) is characteristic. Synchronous air photography gave an idea of the sea surface structure under 4-point rough sea conditions: the size of foam formations, generated at wind wave breaking, were of the order of 3–10 m. By virtue of the small time constant of integration $\tau_0 = 0.05$ s (in flight mode), the amplitudes of signals were recorded virtually completely, the radiothermal contrasts being equal to 3–4 K. The experimental situation corresponds with the condition of $G > S_0$ and, thus, $(T_{BF} - T_{BS})$ will be 80–100 K, which corresponds fully with theoretical estimations (Raizer and Sharkov, 1981; and Chapter 6).

A second interesting example represents synchronous registrations of a scatterometer (3 cm) and radiometers (2 cm and 8 mm) obtained under force 5 and 6 rough sea conditions in performing circular flights (at an altitude of 300–600 m). They are presented in Figures 8.14 and 8.15. The registrations of microwave instruments clearly exhibit the signals from foam structures in the form of "triangles" with a

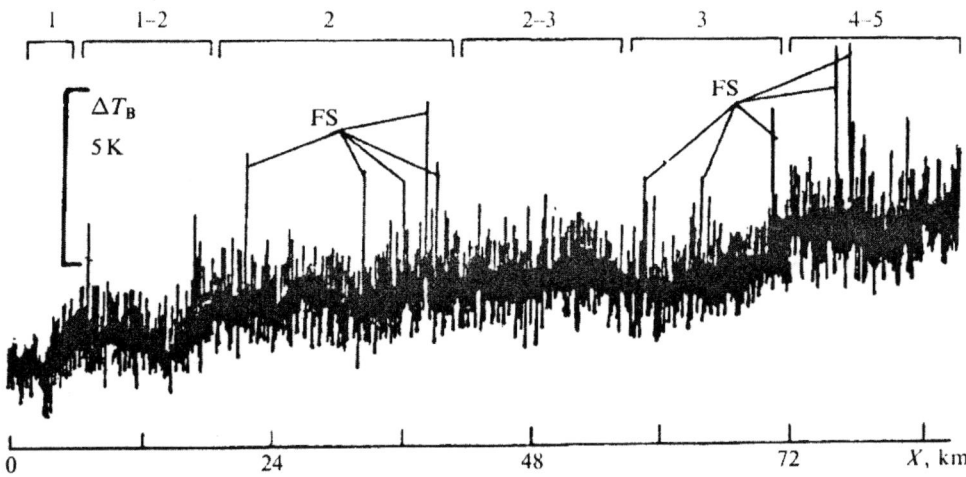

**Figure 8.13.** Fragment of the output signal of an airborne (Russian airplane laboratory IL-18) high-performance radiometer R2 (wavelength 2 cm) versus flight distance $X$ (km). The linear flight took place at a height of 1000 m with the nadir view angle relative to the aircraft as the sea state increased from 1 to 5 points on the Beaufort Scale. The working area is the North Caspian Sea (Russia). The date was November 27, 1977. FS represents the emission signals from foam structures. Numbers show force on the Beaufort Scale of the sea state.

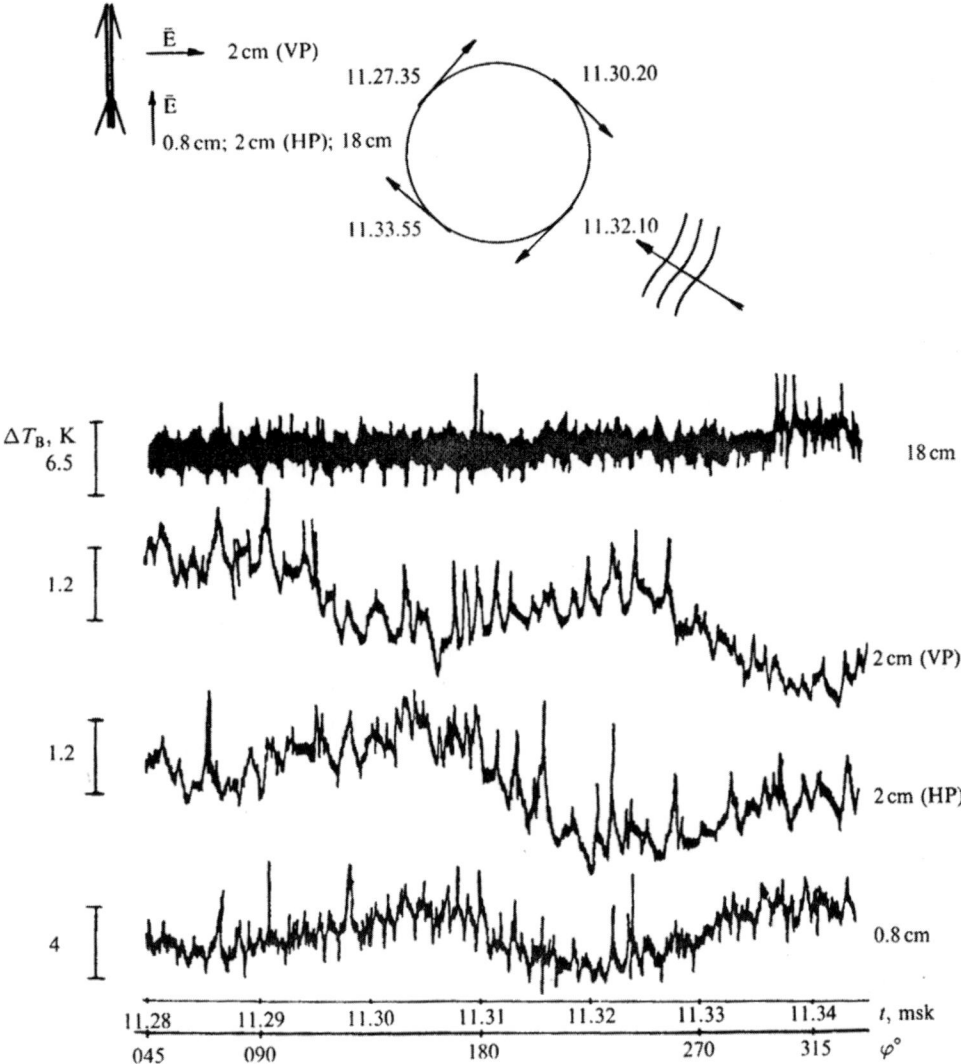

**Figure 8.14.** The first experimental results of flight investigations to make detection of the azimuthal anisotropy effect in sea surface microwave emission possible. The scheme of circular flights and the positions of the linear polarization vectors of the antennas and of surface gravity wavefronts is presented at the top of the figure. The double arrow points along the longitudinal centerline of the research aircraft. Fragments are shown of the output signals of the airborne (Russian airplane laboratory IL-18) radiometer R-18 (wavelength 18 cm), radiometer R2 (wavelength 2 cm, two polarizations: HP, horizontal polarization: VP, vertical polarization), and radiometer R0.8 (wavelength 0.8 cm) when circular flights took place at a height of 300 m and microwave observations were performed at the nadir view angle. The working basin is the North Caspian Sea (Russia). The date was November 28, 1977. Moscow time and brightness temperature intensity scale are shown on the abscissa and on the ordinate, respectively.

contrast amplitude of the order of $T = 2$–$4$ K. Since the experimental situation is such that $G \approx S_0$, but in this case $T_0 > \Delta x/V$, the low values of radiothermal contrasts are clear. Of importance is the fact that the scatterometer channel does not indicate the presence of foam structures because of their small reflectivity (see Chapter 6 and Militskii *et al.*, 1976, 1977). However, subsequent full-scale investigations have shown (Cherny and Sharkov, 1988) that, in general, the situation is more complicated, and the relationship between the values of radiation and backscattered signals depends on the temporal stage of wave breaking and on the appearance of a drop-spray phase against the background of foam structures, the drop-spray phase making a noticeable contribution to the backscattered signal. As a rough sea amplifies—that is, a sea surface state more than 6 points on the Beaufort Scale—the regular azimuthal dependence (Figures 8.13 and 8.14) of the anisotropy in polarization microwave emissivity is violated (Bespalova *et al.*, 1982): the radiothermal field becomes more isotropic in character, since a noticeable contribution to radio emission is made by the drop-spray phase and foam formations, which can be obtained without any azimuthal features regarding wind direction.

Thus, the contribution of foam structures to sea surface radiation in aircraft sensing can have signals of diverse character, and the thermal radiation of individual foam structures can be revealed only by using synchronous air photography at high spatial resolution (10–20 cm), which is obviously not always possible. On the other hand, the use of not such a high-velocity but still more maneuverable carrier (a helicopter) will allow increasing the observation time, which considerably facilitates the experimental problem. In this case it is possible to perform detailed air photography of the studied surface for further identification of the data (see Chapter 3 and Bondur and Sharkov, 1982, 1990; Smith, 1988).

For observation from space carriers, where the linear size of the field of view of the radiothermal instrument's antenna equals 10–50 km, the recorded signal quantity can be written as (Sharkov, 2003):

$$T_B = T_{BS}\left(1 - \sum_{i=1}^{N(t)} S_{Fi}\right) + \sum_{i=1}^{n(t)} T_{BFi} S_{Fi} \tag{8.4}$$

where $N(t)$ is the number of foam structures in a frame corresponding to a displayed "spot" of the antenna directional pattern (ADP) on the surface. In this case it is necessary to know the detailed statistics of spatial fluctuations $N(t)$ in the appropriate spatial frame, as well as of the geometric ($S_{Fi}$) and emissive ($T_{BFi}$) characteristics of foam systems of various types and their "lifetime". These issues have been thoroughly studied in a cycle of works (Pokrovskaya and Sharkov, 1986, 1987a, b, 1994; Zaslavskii and Sharkov, 1987; Bondur and Sharkov, 1982, 1990; Sharkov, 1993a, b, 1994, 1995). Thus, detailed experimental results have shown (Pokrovskaya and Sharkov, 1987, 1991; Sharkov, 1993a, b, 1996a, b) that a stochastic model of large-scale breaking wave fields may be presented as a stochastic spatial wholenumber field of Markov type (the Poisson process) with independent discrete sources (events). The mean value of the specific density of foam structures (the centers of wave energy scattering) has a clear, strong (cubic) tendency to increase with wind velocity and has a "threshold" nature (the stepwise approximation) at a wind velocity of $5\,m\,s^{-1}$ (Bondur and Sharkov, 1982, 1993).

Detailed experimental investigations of the process of breaking and subsequent decaying of foam systems (Sharkov, 1994, 1995) have shown that the process of gravity wave "preparation" to breaking and the breaking itself takes a fraction (about $\frac{1}{3}$) of the wave period. In other words, the nonlinear stage of the breaking process and the formation of a wave crest foam grows at fabulous rates. However, the process of decaying of patch foam structures of a foam mass is characterized by exponential decay with a considerable half-decay time (0.5–10 s), which depends on the physicochemical properties of sea water in this area.

A new specific group of wave breakings, called mesobreakings (MBs), was first detected by Sharkov (1994, 1995). MBs have some particular properties: small lifetime of a foam mass ($\sim$1 s); small characteristic size ($\sim$0.4 m); absence of an accompanying patch foam; faint optical contrast; independence of average size on wave–wind conditions. The instantaneous value of surface foam covering (per unit) for MBs amounts to 0.6%. MBs may be thought of as "noise" fields.

As to radiothermal models of foam systems, the detailed laboratory and theoretical investigations, carried out under the guidance of the author of this book during 1974–1981, have shown that there exist two contrasting classes of foam systems: an emulsion monolayer and a layer of foam of polyhedral structure (Militskii *et al.*, 1978). In this case it was strictly demonstrated (Raizer and Sharkov, 1981) that the electrodynamical models of a homogeneous or inhomogeneous dielectric layer with parameters corresponding to a heterogeneous mixture of water and air (Matveyev, 1971) do not exhibit quantitative agreement with experiment. The same is true for the models of a discrete stratified medium (Rozenkranz and Staelin, 1972; Tang, 1974; Bordonskii *et al.*, 1978) and for the models of a smoothly varying transition dielectric layer (Bordonskii *et al.*, 1978). The electromagnetic properties of coarse disperse media in the microwave band are most completely reproduced by continuously stratified inhomogeneous models with strict allowance for the diffraction properties of hollow spheres and hexagonal structures, with allowance for a blurred interface boundary between the disperse structure and water surface, and with allowance for the altitude inhomogeneity of a dielectric layer (Raizer and Sharkov, 1981). In short, we can say that the matter is as follows. For the electromagnetic range under consideration, hollow water spheres (or hexagonal structures) represent some kind of small blackbodies, which intensively absorb the electromagnetic energy due to inter-bubble diffraction. Complete consideration of these issues is presented in Chapters 2–4 and 6 of the present book. We also recommend Raizer and Sharkov (1981) to the reader for the original literature.

## 8.5 NONLINEAR DYNAMICS OF GRAVITY WAVES IN THE BREAKING WAVE BACKSCATTERING FIELD

As already noted (Chapter 1), in spite of the fact that the properties of wind-driven sea waves have been intensively studied since the 1950s, many physical effects of so-called second-order smallness (in magnitude) relative to features of the energy spectrum still remain insufficiently studied. These effects, however, are of principal significance for understanding the processes of wind wave generation, development, and extinction. They include the effects of weakly nonlinear interactions of near-surface waves, which result in energy redistribution between interacting components and, thus, play a significant part in forming the spectrum of near-surface heavy sea waves (Phillips, 1977). The study of these effects under field conditions meets considerable difficulties, the most important of which lies in the fact that—in a natural rough sea—interacting waves are not isolated. As a result, the same components can be involved in interactions, which resound simultaneously, with the waves of various groups, of various spatiotemporal scales, and of various propagation directions. So, information on a rough sea structure is needed over a considerable basin; or, in other words, a spatiotemporal frame of the wave field needs to be formed. Such information cannot be principally obtained by wave-graphical *in situ* techniques. This is the reason that microwave scatterometry techniques seem to be optimum for studying the effects of weakly nonlinear wave interactions, since they allow us to obtain a spatial picture of wave system distribution over considerable areas. And, at the same time, they allow us to study the temporal dynamics of near-surface waves by analyzing radio images of the sea surface. In their turn, the surface manifestations of wave systems can be most efficiently studied at low grazing angles in the backscattering field of breaking waves (Kalmykov *et al.*, 1976; Phillips *et al.*, 2001; Bulatov *et al.*, 2003, 2004, 2006; Sletten *et al.*, 2003).

In such a methodology the spatial and temporal characteristics of waves can be observed, having presented the radio image of the sea surface in the frame with time–distance coordinates (a "time–distance" image). The radio image in such a format is usually called the $[t, r]$-diagram (Phillips *et al.*, 2001). It can be obtained by sampling the received signal at the range and by recording its variations in time, thus forming a sample for each value at the range at a fixed azimuthal sensing angle. A thus-generated radio image allows us to determine the velocities of motion of sea surface elements, scattering radiowaves, the distances they have traveled in a direction toward the radar or away from it, and to determine their "lifetime" on the sea surface. The two-dimensional Fourier transformation of radar data, specified in "time–distance" format, represents a spatial-frequency spectrum of the intensity of sea surface backscattering in "frequency–wavenumber" format. In such a form, sea surface radar data can be analyzed by studying the fine scattering properties of wave systems resolved by radar, and can also be used for studying the group structure of waves in wave pools and in field experiments, where flows and sea waves in the coastal zone are studied. Bulatov *et al.* (2004, 2006) present the results of a field radar experiment directed at studying the dynamics of the sea surface. Analysis of the data obtained in "time–distance" format allowed us to determine the characteristic velocities and lifetimes of scatterers, which form the radiowave backscattering at vertical and horizontal polarizations, and to explain the obtained results based on a model of interaction of surface waves at three scales. Analysis of the two-dimensional (2D) spectra of radio images of "frequency–wavenumber" type allows us to reveal the effects of the weakly nonlinear resonance interactions of surface waves and to determine the parameters of components satisfying four-wave interaction conditions.

### 8.5.1   Experimental region and wind–wave conditions

The data presented in this section were obtained on September 16, 2002, in the Blue Bay of the Black Sea, near Gelendzhik. Measurements were carried out from a pier that extends 200 m into the sea area. Figure 8.15a shows the shoreline and lines of equal depths, the place of radar location at the pier edge, and the survey zone, corresponding to the position of an antenna beam oriented toward a rough sea, is indicated in the figure as well. Points "A" and "B" correspond to the values of the survey's minimum and maximum range. As follows from the scheme presented in Figure 8.16a, the sensed area of the surface is situated over the underwater trench extending southward from the pier end. As is known, with such a relief of the sea bottom, a surface wave, whose length is commensurable with the sea depth in the trench vicinity, is oriented along its axis. This allows us to simplify the analysis of our measurement results, supposing in advance that the observed wave components are collinear and propagate predominantly in the antenna direction.

Before performing radar measurements there was a calm period, changed by an unstable, gusty southeasterly, and then by steady wind from a southerly direction, whose velocity was $(7.5 \pm 0.5)\,\mathrm{m\,s^{-1}}$ and direction $(180 \pm 10)°$. Estimates of the time taken by the wind to speed up, made according to standard techniques (Roshkov, 1979), showed that at the time of performing radar measurements the sea waves were fully developed, the mean height of waves reached a maximum value for the given wind velocity, which was 1.2 m according to *in situ* measurements.

### 8.5.2   Radar measurement technique

To construct spatiotemporal research diagrams the following sea surface sensing scheme was used (see Figure 8.16b). To carry out radar measurements a two-polarization radar in the 3-cm range was used, which provided resolutions in the range $\Delta r = 7.5$ m. The radar was installed at height $H = 8$ m above sea level. In the azimuth plane the sea surface survey zone was determined by an antenna pattern width equal to $1°$. The boundaries of the survey zone in the range, designated in Figure 8.15b as $R_{\min}$ and $R_{\max}$, were determined by the duration of a sample pulse, which locked the receiver at the instant of emission, and the level of power of the signal received, at which it was possible to ignore the effect of receiver noise on the quality of the radio images received. The minimum and maximum size of the resolution element $(L)$ in the azimuth plane, corresponding to the minimum and maximum ranges, were $L_{\min} = 3.5$ m and $L_{\max} = 17.5$ m. Within survey zone limits the sensing angle relative to nadir varied from 88 to 89.6°. Measurements were carried out at horizontal (HH) and vertical (VV) polarizations alternately—for 10–15 min at each polarization. Sensing pulses of a transmitter of duration 50 ns followed each other at a frequency of 500–2000 Hz. The emitted pulse power was 7.5 kW. Digitizing the analog signal in the eight-digit analog–digital converter and recording its binary

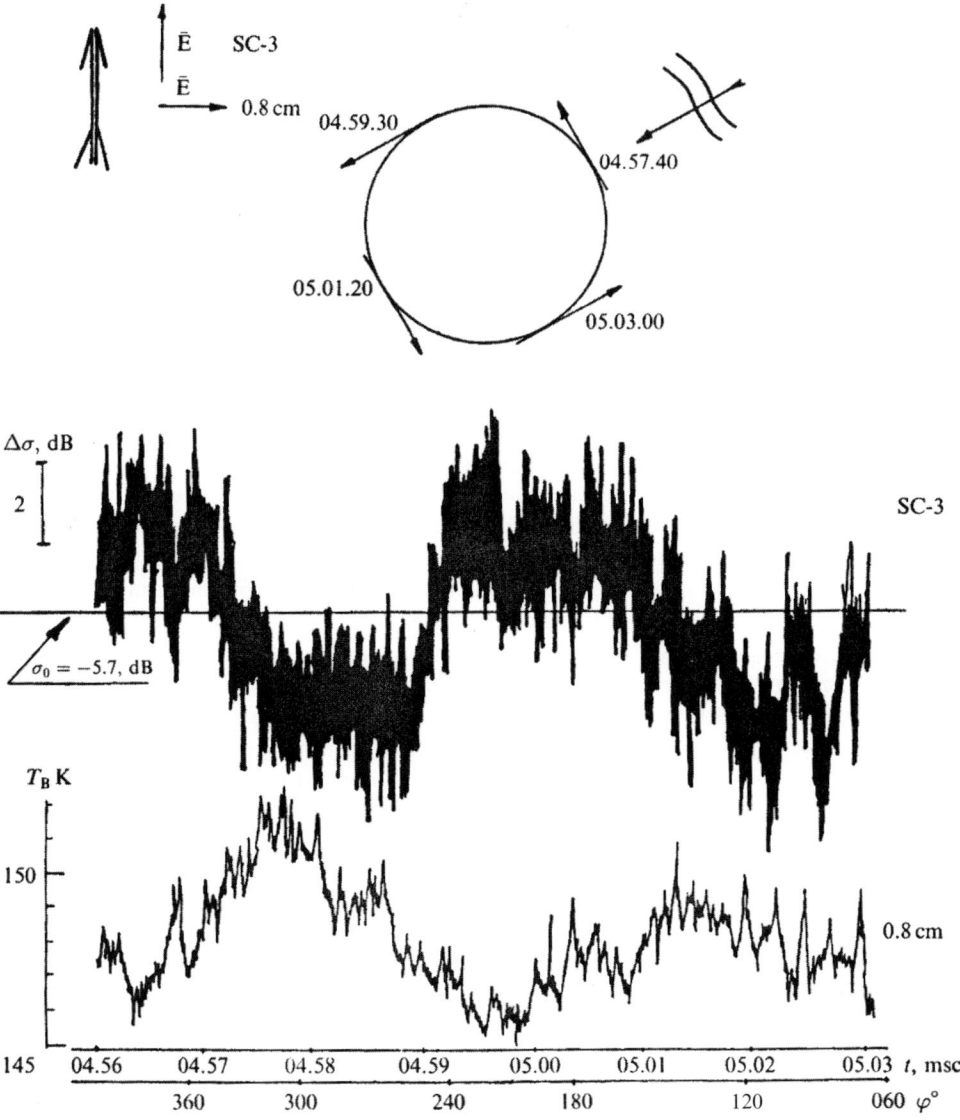

**Figure 8.15.** The experimental results of flight investigations (October 20, 1978 over the Sea of Japan) for the azimuthal anisotropy study on sea surface microwave emission and back-scattering. The scheme of circular flights and the positions of the linear polarization vectors of antennas and of surface gravity wavefronts is presented at the top of the figure. Fragments are shown of the output signals of the airborne (Russian airplane laboratory IL-14) radiometer R0.8 (wavelength 0.8 cm) and of scatterometer SC-3 (wavelength 3 cm) when circular flights took place at a height of 300 m with a 5° angle of slope and microwave observations were carried out at the nadir view angle relative to the aircraft. Sea state was 6 on the Beaufort Scale. Moscow time and the brightness temperature and backscattering intensity scales are shown on the abscissa and on the ordinate.

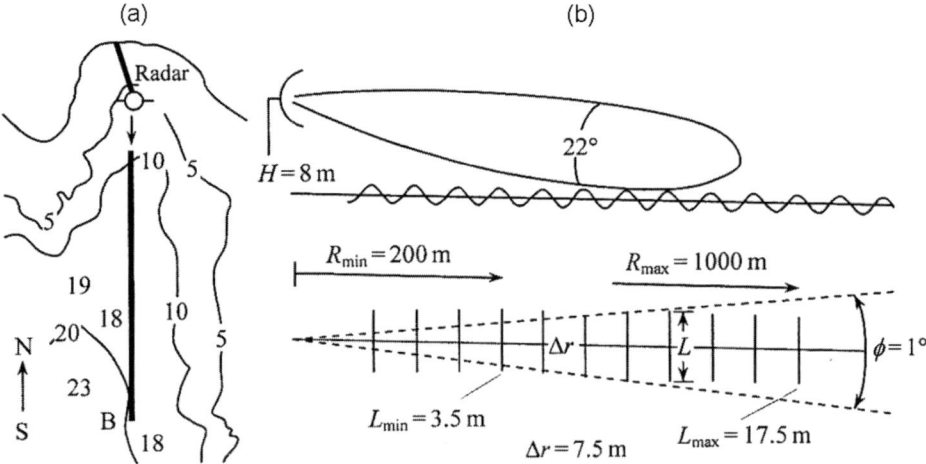

**Figure 8.16.** Sea measurement area and the layout used to view the sea surface during the coastal remote-sensing study. (a) Sea bottom topography and coastline of measurement area; (b) radar viewing of the sea surface using the principal remote-sensing spatial parameters (see text for details).

presentation in the memory of a personal computer took 12.8 µs, which corresponded to a single range scan of length 1280 m consisting of 256 elements. The resulting sample, corresponding to the single range scan, was formed by four-fold averaging of the values obtained after viewing the survey zone once.

### 8.5.3    Analysis of spatiotemporal diagrams

During operation with a motionless antenna, oriented towards the general direction of wave propagation, a sea surface radio image on the time–distance coordinate plane represents a set of points, the brightness of each of which is proportional to the mean strength of a signal scattered by a surface element located at time instant $t$ at distance $R$, whose area is $S = \Delta r \cdot L$. Radio images presented in such a format are usually called RTI-diagrams (range–time–intensity diagrams) (Bulatov *et al.*, 2006). In essence, an RTI-diagram represents a matrix, each line of which consists of a sequence of values of the strength of signals scattered by surface elements located at equal distance from the antenna at various time instants. The maximum number of lines of such a matrix is equal to the number of resolution elements within the survey zone limits $R_{max}/\Delta r$, and the number of columns to the number of time readings of a signal during the observation time. In Bulatov *et al.* (2006) the number of readings in the range was 256, and the time interval between readings was chosen to be about 0.5–1 sec.

Figure 8.17 shows RTI-diagrams obtained at the VV- and HH-polarizations of a sensing signal. Surface elements, such as scattering radiowaves moving along the

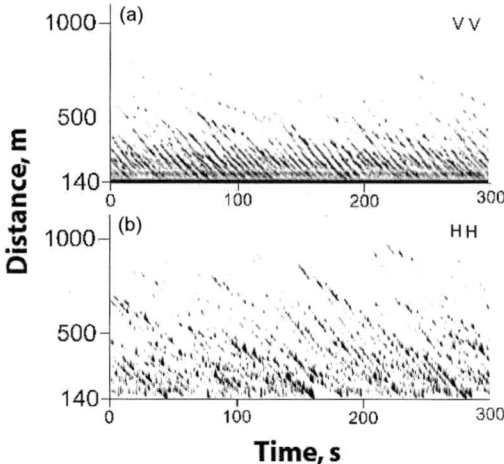

**Figure 8.17.** Time–distance 2D images of the radar backscattering signal from a disturbed sea surface at vertical (VV) and horizontal (HH) polarizations.

antenna beam, are displayed on them as dark strips (i.e., "tracks") whose slope is determined by the velocity of scatterers. Their projections on the range axis correspond to the distances traveled by scatterers toward a radar, and projections on the time axis correspond to the duration of separate acts of scattering (scatterer lifetime).

Comparison of the diagrams in Figure 8.17 shows that the character of tracks is principally different at vertical and horizontal polarizations. Whereas on a diagram obtained at vertical polarization continuous strips are mainly observed, whose extent equals some elements of resolution at the range, at horizontal polarization the track is broken into separate spikes. This distinction is related to the features of electromagnetic wave backscattering at low grazing angles.

Analysis of the histograms of distributions of scatterer lifetimes and velocities, calculated for two sets of tracks, indicates (Bulatov *et al.*, 2004) that at vertical polarization the distribution maximum lies in the velocity range of $7–8 \, \mathrm{m \, s^{-1}}$, whereas at horizontal polarization the upper boundary of the velocity range with a maximum number of tracks equals $10 \, \mathrm{m \, s^{-1}}$. Since the phase velocity of energy-carrying waves in a developed rough sea is virtually equal to wind velocity ($7 \, \mathrm{m \, s^{-1}}$ in our case), the following conclusion can be drawn. Scatterers that form a signal at vertical polarization move at the velocity of crests of energy-carrying waves; in this case, track duration is of the order of one or more wave periods (according to the data of *in situ* measurements, the period of a dominating wave was about 4 s). At horizontal polarization the signal is formed by scatterers, the velocity of a considerable part of which (of about 40%) exceeds the phase velocity of a wave, and their lifetime is essentially shorter than the wave period.

The results obtained can be interpreted both within the framework of a phenomenological two-scale model describing the mechanism of electromagnetic waves backscattering by the sea surface at low grazing angles of sensing (Bulatov *et al.*, 2003), and within the framework of a drop-spray model at gravity wave breaking—see Section 8.2 and Kalmykov *et al.* (1976), and Cherny and Sharkov (1988)—as well as within the framework of a recent model of small-size structures of various types that appeared as waves break (Sletten *et al.*, 2003). Within the framework of the two-scale model, scattering at vertical polarization is determined by the resonance ripple produced by the local wind moving over the crest of a large wave. And scattering amplification at horizontal polarization is associated with the growth of slopes and amplitudes of waves of intermediate scales (of about 1 m and shorter), which gives rise to the appearance of a compelled ripple on the crests of such waves. This ripple moves over the slope of a long wave at the phase velocity of a wave of intermediate scale. In this case the resulting velocity of small-scale scatterers of horizontally polarized electromagnetic waves is equal to the sum of velocities of waves of intermediate scales and of long waves. This explains the different displacements of the central frequency of Doppler spectra measured at vertical and horizontal polarizations. The same effect explains polarization distinctions in the velocity distributions we have obtained. Note here that a similar interpretation can be obtained when using a drop-spray model (see Section 8.2 for more details).

### 8.5.4   Analysis of spatial frequency spectra

Since the tracks on an RTI-diagram are caused by modulation of small-scale scatterers (of various physical nature, generally speaking) by large waves, the two-dimensional Wiener spectrum of the RTI-diagram—also called an $(\omega - \mathbf{k})$-diagram—characterizes the spectrum of modulating surface waves, whose scales are correlated with the spatial resolution of radar. The spectral composition of sea surface disturbance in the region of energy-carrying waves is known to be determined by the wind pumping effect and by fine interwave interactions. So, analysis of $(\omega - \mathbf{k})$-diagrams, obtained by remote techniques under various conditions of rough sea state, represents a good tool to study the complicated wave interactions of such a kind. Note that the traditional techniques of measurements using contact wavegraphs do not in principle allow the possibility of studying such a type of wave system interaction.

Figure 8.18 shows in enlarged scale the fragment of a spatial-frequency spectrum determined in the region of positive values of $(\omega, \mathbf{k})$ variables as a result of two-dimensional Fourier transformation of radio image fragments, presented in Figure 8.16, in the form of RTI-diagrams. Each spectrum was obtained in a spectral window of dimension $512 \times 512$ pixels, whose center was displaced to a distance equal to the window width over the "time-range" diagram when calculating each subsequent spectrum. The levels of brightness on the images in Figure 8.18 are displayed in the logarithmic scale. As analysis has shown, the same wave components are observed on the $(\omega - \mathbf{k})$-diagrams obtained both at vertical and horizontal polarizations. The dark strips, distinctly seen against a gray background in the $(\omega - \mathbf{k})$-plane, are just dispersion curves associated with the frequencies and wavenumbers of long gravity waves, which modulate a signal scattered by the sea surface in the microwave range. After threshold filtration of the original image in an $(\omega - \mathbf{k})$-diagram,

Figure 8.18 displays two sets of regions, whose centers are located approximately along the $\omega^2 = gk$ and $\omega^2 = 2gk$ curves, which correspond in the linear approximation to dispersion relationships for the gravity wave $k$ and its harmonic. The size of regions ("blurring" of dispersion curves) is determined by the final values of amplitudes of real components of the rough sea, as well as by the spatial resolution of radar and by the duration of realization processed in calculating the spectrum. Note also that in a region with high wavenumber values (Figure 8.18) the spots are shifted into a region of high frequencies. This phenomenon may be caused by wave interaction with the surface flow giving rise to a Doppler shift. In this case the dispersion dependence has the following form: $\omega = \sqrt{2gk} + \mathbf{k}\mathbf{U}$, where $\mathbf{U}$ is the flow velocity vector. According to our estimates, the flow velocity corresponding to the observed frequency shift equals $10–15\,\mathrm{cm\,s^{-1}}$, which corresponds, by an order of magnitude, to the wind shifting velocity. We shall now try to relate the results of spectral analysis of the radio images with the data of wave-graphical *in situ* measurements—that is, with the frequency spectra of elevations that have been calculated from wavegrams about 20 min in duration with subsequent four-fold averaging. The frequency spectra recorded at various stages of rough sea development are shown in Figure 8.18. The spectra presented in Figures 8.19a, b were obtained under developing rough sea conditions during 6 hours (Figure 8.19a) and 4 hours (Figure 8.19b) prior to

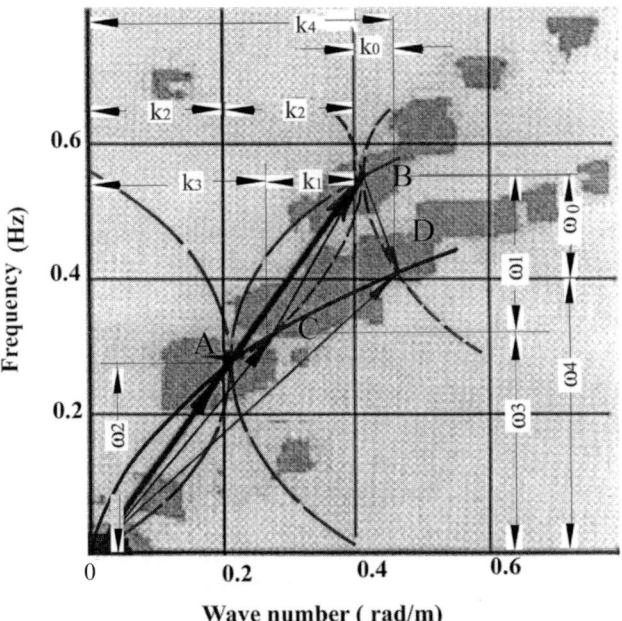

**Figure 8.18.** Fragment of the space-frequency spectrum of a surface gravity wave system determined in the range of positive $\omega$ and $\mathbf{k}$ coordinates. The diagram also shows the interaction of wave components (see the text for details). The shading variations are the result of Fourier transformation (see the text for details).

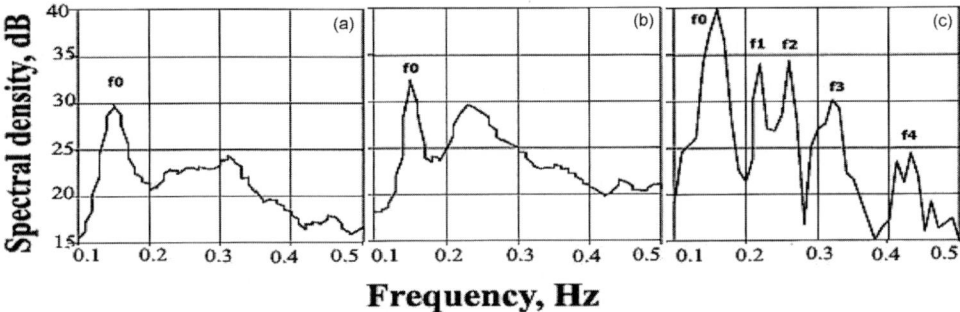

**Figure 8.19.** The experimental frequency spectra of sea surface elevations, measured using a string wave recorder at various stages in the development of roughness development.

the beginning of radar measurements. The spectrum presented in Figure 8.19c was obtained during measurements at the stage of a fully developed rough sea. Note, here, that the low-frequency maximum at frequency $f_0 = 0.16$ Hz observed at all stages of rough sea development is caused—as long-term observations have shown—by the appearance of ripple areas coming into the Blue Bay from other Black Sea areas, and is not associated with local wind. A characteristic feature of the spectrum shown in Figure 8.19c is the appearance of maxima at four frequencies: $f_1 = 0.23$ Hz, $f_2 = 0.27$ Hz, $f_3 = 0.32$ Hz, and $f_4 = 0.41$ Hz. In order to be sure that the presence of these peaks is not caused by statistical processing errors or by the insufficient length of a processed realization, detailed analysis showed that these factors do not explain the appearance of these peaks.

It is known (Phillips, 1977) that the multi-modal structure of a rough sea's frequency spectrum may be caused by nonlinear interactions between gravity waves (resonance interactions of third and higher orders). The angular frequencies and their corresponding wavenumber values can be calculated by the well-known technique of determining the parameters of gravity waves propagating to the sea of finite depth. The angular frequencies of maxima of the spectrum ($\omega_0$–$\omega_4$) are equal to 1.005, 1.44, 1.72, 2.01, and 2.57 rad s$^{-1}$, and the values of their wavenumbers ($k_0$–$k_4$) correspond to 0.078, 0.212, 0.304, 0.412, and 0.66 rad m$^{-1}$. Simple analysis of the values of frequencies and wavenumbers indicates that a set of frequencies $\omega_1$, $\omega_2$, $\omega_3$ satisfies the conditions of frequency synchronism:

$$2\omega_2 = \omega_1 + \omega_3 \tag{8.5a}$$

$$2\mathbf{k}_2 = \mathbf{k}_1 + \mathbf{k}_3 \tag{8.5b}$$

When these conditions are met, a quasi-synchronous component is excited at frequency $\omega_2$. For the other set of frequencies $\omega_0$, $\omega_2$, $\omega_4$ the following synchronism conditions are satisfied:

$$2\omega_2 = \omega_4 + \omega_0 \tag{8.6a}$$

$$2\mathbf{k}_2 = \mathbf{k}_4 - \mathbf{k}_0 \tag{8.6b}$$

which corresponds with wave component excitation at a lateral frequency $\omega_4$. Equalities (8.5) and (8.6) are observed to an accuracy better than 10%, which agrees with the accuracy of real measurements.

To compare these results with radar observations we shall draw a diagram of wavevectors—whose wavenumbers correspond with the peaks of the spectrum in Figure 8.19—on the contours of the spatial spectrum in Figure 8.18, and then we shall start our construction using the graphical method of analysis of nonlinear wave interactions (Bulatov *et al.*, 2004). Having chosen point **O**, corresponding to the zero harmonic of a spatial spectrum, as the coordinate origin, we shall construct vector **OA** with projections $\omega_2$ and $k_2$, which displays wavevector $\mathbf{k}_2$ on the $(\omega - \mathbf{k})$-plane. Having transferred the coordinate origin to point **A** and having repeated a previous construction, we obtain vector **OB**, which displays vector $2\mathbf{k}_2$ on the $(\omega - \mathbf{k})$-plane. The dispersion curves (for all signs of $\omega$ and $k$) passing through point $B$ (dashed lines), while intersecting with the dispersion curve passing through point **O** (solid line), determine the pair of vectors $(\mathbf{BC}, \mathbf{OC})$ and $(\mathbf{BD}, \mathbf{OD})$, which display the obtained spatial-frequency spectra of scattered signal modulation. Analysis of these spectra allowed: (a) determination of the degree of nonlinearity of surface waves and the range of group and phase velocities of surface waves depending on the rough sea development stage, and (b) separation of the spectral components of the modulation spectrum that arise as wave packages form. It is also shown that, by analyzing the spatial-frequency modulation spectra, we can obtain estimates of the nonlinear surface wave amplitude that agree well with the results of *in situ* measurements of the $(\omega - \mathbf{k})$-plane of a pair of wavevectors $(\mathbf{k}_1, \mathbf{k}_3)$ and $(\mathbf{k}_0, \mathbf{k}_4)$. The wavenumbers and frequencies of these vectors correspond to the resonance conditions (8.5) and (8.6) and, at the same time, the values of their projections—determined from the results of constructing a vector diagram and designated by the corresponding indices in Figure 8.18—correspond with the values of frequencies and wavenumbers presented above. Thus, the construction we have carried out above assures us that the spectra presented in Figures 8.18 and 8.19 really do reflect the dispersion properties of two systems of interacting surface gravity waves.

Further investigations in this direction demonstrate the high efficiency of spatial-frequency analysis of the data of active scatterometric remote observations of complicated wave systems in a rough sea. So, in Bulatov *et al.* (2006) the spatial-frequency spectra of modulation of scattered signals were obtained, whose analysis made it possible to determine the degree of nonlinearity of surface waves and the range of group and phase velocities that depend on the rough sea development stage, as well as to separate the spectral components of the modulation spectrum that arise as wave packages form. In addition, it is shown that by analyzing spatial-frequency modulation spectra we can obtain estimates of nonlinear surface wave amplitude, which agree well with the results of *in situ* measurements. Note also that the use of $(\omega - \mathbf{k})$-diagrams, which display the dispersion characteristics of the bound harmonics of a nonlinear wave, allows us to determine the power of modulation spectrum components produced by these harmonics, by excluding the modulation produced by free waves at the same frequencies.

## 8.6  CONCLUSIONS

**1**    The results of specialized field experiments on studying the electromagnetic properties of gravity waves in breaking mode state that remote microwave synchronous measurements, in the radiothermal and scatterometric modes, make it possible to obtain unique information on the breaking process and to separate two temporal phases associated with the appearance of a drop-spray structure over a breaker as it forms and with the formation of a foam strip field. The high-frequency zone of the Doppler spectrum provides information on the field of Doppler velocities of moving scatterers in the drop-spray phase, which stipulates formation of a speckle structure in the field of backscattered signals from a disturbed sea surface.

**2**    In their turn, the surface manifestations of wave systems can be most effectively obtained at low grazing angles in the microwave backscattering field of breaking waves. In this case the information on rough sea structure can be obtained over a considerable water basin at once, or, in other words, the spatiotemporal frame of the wave field can be generated. Such information cannot in principle be obtained by wave-graphical *in situ* techniques. Subsequent investigations in this direction show the high efficiency of a spatial-frequency analysis of the data of active scatterometric remote observations for revealing complicated nonlinear (weak) interactions of the wave systems of heavy seas.

# 9

# Conclusions

The scientific results presented in this book allow us to get the most complete idea, to date, of the modern level of development of microwave and optical remote diagnostics of a rough sea surface when the single-boundedness of the surface is broken and when an intermediate density zone arises, which represents a polydisperse mixture of finite volumes of air and water with highly fluctuating (in space and time) transition characteristics.

The general approach, as well as the methodological and instrumental classification relating to studying the process of the breaking of oceanic waves and the formation of foam-spray structures, are developed in the book. This work principally concerns application of remote optical and microwave techniques, which, along with acoustic techniques, will allow obtention in the near future of the necessary and sufficient information about aerated and disperse layers on the sea surface and in the near-surface layer. It will also facilitate construction of spatial and temporal models to solve the problems of wave dynamics, in addition to models of the energy, mass, and heat exchange in the ocean–atmosphere system.

This work offers and develops experimental remote techniques in detail and presents original results obtained in studying the regularities of the statistical spatial structure and the scale-invariant properties of chaotic fields of breaking gravity waves in the ocean–atmosphere system. Analysis of the results obtained and data on the derivation of existing breaking models have revealed inconsistencies in some generally accepted models and the necessity of developing principally new approaches to describe gravity wave breaking processes.

The original experimental remote results presented in this book are non-traditional in character and were obtained in flight–marine expeditions in the period from 1976 to 1983.

On the basis of detailed theoretical and experimental investigation of the interaction of microwave-range electromagnetic waves with polydisperse, closely packed structures that form in the ocean–atmosphere system under gravity wave breaking conditions, the principal physical mechanism of the natural radiation of disperse structures is established. This mechanism is associated with the diffraction absorption of electromagnetic waves by an emulsion medium's bubbles and by the polyhedral cells of the cellular structure of foam layers. The model approach using the Lorentz–Lorenz and Hulst approximations for a polydisperse medium described in this book can be generalized to a wider class of natural objects. The latter can include sea ice, fresh and thawed snow, water–oil emulsions, wetlands, and other geophysical systems, whose electrodynamic properties essentially depend on the structural characteristics of a disperse medium, as well as on the configuration and packing of its particles.

The results of specialized experiments on studying the electromagnetic properties of concentrated drop fluxes and the possibility of their use in constructing electrodynamic models of the drop-spray phase of breaking waves facilitate the discovery of the boundaries of applicability of the radiation transfer theory to describe the electrodynamic parameters of discrete air–disperse media with a high density of absorbing scatterers. It is experimentally found that, in the electromagnetic interaction of solitary scatterers, the air–disperse discrete system acquires the properties of a continuous turbulent medium with spatial fluctuations in its dielectric characteristics; a qualitatively new electrodynamic model of a discrete medium that takes account of scatterer interaction is offered in the book.

The results of specialized field experiments on studying the electromagnetic properties of gravity waves in breaking mode have shown that remote microwave synchronous measurements in radiothermal and scatterometric modes allow acquisition of unique information on the breaking process and distinction of two temporal phases associated with the appearance of a drop-spray structure over a fully formed breaker and formation of a foam strip field.

In their turn the surface manifestations of wave systems can be most effectively observed at low grazing angles from the microwave backscattering of breaking waves. In this case information on a rough sea structure can be obtained over a considerable basin at once, or, in other words, the spatiotemporal frame of a wave field can be formed. It is principally impossible to obtain such information by wave-graphical *in situ* techniques. Subsequent investigations in this direction have shown the high degree of efficiency of spatial-frequency analysis of the data of active scatterometric remote observations at revealing the complicated nonlinear (weak) interactions of wave systems in a rough sea.

Theoretical and experimental investigations in the immediate future will probably be directed at improving those nonlinear models that describe the structure of a near-surface layer, as well as its electrodynamic properties in the presence of foam systems and spray clouds. The essence of such models is outlined in this book. However, the efficient development and use of theoretical models is only possible on the basis of an extensive complex of experimental remote and *in situ* data; the scope of these data is clearly insufficient at the moment. This is especially true on scales of both the World Ocean as a whole and its regional constituents. Nevertheless, we believe that the experimental, theoretical, and model results stated in the present book will serve as a serious stimulus and basis for such ambitious, potential work.

# References

Abe T. (1957). A supplementary note on the foaming of sea water. *Rec. Oceanogr. Works Jap.*, Vol. 4, No. 1, pp. 1–7.

Abe T. (1962). On the stable foam forming of sea water in seas (preliminary report). *J. Oceanogr. Soc. Japan*, 20th Ann. Vol., pp. 242–250.

Abe T. (1963). In situ formation of stable foam in sea water to cause salty wind damage. *Pap. Meteorol. and Geophys.*, Vol. 14, No. 2, pp. 93–108.

Aden A. L. and Kerker M. (1951). Scattering of electromagnetic waves from two concentric spheres. *J. Appl. Phys.*, Vol. 22, No. 10, pp. 357–361.

Alcock R. K. and Morgan D. G. (1978). Investigation of wind and sea state with respect to the Beaufort Scale. *Weather*, Vol. 33, No. 7, pp. 271–277.

Alpers W. and Hasselmann K. (1982). Spectral signal to clutter and thermal noise properties of ocean wave imaging synthetic aperture radars. *Intl. J. Rem. Sens.*, Vol. 3, No. 4, pp. 432–446.

Atlas D., Ulbrich C.W., Meneghini R. (1984). The multiparameter remote measurement of rainfall. *Radio Sci.*, Vol. 19, No. 1, pp. 3–22.

Au B., Kenney J., Martin, L., and Ross D. (1974) Multifrequency radiometric measurements of foam and monomolecular slicks. *Proc. of the 7th International Symposium Rem. Sens. Env. vol. 3, Michigan*, pp. 1763–1773.

Avanesova G. G., Volayk K. B., Shugan I. V. (1984). The measurement of sea waves characterstics by air-borne side-looking radar. Theory and experiment. *Trudy Phisicheskogo Instituta Akademii Nauk (Transactions of Academy Science Physical Institute)*, Vol. 156, pp. 94–123 [in Russian].

Banner M. L. and Peregrine D. H. (1993). Wave breaking in deep water. *Ann. Rev. Fluid Mech.*, Vol. 25, pp. 373–397.

Banner M. L., and Phillips O.M. (1974). On the incipient breaking of small scale waves. *J. Fluid Mech.*, Vol. 65, pp. 647–656.

Baryshnikova Yu. S., Zaslavsky G. M., Lupian E. A., Moiseev S. S., and Sharkov E. A. (1989). Fractal analysis of the pre-hurricane atmosphere from satellite data. *Adv. Space Research*, Vol. 9, No. 7, pp. 405–408.

Basharinov A. E., Gurvich A. S., Egorov S. T. (1969). The retrieval of geophysical parameters with thermal emission data by Cosmos-243 satellite. *Doklady Acad. Nauk USSR (Trans. of Russian Academy of Sciences/Earth Science Section—Eng. Transl.)*, Vol. 188, No. 6, pp. 1273–1275.

Bass F. G., Braude S. Y., and Kalmykov A. I. (1975). The radiophysical investigations of seas (radiooceanography) developed in Ukrainian SSR Academy of Sciences. Preprint No. 51. Kharkov, IRE of Ukrainian SSR, 53 pp. [in Russian].

Belov D. M. (1978). The technique for experimental study of the ocean–atmosphere energy transfer using drop-spray mechanism. *Izvestia Vsesouznogo Geographicheskogo obschestva (Izvestiya VGO)*, Vol. 110, No. 3, pp. 257–261 ([in Russian].

Bendat J.S. and Piersol A. G. (1966). *Random data and measurements procedures.* Wiley-Interscience, New York, 450 pp.

Bespalova E. A., Veselov V. M., Glotov A. A., Militzkii Y. A., Mirovskii V. G., Pokrovskaya I. V., Popov A. E., Raev M. D., Sharkov E. A., Etkin V. S. (1979). Investigations of wind sea roughness anisotropy with variability of thermal radioemission. *Doklady Acad. Nauk (Trans. of Russian Academy of Sciences/Earth Science Section—Engl. Transl.)*, Vol. 246, No. 6, pp. 1482–1485.

Bespalova E. A., Veselov V. M., Gershenson V. E., Militzkii Y. A., Mirovskii V. G., Pokrovskaya I. V., Raev M. D., Trochimovskii J. G., Semin A. G., Smirnov N. K. *et al.* (1982). On determination of wind speed with polarization anisotropy measurements of thermal and backscattering microwaves. *Earth Research from Space*, No. 1, pp. 87–94.

Bezzabotnov V. S. (1985). Some results of natural measurements of sea foam systems structure. *Izvestia Acad. Nauk. Fizika atmosphery i okeana (Izvestiya, Atmospheric and Oceanic Physics—Engl. Transl.)*, Vol. 21, No. 1, pp. 101–104.

Bharucha-Reid A.T. (1960). *Elements of the theory of Markov processes and their applications.* McGraw-Hill, New York, 360pp.

Bikerman J.J. (1973). *Foams.* Springer-Verlag, New York, 320 pp.

Blanchard D. C. (1963). The electrification of the atmosphere by particles from bubbles in the sea. *Progress in Oceanography*, Vol. 1, pp. 73–202.

Blanchard D. C. and Woodcock A. H. (1957). Bubble formation and modification in the sea and its meteorological significance. *Tellus*, Vol. 9, No. 2, pp. 145–158.

Bohren C. F. and Hoffman D. R. (1983). *Absorption and scattering of light by small particles.* New York, Wiley, 530 pp.

Bondur V. G. and Sharkov E. A. (1982) Statistical properties of whitecaps on a rough sea. *Oceanology*, Vol. 22, No. 3, pp. 274–279.

Bondur V. G. and Sharkov E. A. (1990) Statistical characteristics of linear geometric elements of foam structures on the sea surface for optical sensor data. *Sov. J. Remote Sensing*, Vol. 6, No. 4, pp. 534–550.

Bordonskii G. S., Vasilkova I. B., Veselov V. M., Vorsin N. N., Militskii Y. A., Mirovskii V. G., Nikitin V. V., Raizer V. Y., Khapin Y. B., Sharkov E. A. *et al.* (1978). Spectral characteristics of the emissivity of foam formations. *Izvestia Acad. Nauk. Fizika atmosphery i okeana (Izvestiya, Atmospheric and Oceanic Physics—Engl. Transl.)*, Vol. 14, No. 6, pp. 656–663.

Bordugov V. M., Vereshak A. I., Grodskii S. A. (1986). The investigation of representation for large-scale internal waves parameters on ocean surface. Preprint. Morskoi Hydrophisicheskii Institut (Sea Hydrophysical Institute), Sevastopol, p. 86. [in Russian].

Borisenkov Ye. P. and Kuznetsov M. A. (1976). On the theory of heat and moisture exchange between the atmosphere and ocean under stormy weather conditions. *Meteorologia i hydrologia (Russian Meteorology and Hydrology—Engl. transl.)*, No. 5, pp. 18–26.

Borisenkov Ye. P. and Kuznetsov M. A. (1978). On the parametrization of interaction between the atmosphere and ocean under stormy weather related to the atmosphere circulation models. *Izvestia Acad. Nauk. Fizika atmosphery i okeana (Izvestiya, Atmospheric and Oceanic Physics—Engl. Transl.)*, Vol. 14, No. 5, pp. 510–519.

Born M. and Wolf E. (1999). *Principles of optics: Electromagnetic theory of propagation, interference and diffraction of light* 7th Edition. Cambridge University Press, New York, 952 pp.

Bortkovskiy R. S. (1977). The experimental investigations of droplet fields over wind waves. *Trudy Glavnoi Geophysical Obsevatory (Proceedings of the Main Geophysical Observatory)*, Issue 398, pp. 34–40 [in Russian].

Bortkovskiy R. S. (1983). *Atmosphere–ocean heat and moisture exchange at storm conditions.* Hydrometeoizdat, Leningrad, 158 pp.

Bortkovskiy R. S. (1987). Time–space characteristics of whitecaps and foam patches formed by wave breaking. *Meteorologia i hydrologia (Russian Meteorology and Hydrology—Engl. Transl.)*, No. 5, pp. 68–75.

Bortkovskiy R. S. (2006). Estimation of the oxygen and $CO_2$ mean exchange between the ocean and the atmosphere in key areas of the ocean. *Izvestia Acad. Nauk. Fizika atmosphery i okeana (Izvestiya, Atmospheric and Oceanic Physics—Engl. Transl.)*, Vol. 42, No. 2, pp. 250–257.

Bortkovskiy R. S. and Kuznetsov M. A. (1977). Some results of sea surface condition study. In: *Typhoon 75*, Vol. 1. Hydrometeoizdat, Leningrad, pp. 90–105 [in Russian].

Bortkovskiy R. S. and Timonovsky D. F. (1982). On the microsructure of wind-waves breaking crests. *Izvestia Acad. Nauk. Fizika atmosphery i okeana (Izvestiya, Atmospheric and Oceanic Physics—Engl. Transl.)*, Vol. 18, No. 3, pp. 327–329.

Brekhovskikh L. M. (1957). *Waves in layered media.* USSR Acad. Sciences, Moscow, 502 pp.

Bulatov M. G., Kravtsov Yu. A., Lavrova O. Yu., Litovchenko K. Ts., Mityagina M. I., Raev M. D., Sabinin K. D., Trokhimovskii Yu. G., Churyumov A. N., Shugan I. V. (2003). Physical mechanisms of aerospace radar imaging of the ocean. *Physics-Uspekhi*, Vol. 46, No. 1, pp. 63–79.

Bulatov M. G., Raev M. D., Skvortsov E. I. (2004). Dynamics of sea waves in coastal region according to date of high-resolution radar observation. *Physics of Wave Phenomena*, Vol. 12, No. 1, pp. 18–24.

Bulatov M. G., Raev M. D., Skvortsov E. I. (2006). Study of nonlinear wave dynamics with the spatial-frequency spectra of the sea surface radioimages. *Earth Research from Space*, No. 2, pp. 1–7.

Bunkin F. V. and Gochelashvili K. S. (1968). The overshoots for random scalar field. *Radiophysics and Quantum Electronics*, Vol. 11, No. 12, pp. 1864–1870.

Camps A., Vall-Ilossera M., Villarino R., Reul N., Chapron B., Corbella I., Duffo N., Torres F., Miranda J. J., Sabia R., Monerris A., and Rodriguez R. (2005). The emissivity of foam-covered water surface at L-band: Theoretical modeling and experimental results from the FROG 2003 field experiment. *IEEE Transactions on Geoscience and Remote Sensing*, Vol. 43, No. 3, pp. 925–936.

Carter D. J. (1982). Prediction of wave height and period for a constant wind velocity using the JONSWAP results. *Ocean Eng.*, Vol. 9. No. 1. pp. 17–33.

Cavaleri L. (2006). Wave modeling. Where to go in the future. *Bull. Amer. Meteorological Soc.*, Vol. 87, No. 2, pp. 207–214.

Cherny I. V. (1982). Radiometer-scatterometer in millimeter range for sea surface investigations. Preprint No. Pr-689. Space Research Institute, Moscow, 19 pp. [in Russian].

Cherny I. V. and Sharkov E. A. (1988). Remote radiometry of the sea wave breaking cycle. *Earth Research from Space (Earth Obs. Remote Sensing—Engl. Transl.)*, No. 2, pp. 17–28.

Cherny I. V. and Sharkov E. A. (1991a). Electrodynamics of discrete dispersive concentrated media with absorbing scattering particles. Preprint No. Pr-1753. Space Research Institute, Moscow, 40 pp. [in Russian].

Cherny I. V. and Sharkov E. A. (1991b). Characteristics of backscattering of electromagnetic waves by concentrated air-dispersive media. *Pisma v Zhurnal Tech. Fiziki (Technical Physics Letters—Engl. Transl.)*, Vol. 17, No. 3, pp. 73–77.

Cipriano R. J. and Blanchard D. C. (1981). Bubble and aerosol spectra produced by laboratory breaking waves. *J. Geophysical Research*, Vol. 86, pp. 8085–8092.

Conwell P. R., Barber P. W., Rushforth C. K. (1984). Resonant spectra of dielectric spheres. *J. Opt. Soc. Am. A*, Vol. 1, No. 1. pp. 62–67.

Cramer H. and Leadbetter M. R. (1967). *Stationary and related stochastic processes: Sample function properties and their applications.* Wiley, New York, 398 pp.

Day J. A. (1964). Production of droplets and salt nuclei by bursting of air-bubble films. *Quart. J. Roy. Met. Soc.*, Vol. 90, No. 383, pp. 72–78.

Deane G. B. and Stokes M. D. (2002). Scale dependence of bubble creation mechanisms in breaking waves. *Nature*, Vol. 418, No. 6900, pp. 839–844.

Deirmendjian D. (1969). *Electromagnetic scattering on spherical polydispersions.* American Elsevier, New York, 290 pp.

Dombrovskiy L. A. (1974). Scattering and absorption of light by hollow spherical particles. *Izvestiya, Atmos. Oceanic Phys.*, Vol. 10, No. 7, pp. 720–727.

Dombrovskiy L. A. (1979). Calculation of the thermal radiation emission of foam on the sea surface. *Izvestiya, Atmos. Oceanic Phys.*, Vol. 15, No. 3, pp. 193–198.

Dombrovskiy L. A. (1981). Absorption and scattering of microwave emission by spherical aqueous envelopes. *Izvestiya, Atmos. Oceanic Phys.*, Vol. 17, No. 3, pp. 324–329.

Dombrovskiy L. A. and Raizer V. Y. (1992). Microwave model of a two-phase medium at the ocean surface. *Izvestiya, Atmos. Oceanic Phys.*, Vol. 28, No. 8, pp. 650–656.

Doviak R. J. and Lee J. T. (1985). Radar for storm forecasting and weather hazard warning. *J. Aircraft*, Vol. 22, No. 12, pp. 1059–1064.

Doviak R. J. and Zrnic D. S. (1984). *Doppler radar and weather observation.* Academic Press, Orlando, FL, 458 pp.

Droppleman J. D. (1970). Apparent microwave emissivity of sea foam. *J. Geophysical Research*, Vol. 75, No. 3, pp. 696–698.

Egorov B. P. (1977). The parameters of drop-spray clouds developed during strong surf. *Trudy Glavnoi Geophysical Observatory (Proceedings of the Main Geophysical Observatory)*, Issue 399, pp. 136–145 [in Russian].

Fante R. (1984). Detection of multiscatter targets in K-distributed clutter. *IEEE Trans. on AP*, Vol. 32, No. 12, pp. 1358–1363.

Feller W. (1971). *An introduction to probability theory and its applications*, Vol. II. Wiley, New York, 478 pp.

Frisch U. (1995). *Turbulence: The legacy of A. N. Kolmogorov.* Cambridge University Press, Cambridge, UK, 350 pp.

Frouin R., Iacobelllis S. F., Deschamps P. Y. (2001). Influence of oceanic whitecaps on the global radiation budget. *Geophysical Research Letters*, Vol. 28, No. 8, pp. 1523–1526.

Glazman R. E. (1991a). Statistical problems of wind-generated gravity waves arising in microwave remote sensing of surface winds. *IEEE Trans. on Geoscience and Remote Sensing*, Vol. 29, No. 1, pp. 135–142.

Glazman R. E. (1991b) . Reply. *Journal Geophysical Research*, Vol. 96. No. C3, pp. 4979–4983.

Glazman R. E. and Weichman P. B. (1989). Statistical geometry of a small surface patch in a developed sea. *Journal Geophysical Research*, Vol. 94. No. C4, pp. 4998–5010.

Glazman R. E. and Weichman P. B. (1990). Reply to comments by E. C. Monahan on "Statistical geometry of a small surface patch in a developed sea". *Journal of Geophysical Research*, Vol. 95. No. C2, pp. 1771–1773.

Gradshteyn I. S. and Ryzhik I. M. (2000). *Tables of Integrals, Series, and Products*, 6th Edition (edited by A. Jeffrey and D. Zwillinger). Academic Press, Orlando, FL, 1163 pp.

Grushin V. A., Il'in Yu. A., Lazarev A. A., Lupyan E. A., Malinnikov V. A., Pokrovskaya I. V., Skachkov V. A., Suslov A. I., Stulov A. A., Sharkov E.A. (1990). Simultaneous optical and in-situ studies of the spatio-spectral characteristics of wind-driven waves. *Sov. J. Remote Sensing*, Vol. 6, No. 2, pp. 211–229.

Hayami S. and Toba Y. (1957). Drop production by bursting air bubbles on the sea surface. I: Experiments at still water surface. *J. Oceanogr. Soc. Japan*, Vol. 14, No. 2, pp. 145–150.

Holliger J. P. (1971). Passive microwave measurements of sea surface roughness. *IEEE Trans. Geoscience Electronics*, Vol. 9, No. 3, pp. 165–169.

Horne R. A. (1969). *Marine chemistry: The structure of water and the chemistry of the hydrosphere*. Wiley-Interscience, New York, 398 pp.

Hu J., Gao J., Posner F. L., Zheng Y., Tung W. W. (2006). Target detection within sea clutter: A comparative study by fractal scaling analyses. *Fractals*. Vol. 14, No. 3, pp. 187–204.

Hulst C. H. van de (1981). *Light scattering by small particles*. Dover Publications, New York, 470 pp.

Ishimaru A. (1978). *Wave propagation and scattering in random media*, Vols. I and II. Academic Press, New York, 540 pp.

Ivazyn G. M. (1991). *Propagation of millimeter and submillimeter wavelengths in clouds*. Gidrometeoizdat, Leningrad, 478 pp. [in Russian].

Jameson A. R. (1991). The effect of drop-size distribution variability on radiometric estimates of rainfall rates for frequencies from 3 to 10 GHz. *J. Applied Meteorology*, Vol. 30, No. 7, pp. 1025–1033.

Johnson N. L. and Leone F. C. (1977) *Statistics and experimental design in engineering and the physical sciences*. Wiley, New York, 610 pp.

Kalmykov A. I., Kurekin A. S., Lementa Yu. A., Ostrovskii I. E., Pustovoytenko V.V. (1976). Peculiarities of scattering of microwave radiation by attacked sea waves. *Radiophysics and Quantum Electronics*, Vol. 19, No. 9, pp. 1315–1321.

Kanevsky M. B. (2004). *Theory of radar imaging of the ocean surface*. Institute of Applied Physics, Russian Academy of Sciences, Nizhniy Novgorod, 124 pp.

Karlin S. (1968). *A first course in stochastic processes*. Academic Press, New York, 538 pp.

Kazevich R. S., Tang C. H., Henriksen S. W. (1972). Analysis and optical processing of sea photographs for energy spectra. *IEEE Trans. Geoscience and Electronic*, Vol. 10, No. 1, pp. 51–57.

Kerker M. (1969). *The scattering of light*. Academic Press, New York, 350 pp.

Khusu A. P., Vitenberg Y. R., Palmov V. A. (1975). *The roughness of surfaces (theoretic-probability approach)*. Nauka, Moscow, 344 pp.

Kitaigorodskii S. A. (1973). *Physics of air–sea interaction*. Israel Program for Scientific Translation, Jerusalem, 210 pp.

Kitaigorodskii S. A. (1997). Effect of breaking of wind-generated waves on the local atmosphere–ocean interaction. *Izvestia Acad. Nauk. Fizika atmosphery i okeana (Izvestiya, Atmospheric and Oceanic Physics—Engl. Transl.)*, Vol. 33, No. 6, pp. 828–836.

Kitaigorodskii S. A. (2001). New evidence for the action of the process of nonlinear wind waves breaking on an increase in the kinetic energy dissipation within the sea upper layer.

*Doklady Acad. Nauk (Trans. of Russian Academy of Sciences—Engl. Trans.)*, Vol. 376, No. 4, pp. 539–542.

Koga M. (1981) Direct production of droplets from breaking wind-generated waves: Its observation by a multicolored overlapping exposure photographic technique. *Tellus*, Vol. 33, No. 6, pp. 552–563.

Koga M. (1982). Bubble entrainment in breaking wind-generated waves. *Tellus*, Vol. 34, No. 5, pp. 481–489.

Kokhanovsky A. A. (2004). Spectral reflectance of whitecaps. *J. Geophys. Research*, Vol. 109, C05021, doi:101029/2003JC002177.

Kollias P., Lhermitte R., Albrecht B. A. (1999). Vertical air motion and raindrop size distributions in convective systems using a 94 GHz radar. *Geophysical Research Letters*, Vol. 26, No. 20, pp. 3109–3112.

Kondratyev K. Ya., Rabinovich Yu. I., Nordberg W. (eds.) (1975). *USSR/USA Bering Sea experiment proceedings of the final symposium on the joint results of the joint Soviet–American expedition, Leningrad, May 12–17, 1974.* Gidrometeoizdat, Leningrad, 254 pp.

Korn G. A. and Korn T. M. (1961). *Mathematical handbook for scientists and engineers: Definitions, theorems and formulas for reference and review.* McGraw-Hill, New York, 720 pp.

Krasilnikov N. I. (1987). Dispersion and breaking of gravitation waves in fluid. *Doklady Acad. Nauk SSSR (Trans. of USSR Academy of Sciences—Engl. Transl.)*, Vol. 294, No. 3, pp. 592–594.

Krasilnikov N. I., Lebedev V. B., Khapaev M. M., Gribov B. E. (1986). Computer simulation for sea waves breakings. Preprint No. Pr-1095. Space Research Institute, Moscow, 29 pp. [in Russian].

Krasiuk N. P. and Rosenberg V. I. (1970). *Ship-borne radiolocation and meteorology.* Sudostroenie, Leningrad, 324 pp. [in Russian].

Kutateladze S. S. and Styrikovich M. A. (1976) *Hydrodynamics of gas–liquid systems.* Energiya, Moscow, 296 pp.

Kwoh D. S. and Lake B. M. (1985). The nature of microwave backscattering from water waves. In: *Ocean surface: Wave breaking, turbulent mixing and radio probing.* Dordrecht, The Netherlands, pp. 249–256.

Lai J. R. and Shemdin O.H. (1974). Laboratory study of the generation of spray over water. *J. Geophysical Research*, Vol. 79, No. 21, pp. 3055–3063.

Lakhtakia A., Messier R., Varadan V. V., Varadan V. K. (1987) Fractal dimension from the back-scattering cross section. *J. Phys. A : Math. Gen.*, Vol. 20, pp. 1615–1619.

Landau L. D. and Lifshitz E. M. (1957). *Electrodynamics of continuous media.* Gostechizdat, Moscow, 340 pp. [in Russian] (Engl. Transl.: Pergamon, Oxford, UK, 1960, 350 pp.).

Lappo S. S., Gulev S. K., Rozhdrestvenskii A. E. (1990). *The large scale thermal interaction in the ocean–atmosphere system and the energy-active zones of the World Ocean.* Hydrometeoizdat, Leningrad, 336 pp. [in Russian].

Lavrova N. P. and Stestenko A. F. (1981). *Aerial photography: Aero-photography instruments.* Nedra, Moscow, 297 pp. [in Russian].

Levich V. (1962) *Physicochemical hydrodynamics.* Prentice-Hall, Englewood Cliffs, NJ, 340 pp.

Lewis B. and Olin I. (1980). Experimental study and theoretical model of high-resolution radarbackscatter from the sea. *Radio Science*, Vol. 15, No. 4, pp. 815–828.

Lhermitte R. M. (1988). Cloud and precipitation remote sensing at 94 GHz. *IEEE Trans. Geosci. Remote Sensing*, Vol. 26, No. 3, pp. 207–216.

Liu P. (1993). Estimating breaking wave statistics from wind-wave time series data. *Annales Geophysicae*, Vol. 11. No. 10. pp. 970–972.

Longuet-Higgins M. S. (1969). On wave breaking and the equilibrium spectrum of wind generating waves. *Proc. Roy. Soc. A.*, Vol. 310. No. 1501, pp. 151–159.

Longuet-Higgins M. S. and Turner J.S. (1974). An entraining plume model of a spilling breaker. *J. Fluid Mech.*, Vol. 63, No. 1, pp. 1–20.

Lovejoy S. (1982). Area–perimeter relation for rain and cloud areas. *Science*, Vol. 219, No. 9, pp. 185–187.

Lovejoy S. and Mandelbrot B. B. (1985). Fractal properties of rain and a fractal model. *Tellus*, Vol. 37A, pp. 205–232.

Lovejoy S. and Schertzer D. (1985). Generalized scale invariance in the atmosphere and fractal models of rain. *Water Resources Research*, Vol. 21, No. 8, pp. 1233–1250.

Lubyako L. V. and Parshin V.V. (1986). The use of a 2-mm scatterometer for revealing the statistical characteristics of signals scattered by sea surface. In: *All-union conf. on statistical methods for processing remote sensing data*. Theses of reports, Riga, Latvia, pp. 100 [in Russian].

Lupyan E. A. and Sharkov E. A. (1990 ). Figures of merit for rough sea surface reflectance from optical images. *Sov. J. Remote Sensing*, Vol. 6, No. 2, pp. 230–245.

MacIntyre F. (1972). Flow patterns in breaking bubbles. *J. Geophysical Research*, Vol. 77, No. 27, pp. 5211–5228.

Malinovskii V. V. (1991). Estimation of the interaction between the statistical characteristics of a radar signal scattered by the sea surface at grazing angles and the characteristics of breaking sea waves. *Morskoi Hydrophysical Journal (Physical Oceanography—Engl. Transl.)*, No. 6, pp.32–41 [in Russian].

Mandelbrot B. (1982) *The fractal geometry of nature*. Freeman & Co., New York, 461 pp.

Marmorino G. O. and Smith G. B. (2005). Bright and dark whitecaps observed in the infrared. *Geophysical Research Letters*, Vol. 32, L11604, doi:10.1029/2005GL0231766.

Martsinkevich L. B. and Melentyev V.V. (1975). Model calculations of sea surface thermal emission for a still sea state. *Trudy GGO (Trans. Main Geophys. Observatory)*, No. 331, pp. 73–85 [in Russian].

Matveyev D. T. (1971). On the spectrum of the microwave radiation of the wavy sea surface. *Izvestia Acad. Nauk. Fizika atmosphery i okeana (Izvestiya, Atmospheric and Oceanic Physics—Engl. Transl.)*, Vol.7, No. 10, pp. 1070–1083.

Matveyev D. T. (1978). Analysis of results from radiothermal sounding the sea surface in a storm. *Meteorologia i hydrologia (Russian Meteorology and Hydrology—Engl. Transl.)*, No. 4, pp. 58-66.

Meischner P. (1990). *Cloud dynamics and cloud microphysics by radar measurements*, ESA SR-301. ESA, Noordwijk, The Netherlands, pp. 19-26.

Melnichuk Y. V. and Chernikov A.A. (1971). The spectra of radar signals from sea surface under various radiation reception polarizations. *Izvestia Acad. Nauk. Fizika atmosphery i okeana (Izvestiya, Atmospheric and Oceanic Physics—Engl. Transl.)*, Vol. 7, No. 1, pp. 28–40.

Miyake Y. and Abe T. (1948). A study on the foaming of sea water. *J. Mar. Res.*, Vol. 7, No. 2, pp. 67–73.

Militskii Y. A., Raizer V. Y., Sharkov E. A., Etkin V. S. (1976). On scattering microwave emission by foamy structures. *Pisma v JTF (Journal Technical Physics Letters—Engl. Transl.)*, Vol. 2, No. 18, pp. 851–855.

Militskii Y. A., Raizer V. Y., Sharkov E. A., Etkin V. S. (1977). Scattering microwave radiation by foamy structures. *Radiotechnika i electronica (J. of Communic. Techn. Electronics—Engl. Transl.)*, Vol. 22, No. 11, pp. 2299–2304.

Militskii Y. A., Raizer V. Y., Sharkov E. A., Etkin V. S. (1978). On thermal emission of foamy structures. *Jour. Technicheskoi Fiziki (Journal Technical Physics—Engl. Transl.)*, Vol. 48, No. 5, pp. 1031–1033.

Monahan E. C. (1968). Sea spray as a function of low elevation speed. *J. Geophysical Research*, Vol. 73, No. 4, pp. 1127–1137.

Monahan E. C. (1971). Oceanic whitecaps. *J. Phys. Oceanography*, Vol. 1. No. 2. pp. 139–144.

Monahan E. C. (1990). Comment to "Statistical geometry of a small surface patch in a develoed sea", by R. E. Glazman and P. B. Weichman. *J. Geophys. Research*, Vol. 95, No. C2, 1768–1770.

Monahan E. C. (2001). Whitecaps and foam. In: *Encyclopedia of Ocean Sciences* (edited by J. Steele, S. Thorpe, and K. Turekian). Elsevier, New York, pp. 3213–3219.

Monahan E. C. and Zietlow C. R. (1969). Laboratory comparisons of fresh-water and salt-water whitecaps. *J. Geophysical Research*, Vol. 74, No. 28, 6961–6966.

Monahan E. C., Davidson K. L., and Spiel D. E. (1982). Whitecap aerosol productivity deduced from simulation tank measurements. *J. Geophysical Research*, Vol. 87, No. C11, pp. 8898–8904.

Moor R. and Fung A.K. (1979). Radar determination of wind of sea. *Proc. IEEE*, Vol. 67, No. 11, pp. 1504–1521.

Mouche A. A., Hauser D., Kudryavtsev V. (2006). Radar scattering of the ocean surface and sea-roughness properties: A combined analysis from dual-polarization airborne radar observations and models in C band. *J. Geophysical Research*, Vol. 111, C09004, doi:10.1029/2005JC003166.

Nigmatulin R. I. (1978). *Foundations of mechanics of heterogeneous mixtures*. Nauka, Moscow, 250 pp. [in Russian].

Nordberg W., Conaway J., Ross D. B., Wilheit T. (1971). Measurements of microwave emission from a foam-covered, wind-driven sea. *J. Atmos. Science*, Vol. 28, No. 6, pp. 1971–1978.

Ochi M. and Tsai C. H. (1983). Prediction of occurrence of breaking waves in deep water. *J. Physical Oceanography*, Vol. 13, No. 11, pp. 2008–2019.

Odelevskii V. I. (1951). The dielectric properties of heterogeneous mixtures. *Journal Technicheskoi Fiziki (Journal of Technical Physics)*, Vol. 21, No. 6, pp. 667–673 [in Russian].

Oguchi T. (1983). Electromagnetic wave propagation and scattering in rain and other hydro-meteors. *Proc. IEEE*, Vol. 71, No. 9, pp. 1029–1078.

Okuda S. and Hayami S. (1959). Experiments on evaporation from a wavy water surface. *Rec. Oceanogr. Works in Japan*, Vol. 5, No. 1, 6–13.

Ozisik M. N. (1973). *Radiative transfer and interactions with conduction and convection*. Wiley, New York, 450 pp.

Papadimitrakis Y. A. (2005a). On the probability of wave breaking in deep waters. *Deep-Sea Research*, Part II, Vol. 52, pp. 1246–1269.

Papadimitrakis Y. A. (2005b). Momentum and energy exchange across an air–water interface: Partitioning (into waves and currents) and parameterization. *Deep-Sea Research*, Part II, Vol. 52. pp. 1270–1286.

Pasqualicci F. (1984). Drop size distribution measurements in convective storms with a vertical pointing 35 GHz Doppler radar. *Radio Science*, Vol. 19, No. 1, pp. 177–183.

Phillips O. M. (1977). *The dynamics of the upper ocean*. Cambridge University Press, London, 336 pp.

Phillips O. M. (1988). Radar return from sea surface: Bragg scattering and breaking waves. *J. Phys. Oceanogr.*, Vol. 18, No. 8, pp. 1065–1074.

Phillips O. M., Posner F. L., Hansen J. P. (2001). High range resolution radar measurements of speed distribution of breaking events in wind-generation ocean waves: Surface impulse and wave energy dissipation rates. *J. Phys. Oceanogr.*, Vol. 31, No. 4, pp. 450–460.

Pierson W. J. and Moskowitz L. (1964). A proposed spectral model for fully developed wind seas based on the similarity theory of S. A. Kitaigorodskii. *J. Geophysical Research*, Vol. 69, pp. 5181–5190.

Pokrovskaya I. V. and Sharkov E. A. (1986). Spatio-statistical properties of whitecap fields on sea surface with optical remote sensing. *Earth Research from Space (Sov. J. Remote Sensing—Engl. Transl.)*, No. 5, pp. 18–25.

Pokrovskaya I. V. and Sharkov E. A. (1987a). Foam activity on the sea surface as Markov random process. *Doklady Acad. Nauk SSSR (Trans. of USSR Acad. of Sciences/Earth Science Section—Engl. transl.)*, Vol. 293, No. 5, pp. 1108–1111.

Pokrovskaya I. V. and Sharkov E. A. (1987b). Optical remote sensing study of breaking gravity wave activity with developing sea roughness. *Earth Research from Space (Sov. J. Remote Sensing—Engl. Transl.)*, No. 3, pp. 11–22.

Pokrovskaya I. V. and Sharkov E. A. (1994). Optical remote studies of the azimuth characteristics of the breaking of sea gravity waves. *Sov. J. Remote Sensing*, Vol. 11, No. 2, pp. 311–318.

Preobrazhensky L. Y. (1972). Estimation of droplet concentration in the near-surface atmosphere layer. *Trudy Glavnoi Geophysical Obsevatory (Proceedings of the Main Geophysical Observatory)*, Issue 282, pp. 194–199 [in Russian].

Raizer V. Yu. and Novikov V. M. (1990). The fractal properties of breaking surface waves zones in ocean. *Izvestia Acad. Nauk. Fizika atmosphery i okeana (Izvestiya, Atmospheric and Oceanic Physics—Engl. Transl.)*, Vol. 26, No. 6, pp. 664–668.

Raizer V. Yu. and Sharkov E. A. (1980). On dispersal structure of sea foam. *Izvestia Acad. Nauk. Fizika atmosphery i okeana (Izvestiya, Atmospheric and Oceanic Physics—Engl. Transl.)*, Vol. 16, No. 7, pp. 772–776.

Raizer V. Yu. and Sharkov E. A. (1981). Electrodynamic description of densely packed dispersed media. *Radiophysics and Quantum Electronics*, Vol. 24, No. 7, pp. 553–560.

Raizer V. Yu., Sharkov E. A., Etkin V. S. (1975a). Influence of temperature and salinity on the radioemission of a smooth ocean surface at decimeter and meter bands. *Izvestiya Akad. Nauk USSR, Fizika Atm. Okeana (Izvestya. Atmospheric and Oceanic Physics—Engl. Transl.)*, Vol. 11, No. 6, pp. 652–655.

Raizer V. Yu., Sharkov E. A., Etkin V. S. (1975b). On the thermal radioemission of sea surface with oil pollution. Preprint No. Pr-237. Space Research Institute, Moscow, 15 pp. [in Russian].

Raizer V. Yu., Sharkov E. A., Etkin V. S. (1976). Sea foam. Physico-chemical properties. Emissive and reflective characteristics. Preprint No. Pr-306. Space Research Institute, Moscow, pp. 25. [in Russian].

Raizer V. Yu., Novikov V. M., Bocharova T. Y. (1994). The geometrical and fractal properties of visible radiances associated with breaking waves in the ocean. *Ann. Geophysicae*, Vol. 12, pp. 1229–1233.

Rozenberg G. V. (1958). *The optics of thin-walled coating.* Fizmatgiz, Moscow, 250 pp.

Rozenberg G. V. (1972). *The electromagnetic emission scattering and absorption by atmospheric particles.* Hydrometeoizdat, Leningrad, 270 pp.

Rosenkranz P. W. and Staelin D. (1972). Microwave emissivity of ocean foam and its effect on nadiral radiometric measurements. *J. Geophys. Research*, Vol. 77, No. 33, pp. 6528–6538.

Roshkov V. A. (1979). *Methods of probability analysis for oceanic processes.* Hydrometeoizdat, Leningrad, 279 pp. [in Russian].

Ross D. and Cordon V. (1974) Observations of oceanic whitecaps and their relation to remote measurements of surface wind speed. *J. Geophys. Res.*, Vol. 79. No. 3. pp. 444–452.

Ruben D. L. (1977). A water droplet concentration measuring device for use over the ocean. *AIAA Paper*, No. 305, 5 pp.

Rytov S. M., Kravtsov Yu. A., Tatarskii V. I. (1978). *Introduction to statistical radiophysics*, Part II: *Random fields*. Nauka, Moscow, 462 pp. [in Russian].

Sabinin K. and Serebryany A. (2005). Intense short-period internal waves in the ocean. *J. Marine Research*, Vol. 63, No. 1, pp. 227–261.

Sagdeev R. Z., Stiller H., Ziman Y. L. (eds.) (1980). *Soyuz-22 observes the Earth*. Nauka, Moscow, 231 pp. [in Russian].

Samoilenko V. S., Matveev D. T., Semenchenko B. A. (1974). Materials for quantitative estimation of sea surface coverage by foam. In: *TROPEX-72*. Hydrometeoizdat, Leningrad, pp. 582–559 [in Russian].

Sharkov E. A. (1993a). Spatial features of sea wave breaking fields. *Symposium on the Air–Sea Interface, Marseilles, France, June 24–30, 1993*, Abstracts, pp. 21.

Sharkov E. A. (1993b). Scaling properties of sea wave breaking fields. *Annales Geophysicae*, Part II, Suppl. II to Vol. 11, pp. C310.

Sharkov E. A. (1994). Experimental investigations of lifetimes for the breaking wave disperse zone. *Izvestia Acad. Nauk. Fizika atmosphery i okeana (Izvestiya, Atmospheric and Oceanic Physics—Engl. Transl.)*, Vol. 30, No. 6, pp. 844–847.

Sharkov E. A. (1995). Optical investigations of temporal evolution of foam structures on sea surface. *Earth Obs. Rem. Sens.*, Vol. 12, No. 1, pp. 88–101.

Sharkov E. A. (1996a). The nonlinear evolution of breaking sea gravity waves. *Annales Geophysicae*, Suppl. II to Vol. 14, Part II, pp. C542.

Sharkov E. A. (1996b) Wave breaking as the springs of air–sea gas transfer. *PORSEC 96: Pacific Ocean Remote Sensing Conference, Victoria, Canada*, Abstracts, pp. 123.

Sharkov E. A. (1998) *Remote Sensing of Tropical Regions*. Wiley/Praxis, Chichester, UK, 320 pp. (ISBN 0-471-97171-5).

Sharkov E. A. (2000). *Global Tropical Cyclogenesis*. Springer/Praxis, Chichester, UK, 370 pp.

Sharkov E. A. (2003) *Passive Microwave Remote Sensing of the Earth: Physical Foundations*. Springer/Praxis, Chichester, UK, 613 pp. (ISBN 3-540-43946-3).

Sharkov E. A. and Bondur V. G. (1993). Statistical characteristics of linear and area geometry of foam structures on a disturbed sea surface. *Symposium on the Air–Sea Interface, Marseilles, France, June 24–30, 1993*, Abstracts, pp. 177.

Shibata A., Uji T., Isozaki I. (1985). Doppler spectra of microwave radar echo returned from calm and rough sea surface. In: *Ocean surface: Wave breaking, turbulent mixing and radio probing*. Dordrecht, The Netherlands, pp. 263–268.

Shiotsuki Y. (1976). An estimation of dropsize distribution in the severe rainfall. *J. Met. Soc. Japan*, Vol. 54, No. 4, pp. 259–263.

Shlaychin V. M. (1987). The probability models of non-Rayleigh fluctuations for radar signals. *Radiotechnika i electronica (J. of Communic. Techn. Electronics—Engl. Transl.)*, Vol. 32, No. 9, pp. 1793–1817.

Shulgina E. M. (1972). Calculation of emissivity of disturbed sea surface at microwave band. *Izvestia AN. Fizika atmosphery i okeana (Izvestiya. Atmospheric and Oceanic Physics—Engl. Transl.)*, Vol. 8, No. 7, pp. 773–776.

Skolnik M. I. (1980). *Introduction to radar systems*. McCraw-Hill, New York, 450 pp.

Sletten M. A., West J. C., Liu X., and Duncan J. H. (2003). Radar investigations of breaking water waves at low grazing angles with simultaneous high-speed optical imagery. *Radio Science*, Vol. 38, No. 6, p. 1110, doi:10.1029/2002RS0027166.

Snyder R. L. and Kennedy R. M. (1983). On the formation of whitecaps by a threshold mechanism. Part I: Basic formalism. *J. Phys. Oceanography*, Vol. 13, No. 8, pp. 1482–1492.

Smith P. M. (1988). The emissivity of sea foam at 19 and 37 GHz. *IEEE Trans. Geoscience and Remote Sensing.*, Vol. 26, No. 5, pp. 541–547.

Spiridonov Y. G. and Pichugin A. P. (1984). The influence of meteosituations on characteristic radar images of the terrestrial surface from space. *Earth Obs. Rem. Sens.*, No. 6, pp. 21–27.

Stogryn A. (1972). The emissivity of sea foam at microwave frequencies. *J. Geophys. Research*, Vol. 77, No. 9, pp. 1658–1666.

Stratton J. A. (1941). *Electromagnetic theory*. McGraw-Hill, New York, 615 pp.

Takahashi T. (1978). Raindrop size distribution with collision breakup in an axisymmetric warm cloud model. *J. Atmosph. Science*, Vol. 35, No. 8, pp. 1549–1553.

Tang C. C. H. (1974) The effect of droplets in the air–sea transition zone on the sea brightness temperature. *J. Physical Oceanography*, Vol. 4, No. 11, 579–593.

Tedesco R. and Blanchard D. C. (1979). Dynamics of small bubble motion and bursting in freshwater. *J. Rech. Atmos.*, Vol. 13, No. 3, pp. 215–226.

Teich M. C. and Diament P. (1989). Multiple stochastic representations for K distributions and their Poisson transforms. *J. Opt. Soc. Am. A*, Vol. 6, No. 1, pp. 80–91.

Thorpe S. A. (1982). On the cloud of bubbles formed by breaking wind-generated waves in deep water and their role in air–sea gas transfer. *Phil. Trans. Roy. Soc.*, Vol. 304A, No. 1483, pp. 155–251.

Thorpe S. A. and Humphries P. N. (1980). Bubbles and breaking waves. *Nature*, Vol. 283, No. 57746, pp. 463–465.

Tikhomirov V. K. (1975). *Foams: Theory and practice of their production and destruction*. Khimia, Moscow, 320 pp. [in Russian].

Tikhonov V. I. (1970). *The overshoots in random processes*. Nauka, Moscow, 392 pp. [in Russian].

Timofeev P. V. and Sharkov E.A. (1992). Field optical measurements of the dispersive zone of sea wave breaking. Preprint No. Pr-1841. Space Research Institute, Moscow, pp. 34. [in Russian].

Toba Y. (1962). Drop production by the bursting of air bubbles on the sea surface. III: Study by use of a wind plume. *J. Met. Soc. Japan*, Vol. 40, No. 1, pp. 13–17.

Varadan V. K., Bringi V. N., Varadan V. V., Ishimaru A. (1983). Multiple scattering theory for waves in discrete random media and comparision with experiments. *Radio Science*, Vol. 18, No. 3, pp. 321–327.

Veselov V. M., Davydov A. A., Skachkov V. A., Chernyi I.V., Volyak K. I. (1984). Ship-board remote microwave measurements of internal waves. *Izvestia Acad. Nauk. Fizika atmosphery i okeana (Izvestiya, Atmospheric and Oceanic Physics—Engl. Transl.)*, Vol. 20, No. 3, pp. 308–317.

Vinje T. and Brevig P. (1981). Numerical simulation of breaking waves. *Advanced Water Resources*, Vol. 4, No. 6, pp. 77–82.

Volkov Y. A. (1968) The analysis of spectra for sea waves developed by turbulent wind. *Izvestia Acad. Nauk. Fizika atmosphery i okeana (Izvestiya, Atmospheric and Oceanic Physics— Engl. Transl.)*, Vol. 4, No. 9, pp. 968–987.

Vorsin N. N., Glotov A. A., Mirovskii V. G., Raizer V. Y., Troizkii I. A., Sharkov E.A., Etkin V. S. (1984). Natural radioemissive measurements of sea foam structures. *Sov. J. Remote Sensing*, Vol. 2, No. 3, pp. 520–525.

Wallis G. B. (1969). *One-dimensional two-phase flow*. McGraw-Hill, New York, 440 pp.

Wang C. S. and Street R. L. (1978). Transfers across an air–water surface of high wind speeds: the effect of spray. *J. Geophys. Res.*, Vol. 83, No. C6, 2959–2969.

Weaire D. and Hutzler S. (2000). *The physics of foams.* Oxford University Press, Oxford, UK, 224 pp.

Webster W. J., Wilheit T. T., Ross D. B., Gloersen P. (1976). Spectral characteristtics of the microwave emission from a wind-driven foam-covered sea. *J. Geophysical Research*, Vol. 81, No. 18, pp. 3095–3099.

Williams G.F. (1969). Microwave radiometry of the ocean and the possibility of marine wind velocity determination from satellite observations. *J. Geophysical Researh*, Vol. 74, No. 18, pp. 4591–4610.

Wu J. (1973). Spray in the atmospheric surface layer. *J. Geophysical Research*, Vol. 78, No. 3, pp. 511–519.

Wu J. (1979). Spray in the atmospheric surface layer: Review and analysis of laboratory and oceanic results. *J. Geophysical Research*, Vol. 84, No. 4, pp. 1693–1704.

Wu J., Murray J., Lai R. (1984). Production and distribution of sea spray. *J. Geophysical Research*, Vol. 89, No. C5, pp. 8163–8169.

Young I. R. and Babanin A. V. (2006). Spectral distribution of energy dissipation of wind-generated waves due to dominant wave breaking. *J. Physical Oceanography*, Vol. 36, No. 3, pp. 376–394.

Zacharov V. E. and Zaslavskii M.M. (1982). The kinetic equation and Kolmogorov's spectra in the weak wind waves turbulence theory of wind waves. *Izvestia Acad. Nauk. Fizika atmosphery i okeana (Izvestiya Acad.Sci. USSR. Atmospheric and Oceanic Physics—Engl. Transl.)*, Vol. 18, No. 9, pp. 970–979.

Zaitsev Y. P. (1970). *Sea Biology.* Naukova Dumka, Kiev, 230 pp. [in Russian].

Zappa C. J., Asher W. E., Jessup A. T., Klinke J., Long S. R. (2004). Microbreaking and the enhancement of air–water transfer velocity. *J. Geophysical Research*, Vol. 109, C08S16, doi:10.1029/2003JC001897.

Zaslavskii G. M. and Sagdeev R. Z. (1988). *Introduction to nonlinear physics.* Nauka, Moscow, 368 pp.

Zaslavskii G. M. and Sharkov E. A. (1987). Fractal features in breaking wave areas on sea surface. *Doklady Acad. Nauk SSSR (Trans. of USSR Academy of Sciences—Engl. Trans.)*, Vol. 294, No. 6, pp. 1362–1366.

Zhang W., Perrie W., Li W. (2006). Impacts of waves and sea spray on midlatitude storm structure and intensity. *Monthly Weather Review*, Vol. 134, No. 9, pp. 2418–2442.

Zilitinkevich S. S., Monin A. S., Chalikov D. V. (1978). Interaction between the ocean and atmosphere. In: *Ocean physics*, Vol. 1.: *Hydrophysics of the ocean*. Nauka, Moscow. pp. 208–339. [in Russian].

# Index

Printing: Mercedes-Druck, Berlin
Binding: Stein+Lehmann, Berlin